TECHNOLOGY AND THE GOOD LIFE?

ACKNOWLEDGMENTS

This project had a twin birth. The first one occurred on May 23, 1994, during a session on sustainable technology (featuring papers by Eric Katz and Aidan Davison) at a Society for Philosophy and Technology international conference held in Peniscola, Spain. During the discussion that followed the session (which touched on Borgmann's work), Andrew Light realized that while it had been ten years since the publication of Borgmann's *Technology and the Character of Contemporary Life*, there had been only scattered discussion of the book, and no attempt at a sustained and careful consideration of Borgmann's device paradigm by the community of philosophers of technology as a whole. Light passed a note to Eric Higgs, also in attendance at the session, suggesting a ten-year retrospective on Borgmann's 1984 book. A workshop could, in addition to serving as a basis for a reevaluation of Borgmann's philosophy, also help to focus the field of philosophy of technology. Here was an opportunity to advance the cause of philosophy of technology by centering people's attention on the theory of the device paradigm, a theory that has impressive depth and reach and has inspired the work of people in many disciplines.

This idea was further developed in November 1994 when Albert Borgmann came to the University of Alberta for a stint as a Distinguished Visitor. As part of this two-week series of lectures and seminars, Eric Higgs organized a one-day workshop centered on Borgmann's (then unpublished) essay, "The Nature of Reality and the Reality of Nature." Andrew Light, Jonathan Perry (then a graduate student in geography at the University of Wisconsin, Madison), and Mary Richardson (a philosopher at Athabasca University) provided commentaries. Light's paper from this workshop, "Three Questions on Hyperreality," was later published along with a response by Borgmann in *Research in Philosophy and Technology*, volume 15 (1995). After the workshop, Light and Higgs began planning a workshop on *Technology and the Character of Contemporary Life*.

The other birth took place in the summer of 1994. David Strong and Jesse Tatum, sensing the need for a collected volume addressing Borgmann's work, had begun independently soliciting written contributions for such a volume as they attended the National Endowment for the Humanities Summer Seminar "Rethinking Technology" at Penn State. Realizing our similar ambitions, we put our heads together, and with the help of the University of Alberta's conference fund, the Eco-Research Chair in Environmental Risk Management, the Department of Anthropology, and the generosity of participants willing to travel at their own expense, a workshop, "Technology and the Character of Contemporary Life," was held in Jasper, Alberta, from September 30 through October 2, 1995. The response to our invitations was remarkable: over thirty people attended from across North America and Europe. In addition to most of the contributors to this volume, participants included Robert Burch, Irene Klaver, Mary Ann McClure, David Rothenberg, and Pieter Tijmes. We bore two principles in mind in organizing the workshop. First, at Borgmann's insistence, we ensured that the conversation focused on *Technology and the Character of Contemporary Life;* in other words, it was a reflection on the work and not the scholar. Second, as much room as possible would be given over to discussion and open-ended conversation. The mountains and the words made this workshop memorable.

Since that time a variety of friends and colleagues have helped to make possible the transformation from the conference to the present volume. We are grateful to Albert and Nancy Borgmann, Amaya Garcia, Carl Mitcham, Sheila Gallagher, Ginger Gibson, and Dorit Naaman for help and encouragement throughout this process. Susan Abrams, our editor at the University of Chicago Press, has encouraged us from the beginning of the life of the project. Leslie Keros admirably kept us on track through the final stages of production, and Michael Koplow provided what we all agree was a very thorough copyediting job.

CONTRIBUTORS

ALBERT BORGMANN is Regents Professor of Philosophy at the University of Montana, where he has taught since 1970. His area of research is the philosophy of society and culture. His publications include *Technology and the Character of Contemporary Life* (1984), *Crossing the Postmodern Divide* (1992), and *Holding On to Reality: The Nature of Information at the Turn of the Millennium* (1999).

GORDON G. BRITTAN JR. is Regents Professor of Philosophy at Montana State University. He is author of *An Introduction to the Philosophy of Science* (1970) and *Kant's Theory of Science* (1978, with Karel Lambert). He is currently working on a book on metaphysics.

MORA CAMPBELL is associate professor in the Faculty of Environmental Studies at York University in Canada. She has previously published work on time and technology, gender, sustainable agriculture, and food security. She is currently studying embodied knowing by looking at relationships between yogic, martial, and environmental philosophies and practices.

PAUL T. DURBIN is a professor in the Department of Philosophy, the Medical Scholars Program, and the Center for Energy and Environmental Policy at the University of Delaware. He is author of *Social Responsibility in Science, Technology, and Medicine* (1992). He is currently working on another manuscript, "Activist Philosophy of Technology." He has edited many volumes for the Society for Philosophy and Technology, most recently the first four volumes of the society's quarterly electronic journal.

PHILLIP R. FANDOZZI is professor of humanities and philosophy and director of the Liberal Studies Program at the University of Montana. He is author of *Nihilism and Technology* (1982) and writes and teaches on philosophy and film.

Andrew Feenberg is professor of philosophy at San Diego State University. His interests include Critical Theory, computer-mediated communication, and Japanese philosophy. In addition to numerous articles in these fields, he has published *Lukács, Marx, and the Sources of Critical Theory* (1981), *Critical Theory of Technology* (1991), *Alternative Modernity* (1995), and *Questioning Technology* (1999).

Lawrence Haworth is Distinguished Professor Emeritus at the University of Waterloo and a Fellow of the Royal Society of Canada. He is author of *The Good City* (1963), *Decadence and Objectivity: Ideals for Work in Post-consumer Society* (1977), and *Autonomy: An Essay in Philosophical Psychology and Ethics* (1986), and coauthor of *Value Assumptions in Risk Assessment: A Case Study of the Alachlor Controversy* (1991, with Conrad G. Brunk and Brenda Lee) and *A Textured Life: Empowerment and Adults with Developmental Disabilities* (1999, with Alison Pedlar and Peggy Hutchison).

Larry Hickman is director of the Center for Dewey Studies and professor of philosophy at Southern Illinois University, Carbondale. He is author of *Modern Theories of Higher Level Predicates* (1981) and *John Dewey's Pragmatic Technology* (1990). He is also editor of *Technology as a Human Affair* (1990); *Reading Dewey: Interpretations for a Postmodern Generation* (1998); two volumes of *The Essential Dewey* (1998, with Thomas Alexander); and *The Correspondence of John Dewey,* vol. 1, *1871–1918* (1999).

Eric Higgs is associate professor of anthropology (adjunct in sociology) at the University of Alberta. He has interests in technology studies, environmental anthropology and policy, and ecological restoration. He is presently secretary of the Society for Ecological Restoration and completing a book on ecological restoration, "Nature by Design."

Douglas Kellner holds the George F. Kneller Chair in Philosophy of Education at UCLA. He has authored, edited, or coedited over twenty books, including most recently *Media Culture: Cultural Studies, Identity, and Politics between the Modern and the Postmodern* (1995) and *The Postmodern Turn* (with Steven Best, 1997). Kellner is also the general editor of the unpublished and uncollected works of Herbert Marcuse, with the first volume, *Technology, War, and Fascism,* appearing in 1998.

Andrew Light is assistant professor of environmental philosophy and director of the graduate program in environmental conservation education at New York University. He is author of over forty articles and book chapters on environmental ethics, philosophy of technology, and philosophy of film and has edited or coedited ten books, including *Environmental Pragmatism* (1996, with Eric Katz). He has just completed a new monograph on environmental ethics and policy.

DIANE P. MICHELFELDER is professor of philosophy and head of the Department of Languages and Philosophy at Utah State University. Her research focus is on Heidegger, hermeneutics, and issues in ethics and technological change. Her publications include *Dialogue and Deconstruction: The Gadamer-Derrida Encounter* (1989, coedited with Richard Palmer) and an anthology, *Applied Ethics in American Society* (1997, with William Wilcox). She is currently writing a book entitled *Ethical Challenges in Cyberspace.*

CARL MITCHAM is professor of liberal arts and international studies at the Colorado School of Mines. His publications include *Thinking through Technology: The Path between Engineering and Philosophy* (1994) and *Engineering Ethics* (1999). An edited volume, *Visions of STS,* is forthcoming. He is also editor of the annual series *Research in Philosophy and Technology.*

THOMAS MICHAEL POWER is professor and chair in the Department of Economics at the University of Montana. His most recent books are *Lost Landscapes and Failed Economies: The Search for a Value of Place* (1996) and *Environmental Protection and Economic Well-Being: The Economic Pursuit of Quality* (1996). Among his current projects is a book analyzing the impact on workers and communities of the shift of rural areas to "post–natural resource" economies.

DAVID STRONG is associate professor of philosophy at Rocky Mountain College. He is author of *Crazy Mountains: Learning from Wilderness to Weigh Technology* (1995) and of numerous articles relating the philosophy of technology to environmental philosophy and nature writing. Strong is currently developing his idea of philosophy in the service of things.

JESSE S. TATUM is lecturer in science and technology studies at Rensselaer Polytechnic Institute. He is author of *Energy Possibilities: Rethinking Alternatives and the Choice-Making Process* (1995) and *Muted Voices: The Recovery of Democracy in the Shaping of Technology* (1999).

PAUL B. THOMPSON is Distinguished Professor of Philosophy at Purdue

University, where he holds the Joyce and Edward E. Brewer Chair in Applied Ethics. He has published seven books and more than sixty papers on applied ethics, farming, and technology, including *Spirit of the Soil: Agriculture and Environmental Ethics* (1995) and *Food Biotechnology in Ethical Perspective* (1997).

Eric Higgs, Andrew Light, and David Strong

To write that our lives are radically altered by contemporary technology is at once to admit everything and nothing. For example, between the 1980s and the 1990s, electronically encoded identification cards have become widespread, admitting a host of new and increasingly integrated services. University students in some places can now use a single card to withdraw money from a bank, pay for books, make photocopies, check out materials from the library, buy lunch, and gain access to the pub. The convenience, power, and integration of such technologies make them irresistible, and their rate of diffusion in North America and other developed countries has been impressive. The same holds for e-mail, the Internet, cellular telephones, computer diagnostics of automobiles, personal computers, household appliances, and on down a nearly endless list.

Beneath each of these developments lies a complicated set of social, economic, political, and ecological concerns. For instance, on the flip side of convenient identification is surveillance, a practice that is growing exponentially as databases become interconnected and tracking procedures more sophisticated. Purchasing habits are monitored and analyzed, resulting in well-massaged direct-marketing campaigns. The convenience of these cards promotes habitual use, which increases consumption, further embedding the practice in everyday life. Dependency grows, a feature made manifest when a card is misplaced. As we become adroit with the use of identification cards, life before such a device becomes obscure and difficult to imagine. The pervasiveness of such practices numbs us to experience, displaces our critical capacities, and obscures reality.

Growing numbers of people are posing more difficult questions about our advanced technological way of life: Are we allowing time for a genuinely good life? Are we too distracted by technology to realize a good life? The growth of technology seems explosive and our lives are reshaped by the results. Technology in this light appears as a blessing and as a curse, and the line separating these values can be excruciatingly thin. The tendency is to accept this ambivalence—everything and nothing. Yet acceptance permits us to avoid discriminating judgments about the consequences of technology. Typically, this ambivalence gives birth to cynicism: nothing can be done to change our relationships with technology, so why bother trying?

The hunch that motivates this book is that we have not yet developed a philosophy of technology adequate to allow us to make discriminating judgments at personal, social, and political levels on issues like these. It is not that we are bound to the will of technology as some strong technological determinists would argue, but that we have not yet evolved theories that guide us toward a critical rather than a passive engagement with technology and its effects in our lives. At a broader public level, technology remains an alluring and only occasionally disturbing presence. The manifold consequences of technology, especially the way in which it remakes reality, are mostly pushed aside.

An investigation like this one does have its precedents. Following World War II, the French sociologist Jacques Ellul, the American cultural critic Lewis Mumford, and the German philosopher Martin Heidegger urged that we think of technology, in Ellul's words, as "the stake of the century." They argued that modern technology, although it may seem to be just a more efficient means of doing what humans have always done, confronts humanity with issues that go to the very core of who we are and how we live. In this book, we want to reenergize debate about the role and significance of technology in our lives, and in so doing begin again the perennial philosophical quest to find and comprehend a good life. We are philosophers by training and practice, as are most of the contributors. However, we want to bring philosophy out of the cloisters to address matters of social and cultural gravity; technology is arguably the most pressing issue of our age.

To meet this challenge, the business of a philosophy of technology is to give technology the quality of reflection it deserves. But where is the current state of this discussion among philosophers? What is philosophy of technology today? We believe this book pushes the conversation on philosophy and technology further than it has gone to date, and closer to the heart of both broader philosophical discussions and work in technology studies more generally. This collection marks the first time that an interdisciplinary group of scholars working in the philosophy of technology have come together to investigate the field through careful attention to one set of ideas, those promulgated by our colleague, Albert Borgmann. We have gathered a critical mass of sixteen scholars working in technology studies to investigate Borgmann's theory of the "device paradigm," a controversial and powerful theory that attempts to explain the operation and implications of technology. Attention has been given to Borgmann's work in a variety of reviews, articles, and books, but there has not been a systematic attempt to evaluate its claims. But this is not unusual in the fledgling field of philosophy of technology, where more focus has been drawn in the past simply to articulating the variety of positions in the field. The attempt here is to determine whether the arguments and distinctions that make up one of the most original, powerful, and comprehensive theories in the philosophy of technology today are a good place from which to orient the discipline.

In introducing this book we will first survey the field of philosophy of technology and then more precisely describe Borgmann's place in it. We hope to justify our choice of Borgmann's work as a lightning rod for other theories in the field to begin to advance the conversation concerning the philosophical implications of technology.

THE PROBLEM WITH PHILOSOPHY OF TECHNOLOGY TODAY

Many of the philosophical forerunners of what we now call philosophy of technology, such as Heidegger (1954) and Ellul (1964), presented radically disquieting views of technology. They did so arguing that technology was a pernicious and nearly all-determining force, leading to troubling social, political, and economic conditions. To those concerned about the conse-quences of technology, such technological determinism was gripping, and fanned the flames of criticism and dissent (Paul Durbin offers a retrospective view of these developments in his chapter). In part, the notoriety of these views was explained because they went far beyond traditional accounts of technology as socially, culturally, morally, and politically neutral. Ellul's magnum opus, *The Technological Society,* inspired a generation of political activists and technology students with its dark message about loss of human autonomy in the service of machines. He claimed that all modern technology (which he called "*technique*") shares an overarching descriptive-normative dimension, one that is irretrievably harmful to humans. By making such powerful and far-reaching claims, thinkers such as Ellul and Heidegger made a case for something other than a simple instrumental view of technology, a view that sees technologies as only a means to an end. By making these claims as strong as they did—for example, by seeming to argue that the effects of technology on society were absolutely determinant and pernicious—their case was made even more compelling.[1]

1. Our views of Ellul's and Heidegger's theories have evolved too. For Ellul, see Paul Durbin's chapter 2 in this volume. For Heidegger, see Borgmann 1984, 40; Borgmann

More recent critics, notably Langdon Winner in his *Autonomous Technology* (1977), have been awake to the power and hazards of technological determinism. True, technology exerts far-reaching and disturbing effects on contemporary society, but this has not erased the capacity for resistance and change to the supposed ends of technology. Technological determinism fell from favor because the arguments advanced to support it seemed crude. It was during this period, in the 1970s and later, that philosophers of technology, as well as scholars in other disciplines, turned their attention to advancing a richer understanding of technology, one informed by a wide range of traditions and perspectives. Not surprisingly, technological determinism has made a comeback recently (Marx and Smith 1995), no doubt as a result of the remarkable combination and convergence of devices associated with information technology. The arrival of cars that talk and computers that listen is bound to cause unease, and certainly to convey a sense that measurable amounts of human control are being lost to technology. However, the conversation about the social effects of technology, thirty years later, is much more finely honed and shows the results of careful scholarship. But the development of a philosophy of technology has been gradual and behind the scenes for most people.

Philosophy of technology had its organized professional start in North America in the mid-1970s with the formation of the Society for Philosophy and Technology. Over the last quarter century, philosophy of technology has become a recognizable subdiscipline in the Americas and Europe. There are journals, societies, and international meetings devoted to the subject.[2] At this time, identifiable theoretical positions have emerged, including schools of thought emanating from a wide range of philosophical traditions including phenomenology, pragmatism, critical theory, and analytic epistemology and philosophy of science, and influenced by a broad array of historical and contemporary thinkers (more detail is provided in Durbin's chapter). Carl Mitcham has produced the most exhaustive historical survey of the field, finding its roots in thinking well before that outlined by Ellul, Mumford, and Heidegger (Mitcham 1994).

The Society for Philosophy and Technology is a good organization for

---

1987. The views of the authors of this introduction on the determinism at work in Ellul and Heidegger are, however, divergent.

2. Since 1978 a hardcover annual, *Research in Philosophy and Technology,* published by J.A.I. Press, has served as the primary journal in the field. Another book series, Philosophy and Technology, was published in hardcover from 1981 to 1994 by Kluwer Academic Publishers. The Society for Philosophy and Technology has met continuously since 1977, holding international conferences every other year since 1981 and meeting regularly at the divisional meetings of the American Philosophical Association. The society now maintains a quarterly electronic journal, *Techné,* at http://scholar.lib.vt.edu/ejournals/SPT/spt.html.

promoting the interests of those concerned with philosophical approaches to technology. But it has not, at least yet, achieved the professional clout or numbers of the other professional societies concerned with the study of technology such as 4S (the Society for the Social Study of Science) or SHOT (the Society for the History of Technology). And viewed in relation to the larger profession of philosophy, the Society for Philosophy and Technology, indeed philosophy of technology generally, remains curiously on the sidelines. Why? Arguably, the explanation is that there is a lack of focus to the field. Philosophy of technology suffers from a lack of clarity regarding questions that define it as a philosophical discipline, central questions whose resolution will drive the discipline forward. Certainly there are many views now on the social effects of technology and how we are to evaluate those effects, but the field nonetheless lacks a critical discussion of those competing views of the kind that characterizes most philosophical subfields. We have many methods for evaluating the social effects of technology now but precious little by way of an attempt to figure out who, if anyone, has got a better method than anyone else's and on what grounds.[3]

Something similar to this concern was voiced over ten years ago by Elisabeth Stöker in one of the first anthologies assembled on philosophy of technology. Stöker even came up with a name for this problem, the "paradox of continual beginning":

> One is still confronted at present [in philosophy of technology] only with a multitude of detailed philosophical studies each of which can be characterized as merely a sketch or an attempt . . . Furthermore, although it occurs quite often that authors refer to each others' [*sic*] work—even if they disagree in their views—it is only in rare cases that a critical discussion of the same subject is carried out . . . Systematic elaborations and assessments have yet to be entered upon. (Stöker 1983, 333–34)

Stöker's solution to the problem involves the adoption of a more rigorous interdisciplinary connection between philosophers studying technology and the actual science involved in the production of new technologies. More importantly, Stöker advocates the pursuit of a richer history of technology. Regardless of the validity of her prescription, the motivation that drives her analysis is one that we share: "only as a systematic discipline can [philosophy of technology] take on the rank among other philosophical disciplines that

---

3. The view presented here about early philosophy of technology, the lack of critical exchange in the field, and the following analysis of Stöker's remarks is more substantively defended in Light and Roberts forthcoming.

it deserves (today more than ever) if one judges by the importance of its tasks" (Stöker 1983, 334).

Stöker's worries have been mitigated in the years since she identified the paradox of continual beginning. After all, many theorists including Albert Borgmann, Paul Durbin, Andrew Feenberg, Frederick Ferré, Larry Hickman, Don Ihde, Carl Mitcham, Kristin Shrader-Frechette, Caroline Whitbeck, and Langdon Winner have gone well beyond mere "sketches" of a theory of technology. Yet there has not been a concerted attempt to test our understanding of technology through a cohesive theory, focusing the central issues and questions in the field that must be addressed in order for it to progress. What is needed is a redirection toward a set of questions that may reorient us and help us to decide the grounds upon which we can determine if the field is progressing apace.

If our reasoning up until now is correct then there are, of course, many different directions in which we could proceed, many different theories in the field around which we could focus our critical energies in order to move the field forward. While there are many reasons to focus on Borgmann's work in a reassessment of the field, one critical concern of his work that draws our attention is the central importance in his philosophy of technology on the relationship between technology and the good life. Borgmann challenges the notion that our lives are necessarily made better, and our communities stronger, by technological innovations that appear to create more leisure time, social mobility, and entrepreneurial efficiency. If philosophy of technology is to reorient itself, then focusing on issues of the good life will remind us of the importance of keeping the field connected to the concerns of most citizens rather than digressing into another intramural conversation, potentially of interest only to other philosophers.

Central to this volume, then, is a philosophical discussion of technology and what constitutes the good life. Our point of access to this discussion is Borgmann's important distinction between "things" and "devices." The term "device," uniquely Borgmann's idea as we will see in chapter 1, refers very generally to the mass-produced artifacts that shape so much of contemporary life. In contrast, for now we can think of "things" as older, traditional technologies (though Borgmann's is not ultimately a historical distinction) that reflect their surrounding natural and communal context and require developed skill and attention in use. Influenced by Heidegger's (1971) discussion of the etymological meaning of "thing" as "gathering," Borgmann calls these things "focal things." The use or engagement with a thing helps to, in some sense, focus who we are by the skill we employ in its use and by grounding us in a particular place. It is Borgmann's contention that the limitless pursuit of devices has damaged and displaced these focal things, thereby significantly impoverishing the quality of our lives. This view, as we shall see, is not without its problems or detractors. But what we all agree on is that Borgmann's work can serve as a flashpoint for revitalizing philosophy of technology so that it can better address the challenges briefly laid out here and fulfill its potential as both a philosophical subfield and a point of departure for social criticism.

Broadening the Scope of Philosophy of Technology

The set of questions a philosophy of technology should address in order to fulfill its promise are often at the intersection of it and other fields. In our opinion, philosophy of technology at its best should appeal to a very wide audience partly because it illuminates our shared, ordinary everyday life, such as with things and devices, and partly because the issues it probes cut across the full range of disciplines. Many of these issues are already vital matters of concern for these disciplines, such as ethics, social and political philosophy, aesthetics, art history, architecture, music, anthropology, religion, history, history of science and technology, cultural studies, sociology, political science, economics, linguistics, literary criticism, visual culture, and the hard sciences. For example, one of the interdisciplinary successes that philosophy of technology has had is with environmental ethics. Issues that join both fields are addressed in journals regularly, and numerous books have appeared. However, this kind of success should be occurring with other fields as well. What does philosophy of technology have to offer other disciplines? In the view of some of our contributors, traditional philosophical approaches may not be capable of questioning and challenging technology in a sufficiently radical manner. Nevertheless, we can show the kinds of questions a robust philosophy of technology can raise and address. It offers an important and badly needed voice as a participant in ongoing conversations in other fields. In what follows in this section we illustrate with television, and other everyday devices discussed by authors in this collection, how some of these questions and connections would flourish with more philosophical examination.

The development of television illuminates key points of an active political debate concerning science, technology, and society issues. Public funding for basic research in the sciences has typically been justified according to the Vannevar Bush model, developed in the United States in 1945. According to this model, which is currently the subject of much debate, the body of knowledge acquired through basic scientific research would serve as a reservoir that could be tapped and applied by engineers and others to generate technological advances for meeting society's wants and needs. For example, the cathode ray tube was first developed so that physicists could study the behavior of electrons under ideal conditions. A direct descendent of the cathode ray tube is television.

Despite the progress inherent in such a process for the development of new technologies, just when a need (or whose need) has been met, or when a society has benefited from a new development, is left entirely unclear and unpredictable. Policy analyst and geologist Daniel Sarewitz put the case with respect to television in *Frontiers of Illusion: Science, Technology, and the Politics of Progress:*

> When a new process or product emerges from the laboratory, it undergoes a profound transition—from well-behaved, insular idea or object to dynamic component of a complex interactive social system. Once imbedded in that social system, the new idea or innovation may produce effects that are completely surprising. When a television is turned on, a series of intrinsically predictable electromagnetic processes occurs inside the television that always leads to the generation of a visual image on the screen. But nothing else about the television is predictable or immanent because all of its other attributes derive not from the physical laws that allow it to operate but from the context in which it is used: when, where, and by whom it is turned on; what is being broadcast; how the viewer is affected by the program; what activities the viewer chose to forego in making the decision to watch; how this decision affected others who interact with the viewer (a sports-hating spouse on Superbowl Sunday, for example); how the total number of viewers influences the economic prospects of companies that are advertising at that particular time. (Sarewitz 1996, 9)

As a result of such unexpected consequences (though whether such consequences are unpredictable is still a valid concern), the connection the Bush model assumes between basic research and wider social and environmental contexts within which this research is embedded can no longer be taken for granted.

Because of problems arising from this lack of an explicit and just connection between science and its context, critics have called for a new social contract between science and society (Piekle and Byerly 1998). Which members of society benefit from the kinds of basic research that now get funded? But even under more socially just conditions, science could still be in the service of consumption and technology as a way of life. So aside from questions of social justice, there are questions that can be raised by a more robust philosophy of technology about the character of the society science ought to better serve. Should it serve a consumption-driven technological society that has never really paused to reflect on its way of life? (Can there be a science in the service of things rather than in the service of consumption and technology?) If some other way of life is open to our communities and larger society, then how differently would the sciences receive funding in order better to serve this kind of society? Such questions are not limited to the discipline of philosophy of technology, but they have crucial philosophical dimensions that demand attention, and which a philosophy of technology can lead the way in helping to answer.

Turning now to other disciplines, similar questions can be asked of engineers and those involved in technological design. Should we always pursue narrowly defined ends as efficiently as possible? How would engineering and design themselves change if intimacy and communal and political relations were folded into the mix of what our society cares about? For those inspired by the thought of political theorist Langdon Winner, such as Jesse Tatum in the present volume, not only might our society opt for more wind power over nuclear power, for example, we might also be more concerned about the political implications of various designs of the very wind machines themselves.

Then, too, we must also consider what activities we have chosen to forgo when we watch television or when we turn to other aspects of our technologically mediated lifestyle. Traditional religions and the wisdom traditions of philosophy have often been concerned to articulate a sense of existence beyond human making and control. Gadgets, like television, transform our power relationships, giving us more control over whatever we want, whenever we want it, wherever we are. Does spending so much time with these gadgets color the character of human existence for us, individually or as a culture? Does this mean, as some of our contributors discuss, that traditional religious views are less relevant to today's technologically mediated world, or potentially more relevant?

Different disciplines offer complications to conventional answers to such questions. For example, many anthropologists eschew television and decry the effects it has on traditional ways of life. However, these same anthropologists are much more likely to observe local variations that contradict or challenge expected norms. For instance, the introduction of television in Arctic regions of northern Canada has been sharply scrutinized by a small host of southern critics, who argue that it will displace and deaden northern traditions. This is the position adopted by media critic Jerry Mander in his contentious and widely-read book, *In the Absence of the Sacred: The Failure of Technology and the Survival of the Indian Nations* (1991). However, other more recent observers, anthropologists in particular, have noted how northern native peoples appropriate television in specific ways that are quite different than what occurs in the south.

Consistent with the work of Andrew Feenberg (1995), who also appears in this volume, native peoples have in some sense subverted the rationale of these devices to suit their own particular ends. There is a case to be made for understanding cultural differences and the way in which cultural practices and beliefs shape the character of television viewing and other practices involved with making and using technology. Such concerns open some fascinating challenges to the work taken up by some philosophers of technology.

Technology, of course, enriches us materially. In an obvious way, the material world is brought from afar to our screens. Microsoft Art Gallery, for example, displays instantly and vividly an artwork in London to our screens at home. Surely, reproductions of paintings on posters and calendars have always served the same purpose, but in another way the material world is uniquely left behind in the virtual gallery where the presentation of paintings is offered as practically substitutable for the real experience. This is not just one poster, but an entire collection. The aura of an actual painting grabs our attention when we stand before it, but the full texture of the painting is lost on the screen. One cannot move toward the virtual painting, back up, tilt one's head, view it from this angle now, all in accordance with what seems to be elicited by the actual painting when we are in its presence. The museum's character and the sense of being in it, the nearby parks, buildings, and monuments, and then the larger setting of the city and country themselves do not appear on the screen. How important to the artwork's lasting value for us is this material world that is left behind? Are the actual textures of the painting and its larger context necessary for bringing us to a full appreciation of an artwork? More generally, is the substance of virtual reality mostly borrowed from the material world, parasitic on it and never able to reproduce the experience of the world yet seductive in its attempt to do so? These interdisciplinary questions call for philosophically informed and articulate discussions.

From the standpoint of literature and literacy, what about the impact of technology on the act of reading itself? Has the electronic age made much of a difference to the experience of reading? According to literary critic Sven Birkerts new media technologies have made all the difference in the world for reading, literary culture, and our inner life. The book itself and the entire familiar tradition surrounding it are being exchanged for the video monitor and hypertext technologies. Says Birkerts, "[T]he world we have known, the world of myths and references and shared assumptions, is being changed by a powerful, if often intangible, set of forces. We are living in the midst of a momentous paradigm shift" (Birkerts 1994, 18). If true, where do we turn for the maintenance of literary culture, since turning back is neither desirable nor possible? Can we hope with Birkerts for "a genuine resurgence of the arts, and literature in particular" as we grapple with the vacuity of a "crisis of meaning" (Birkerts 1994, 196–97)? Is such speculation merely an elitist holdover from the old divide between highbrow and lowbrow culture or is there something more peculiar going on here? Philosophy of technology at its best ought to be able to delve to the root of such questions, envision alternative futures, and help locate the decisive steps that would begin to move us toward (or away from) them.

WHY BORGMANN'S PHILOSOPHY OF TECHNOLOGY?

Albert Borgmann's work is a good candidate to begin such a rethinking of philosophy of technology so that it is better prepared to answer the challenges laid before it, such as those just mentioned. His work falls in the tradition of the kind of substantive philosophy of technology initiated by Heidegger, Ellul, and Mumford. As a philosophy of technology it is far more comprehensive and ambitious than earlier philosophy of technology, setting its sights on larger issues of social criticism while simultaneously meeting scholarly demands already established in the field by previous works. Specifically, there are four chief reasons why Borgmann's work deserves a central place in advancing the philosophical study of technology.

The first reason is that Borgmann builds his theory from a descriptive phenomenological account. He takes up his field of inquiry with a description of the shift from "things" to "devices," from fireplaces to central heating, from candles to sophisticated lighting systems, from wooden tables to Formica, from traditional foods and drinks to Lite versions, from shoelaces to Velcro, from craftwork to automation, from traditional performances and physical activities to home entertainment centers. For Borgmann these substitutions constitute a repeated pattern that can be described, a pattern that Borgmann claims also has repeated consequences (which can be similarly described) for our relationships to our physical surroundings, our relationships to ourselves and others. Discussing whether Borgmann's characterizations are accurate is a fruitful beginning for a discussion of how technology effects our assessment of the good life.

Consider for example the practice—what we will later understand as a focal practice—of the family and friends sitting down on Friday night to a traditional Jewish Shabbat meal. We start with the *brachot*—the blessings over the candles, the wine, and the *challah* bread—and move to a dinner that in part calls on us to take a moment out of time, steeped in tradition, to reflect on the connections between those gathered at the table, and perhaps also including a vigorous debate and discussion of things that matter to all present. Whether any of the participants at the meal be theists or not (perhaps only embracing a cultural recognition of their heritage and a desire to share it with friends or the next generation) the culture of the Shabbat table focuses the character of each person present in some context, and binds the participants in some relation—relations, we might add, that can so easily be absent with a fast-food, drive-through window experience.

Second are the diagnostic aspects of Borgmann's philosophy. Borgmann locates the problem of technology in relationships. His critique considers the adverse effects technology has on our relationships to our physical surroundings, and our human relationships in their political, social, and aesthetic dimensions. In this sense, the focus of Borgmann's work is not simply technology itself as an object of study, but more thoroughly human relationships and our relationships to our surroundings as they are inevitably effected by technology.

Third, considered prescriptively and on the basis of his diagnosis, Borgmann argues that these relationships can be reconfigured into a socially reconstructive program. In fact, Borgmann's theory, along with others such as those of Andrew Feenberg and Langdon Winner, is one of the few attempts at developing a comprehensive series of reform proposals for technology. It also addresses questions of nature and environment, rather than restricting reform of technology to built space and artifacts, thus exceeding the traditional purview of the field. Focusing on Borgmann's work in conversation with and divergence from these other reform proposals will help to move the field forward.

From another standpoint, Borgmann calls for a philosophical reassessment of social life that challenges received notions of what constitutes the good life. While many moral theorists of late have followed the charge of the communitarians to expand moral discourse beyond a thin assessment of the good, Borgmann adds a call for attention to the material and artifactual foundations of a thicker reconception of the good.[4]

Borgmann's is not an abstract theoretical contribution to an assessment of the good life but a grounding and practical means to create a context and a language whereby our material world can be normatively assessed as part of a more robust moral ontology. Borgmann puts it this way after briefly acknowledging a debt to Heidegger in formulating the wider contours of these views:

> Heidegger says, broadly paraphrased, that the orienting power of simple things will come to the fore only after the rule of technology is raised from its anonymity, is disclosed as the orthodoxy that

4. Note that Charles Taylor accepts Borgmann's description of the device paradigm as part of the ails of modernity. See Taylor 1989, 501; and Taylor 1991, 7.

heretofore has been taken for granted and allowed to remain invisible. As long as we overlook the tightly patterned character of technology and believe that we live in a world of endlessly open and rich opportunities, as long as we ignore the definite ways in which we, acting technologically, have worked out the promise of technology and remain vaguely enthralled by that promise, so long simple things and practices will seem burdensome, confining, and drab. But if we recognize the central vacuity of advanced technology, that emptiness can become the opening for focal things. It works both ways, of course. When we see a focal concern of ours threatened by technology, our sight for the liabilities of mature technology is sharpened. (Borgmann 1984, 199)

And finally, fourth, Borgmann's work is important because of the depth and breadth of his diagnosis and his prescriptions. Borgmann's reform program advocates a set of issues that any political system must address if it is to be effective in a social sphere dominated by technology. The work is therefore potentially of interest to a great variety of political positions and not simply an appeal to the most effective program for the reform of technology by a particular ideological persuasion. But in order to appreciate these four points better, we need to introduce Borgmann's ideas and how they come together in his most lasting description of our current technological age: the device paradigm. A detailed introduction to his theories, especially the device paradigm, is presented in chapter 1.

OUTLINE OF THE BOOK

In order to better assess Borgmann's approach to technology, we have divided the book into four parts. The parts within the book and the chapters within the parts are arranged to give readers who are unfamiliar with Borgmann's books a solid grasp of the essentials of his work before moving on to more detailed commentary.

Part 1, "Philosophy of Technology Today," lays out a thorough and detailed summary of Borgmann's philosophy of technology and then continues with a brief history of contemporary philosophy of technology and looks toward the issues that will shape its future. Part 2, "Evaluating Focal Things," presents both sympathetic and critical reconstructions of Borgmann's work. The chapters in this part move from those that evaluate Borgmann's work on its own terms to those that consider implications for the field as a whole. The next part, "Theory in the Service of Practice," takes Borgmann's device paradigm and applies it to areas that were not originally considered by its author. Here, scholars of technology think about how the idea of a focal thing is important in a serious analysis of film, agriculture, design technology, and the restoration of damaged ecosystems. Authors in the final part, "Extensions and Controversies," challenge the orthodox meaning of focal things and practices and look at the ground that still needs careful examination both from the perspective of Borgmann's work and from the field as a whole. Closing the volume is Borgmann's reappraisal of his work in light of the comments and criticisms offered by his colleagues. His chapter is written with an eye to advancing not only his work, but also toward the broader goal of orienting the philosophy of technology and improving its relevance to larger social, political, and environmental concerns.

Borgmann succeeds in opening up his own views, and philosophy of technology in general, to much broader philosophical debates. While we laud his work, even though we differ in our assessments of it, our only worry is that philosophers of technology may forget to talk to each other and to critically evaluate each other's work. Accordingly, the contributors to this volume are engaged in moving to a more robust philosophy of technology by developing their respective ideas, arguments, and criticisms with and against the big picture presented by Borgmann. It is healthy disagreement we wish to encourage with this book, even and especially if Borgmann's diagnoses and prescriptions are pushed to their limits. Such conversation may even question whether technology is an issue of the magnitude we, along with Heidegger, Mumford, and Ellul, believe it is. These kinds of controversies are the best tonic for vitalizing philosophy of technology.

References

Birkerts, Sven 1994. *The Gutenberg Elegies: The Fate of Reading in an Electronic Age.* New York: Fawcett Columbine.

Borgmann, Albert. 1984. *Technology and the Character of Contemporary Life: A Philosophical Inquiry.* Chicago: University of Chicago Press.

Borgmann, Albert, and Carl Mitcham. 1987. "The Question of Heidegger and Technology: A Critical Review of the Literature." *Philosophy Today* 31:99–194.

Ellul, Jacques. 1964. *The Technological Society.* Trans. J. Wilkinson. New York: Vintage.

Feenberg, Andrew. 1995. *Alternative Modernity: The Technical Turn in Philosophy and Social Theory.* Berkeley and Los Angeles: University of California Press.

Heidegger, Martin. 1954. "Die Frage nach der Technik." In *Vorträge und Aufsätze.* Pfullinger: Günther Neske.

———. 1971. "The Thing." In *Poetry, Language, Thought.* Trans. Albert Hofstadter. New York: Harper & Row.

Light, Andrew, and David Roberts. Forthcoming. "Toward New Foundations in Philosophy of Technology: Mitcham and Wittgenstein on Descriptions." *Research in Philosophy and Technology* 19.

Mander, Jerry. 1991. *In the Absence of the Sacred: The Failure of Technology and the Survival of the Indian Nations.* San Francisco: Sierra Club Books.

Marx, Leo, and Merritt Roe Smith. 1995. *Does Technology Drive History?* Cambridge: MIT Press.

Mitcham, Carl. 1994. *Thinking through Technology: The Path between Engineering and Philosophy.* Chicago: University of Chicago Press.

Piekle, Roger A., Jr., and Radford Byerly Jr. 1998. "Beyond Basic and Applied." *Physics Today,* February, 42–46.

Sarewitz, Daniel 1996. *Frontiers of Illusion: Science, Technology, and the Politics of Progress.* Philadelphia: Temple University Press.

Stöker, Elisabeth. 1983. "Philosophy of Technology: Problems of a Philosophical Discipline." In *Philosophy and Technology.* Ed. Paul Durbin and Friedrich Rapp. Dordrecht: Reidel.

Taylor, Charles. 1989. *Sources of the Self.* Cambridge: Harvard University Press.

———. 1991. *The Ethics of Authenticity.* Cambridge: Harvard University Press.

Winner, Langdon. 1977. *Autonomous Technology.* Cambridge: MIT Press.

PART ONE

Philosophy of Technology Today

Part 1 exemplifies the twofold purpose of the volume as a whole. The ultimate purpose is to help revitalize the important conversation we ought to be having about the philosophy of technology. Its current shortcoming, we believe, has largely to do with its lack of focus and orientation. By addressing in particular Borgmann's comprehensive work (the second purpose) as a springboard to this ultimate purpose, we hope that this volume contributes to revitalizing this conversation with renewed focus and energy. In chapter 1, "Borgmann's Philosophy of Technology," David Strong and Eric Higgs initiate this discussion by describing the public character of Borgmann's philosophy of technology and presenting an overview of the device paradigm. In chapter 2, "Philosophy of Technology: Retrospective and Prospective Views," Paul Durbin, one of the founders of the Society for Philosophy and Technology and arguably its principal overseer, takes up this twofold task by first reviewing the two decades, 1965–85, that saw the formation of the philosophy of technology in North America. From his explicit pragmatist standpoint, he finds that philosophers were originally reaching to the philosophy of technology in hopes of better understanding and addressing our "major technosocial disasters." Durbin urges today's philosophers of technology to remember again these original concerns and to take up this struggle with particular pernicious technosocial problems one at a time, even though addressing these problems at a philosophical level will be very difficult, he maintains, in the coming technological culture. Within this frame he evaluates several possible readings of Borgmann's work.

Borgmann's Philosophy of Technology

David Strong and Eric Higgs

The book [Technology and the Character of Contemporary Life] *that helped me find a voice, however, also opened up a troubling rift between the instruments of thought and what matters in thought, between the discipline of philosophy and the task of philosophy. The latter, I thought, was a matter of helping people become more conscious of the distractions of the culture of technology and more confident of the focal things and practices that can center one's life. I had engaged the arguments of the discipline to advance the task. The clanging and grinding of the disciplinary machinery was music to some colleagues, an ordeal to others, and incomprehensible noise to all non-philosophers. Where, then, is philosophy to be found, in the discipline, or in the task?*
—Albert Borgmann, "Finding Philosophy"

In modern life we swim deep in a sea of technology, surrounded by artifacts and patterns of our own making. These artifacts and patterns, like water, are often transparent to us. They are everywhere and nowhere to be seen as we fin our way along chasing after whatever is new, stylizing and restyling our lives. Yet something feels wrong. Leisure leaves us stressed. Time saving leaves us with no time. Freedom amounts to deciding where to plug into the system. Nature is pushed aside. Even our sense of who we are is transformed in relation to this surrounding sea. So we dart anxiously here and there trying one technological fix after another. It has not occurred to us yet that, like fish in polluted water, what may be wrong lies closest to us. Philosophers of technology along with political and social theorists and others have made insightful attempts to understand the problem or problems of what is in the water. Nevertheless, despite debates within and among these disciplines, scholarly sophistication here remains at an early stage. In addition to this disciplinary challenge, the task of reaching beyond academic specialists to get other people to realize that a significant problem is hidden in their transparent surroundings has scarcely made a dent (Noble 1997).

These two challenges, both the challenge of what Albert Borgmann calls the discipline of philosophy and the task of philosophy, are important. As professional members of the discipline of philosophy, we are concerned with the "instruments of thought." We take theories, distinctions, and arguments, such as those advanced by Borgmann in *Technology and the Character of Contemporary Life* (or *TCCL*), as a point of departure for testing, revising, or forwarding alternative theories. From this perspective, we are the ones who find music, or at least hours well spent, in working through the complex turns of thought in Borgmann's book. However, we also share Borgmann's disappointment that genuine achievements in the discipline of philosophy often fail at the *task* of philosophy, that is, to engage the public more broadly in a reflective conversation about matters of great concern to all. We feel this shortcoming poignantly when these philosophies could have something decisive to contribute to a conversation about the quality of our lives by uncovering what is hidden yet harmful in our surroundings and by helping us to understand what can be done about it. From the particular standpoint of Borgmann's theory of the device paradigm, as we will see in detail later, this means "helping people become more conscious of the distractions of the culture of technology and more confident of the focal things and practices that can center one's life." Such philosophies ought to be, but often are not, part of the mix of a widespread public conversation.

A public and philosophical conversation about technology, in particular, is urgent, as Ellul, Heidegger, Mumford, and many others have pointed out. For such a public conversation to develop in a meaningful sense, it must be much more widespread than a debate among a handful of academic specialists. If the philosophical ramifications of technology remain little discussed by the larger society, then no matter how successful philosophers are at articulating and debating these ramifications, they will have failed at the greater task.

An example of a philosophical and technological issue that may open up a public conversation in this fashion—in fact an underlying contention for a number of chapters that follow—is the challenge posed by rethinking, in our contemporary technological context, the general relationship between the useful and the good. According to *TCCL,* technology has produced extraordinarily useful things and successfully taken on the ancient scourges of hunger, disease, and confinement. It did so, however, in following a particular pattern, the device paradigm. In following that pattern, we have been inattentive to the distinction between two kinds of burdens: the odious burdens of hunger, disease, etc. and the ennobling burdens exacted by the demands of community and of human excellence.[1] Rather we seek relief from all burdens whatsoever. Hence we have under the banner of usefulness reduced our devotion to community and excellence. One the one hand, ambulances save lives and so are eminently useful; on the other hand, cars save us bodily exertion and the annoyances of fellow pedestrians or passengers and are thus, at least in part, a threat to the goods of community and our physical health in the form of exercise. Hence, according to this particular philosophy of technology, we need to focus on those specific goods that are both irreplaceably good (viz., focal things and practices) and threatened by the thoughtless employment of technology.[2]

BORGMANN AND THE TASK OF PHILOSOPHY

The twofold task of philosophy is to engage philosophy with issues that matter and to involve the public in a philosophical conversation about these matters. For Borgmann, the task of philosophy is to engage "the things that matter" quite literally. The distinction between "focal things" and "devices" has proven to be valuable to philosophers and laypeople alike: Borgmann's account of the difference between things and devices is easy to grasp intuitively, as we will see shortly. Moreover, it helps people to become aware of the otherwise invisible water we are immersed in, Borgmann believes, by making them conscious of the significance of technological change as it impinges on important centers of their lives. On his view, it helps people to identify and guard these centers against corrosive forms of technology. To see these latter advantages, we need to understand the distinction itself first.

What are these focal things and devices? In general, Borgmann characterizes "focal things," "centering things," or what he sometimes calls "focal reality" as simply different placeholders "for encounters each of us has with things that of themselves have engaged mind and body and centered our lives. Commanding presence, continuity with the world and centering power are the signs of focal things. They are not warrants, however. To present them is never more than to recall them" (Borgmann 1992, 119–

1. Excellence, for Borgmann, arises out of focal practices that require our exertion, attention, patience and so on. For example, backpackers literally burden themselves with packs and walking in order to encounter the wild on its own terms. What he means by excellence can often be summarized in terms of engagement, skill, discipline, fidelity, resolve, celebration, and, as Gordon Brittan will call attention to in this volume, traditional excellence.

2. As we will see, Borgmann argues that we are already assume the device paradigm when we separate means and ends. Special care, then, may need to be taken if we are not to become subverted by the device paradigm when we reconsider what is useful and what is good.

20). Before developing this terminology, it is helpful to think of the device for now (although a more rigorous account will be offered later) as referring to a descriptive characterization of most of the mass-produced artifacts around us as well as our commonly employed procedures. Devices are just the opposite of focal things.[3] Devices are disposable, discontinuous with their larger context, and glamorous in their appeal.

To give content to these characterizations of things and devices, let us take an example that figures prominently in Borgmann's works: running. Among focal things for runners are an ocean road that George Sheehan runs almost daily; a path along Rattlesnake Creek for a runner in Missoula, Montana; the course that the New York City marathon takes for Peter Wood. Like other focal things, these things often lie inconspicuous until runners disclose them, bringing what Borgmann calls their eloquence into relief.

The focal thing's *commanding presence,* in part, is its capacity to make demands on us. It takes getting in shape and staying in shape to be equal to the six miles of ocean road. Nor can one simply push a button and step off the device, disposing of the run, if one does not feel like running that day. Focal things demand patience, endurance, skill, and the resoluteness of regular practice—a *focal practice.* Even a certain character, that of a runner, is developed in order to become a match for the thing. Commanding presence, too, in part, has to do with the thing's attractions. The sights and sounds, the events of the run, the uniqueness of a particular run, or the harmony one feels with the surroundings cannot be instantly replayed at our disposal. These demands and attractions of the focal thing's commanding presence make things engaging for mind and body, serving to unify them. Commanding presence, then, describes (Borgmann excels at descriptions on a general level) the characteristics of focal things that contrast with the *disposability* of the device. Most of these devices are designed to be under our control. A computer software advertisement, picturing a mouse, brags about how much can be done "without lifting a finger." Following the trajectory of disposability, it's hardly surprising to find the recent development of a wireless mouse—it makes control even easier. In the process of lifting all these burdens from us, however, these devices are often disengaging. Most devices require little in terms of skill, patience, effort, or attention.

Unlike exercise in front of a video in the controlled environment of a health club, runners experience a *telling continuity* between their focal thing, say the Rattlesnake Creek path, and the weather overhead, between the

---

3. It should be noted as well that Borgmann thinks of things and devices as being on a continuum: between the clear examples of things and devices are many degrees of variation.

high, roily waters of the creek, the month of May, the receding snowfields, and the previous winter's snowfall. The office window from which one sees the mountains still capped with snow, the home where one lives, and the conversations one has with other members of the community are of one piece with this focal thing. The hour spent in the club exercising in front of a video, as we will later see with devices generally, is *discontinuous* with this larger context of one's life, community, and place. While the function of a device captures one or a few aspects of the original thing, such as the exercise of muscles, devices sever most other relationships. At the health club, one might be reading a book, riding a stationary bicycle, and listening to music with headphones. Mind, body, and world are all dissociated from one another. In general contrast, then, a focal thing is not an isolated entity; it exists as a material center in a complicated network of human relationships and relationships to its natural and cultural setting.

Focal things gather this complicated web of relations in a way analogous to how a grizzly bear concentrates the web of ecological relationships dispersed throughout an ecosystem large enough to support the species. But mere contact with the thing, the material center, does not guarantee that this web of relations will be brought home to us automatically. A six-mile run along Rattlesnake Creek can be boring or a mere relentless chore. Although mind, body, and world may not be quite as dissociated as in a health club, runners do feel this discord and find themselves to be out of touch at such times. Presumably, runners would not be runners were it not for better days. On the good days, runners come away appreciating these *centering powers* of the thing. They come away invigorated, knowing that "this is where I want to be and what I want to be doing." Through focal things and practices they affirm the place where they live and the direction of their lives. On such days they have had a *centering experience.* These centering powers of focal things contrast with the short-lived but admittedly alluring *glamorous* appeal and thrill of many devices.

These centering powers of a thing can also be thought of as its unifying powers. As we will see better later, devices separate means and ends. With automobiles, for example, we "cash in prior labor for present motion . . . my achievement lies in the past, my enjoyment in the present" (Borgmann 1984, 202). Things, through their centering powers, unify means and ends, achievement and enjoyment, competence and consummation, mind and body, body and world, individual and community, present and tradition, culture and nature. These latter can be seen more clearly by summarizing our account with an example of a focal thing drawn from music. A fine violin, for instance, is brought to life in the hands of a caring and gifted performer, and simultaneously the life of the performer is enriched in relation to the violin. This relationship between the artifact—the violin— and the performer requires skill, and as such helps to create the character of the performer in relation to the artifact, here, the focal thing. Communal ties can be forged when focal thing, performer, and audience come together in a performance that offers a communal celebration.

As are many of the focal things that are likely to come first to mind, violins are an older technology and somewhat "a thing of the past." Most musicians, whether with violins or mandolins, cannot make a living at their focal practice. Many are forced to give up, choosing a different occupation altogether, often leaving their instruments behind. Devices, such as sound systems in one way and televisions in another, have displaced performance both by individuals and within the community. From a historical perspective, these things of the past and their world are all but gone; devices have come to replace them. On Borgmann's account, the destruction of things and the reconstitution of them into devices continues to this day, perhaps even more rapidly, with newer forms of sophisticated technology, such as information technology. In this rising tide of technological devices, disposability supersedes commanding presence, discontinuity wins over continuity, and glamorous thrills trump centering experiences. The pervasive presence of these devices and these experiences, Borgmann finds, tends to contribute to a life that lacks a center and that is missing a rich social and ecological context. Thus, if Borgmann's theory is right, there exists a profound conflict between the expansion of technological devices and the focal things and practices these devices displace. On that basis, he appeals to readers, for the sake of the quality of their lives, not to let devices completely overrun these things and practices. Such protection of centering things can occur, he argues, only if prescribed steps are taken to make room for them in our individual lives, communities, and culture.

Understandably, Borgmann thinks of the task of philosophy as making these points about things, devices, and the quality of our lives not only for other philosophers and other specialists but also for all his fellow citizens. No contemporary philosopher has drawn more attention to these "things that finally matter" than Borgmann, and for that reason, his philosophy has received widespread attention beyond the disciplines of philosophy of technology and technology studies. If more successful with this conversation about technology and the quality of our lives, it certainly would help to spur and revitalize philosophy generally and the philosophy of technology in particular. Significantly, for our purposes too, these focal things yield a standpoint from which Borgmann's theory as a whole can be evaluated.

*

We can get clearer about this task of philosophy and tactics for widening this philosophical conversation with a brief look at the several traditions that Borgmann's philosophy of technology is rooted in. In one classification system Borgmann belongs to the humanities as opposed to engineering tradition in the philosophy of technology. Philosophers in the engineering tradition tend to take a narrow view of the philosophy of technology, thinking of it as a field aimed at examining mostly technical philosophical problems arising out of applied science and engineering and taught most relevantly at technical universities. In the humanities tradition by contrast the task is a broader one of reflecting on the world and technology, and here is where Borgmann's work is clearly situated (see Mitcham 1994). Technology, for him, is not only applied science and engineering; technology is the larger context, the way we "take up with the world." The world within which we exist with modern technology has its own special features and patterns. To bring out these special and decisive features of our age, that is, to evaluate the *significance* of technology is to do, on this view, philosophy of technology.

This reflection on the world accords, too, with Heidegger's influence on Borgmann. No doubt Borgmann, in some sense, is a neo-Heideggerian. Readers of Borgmann will continue to find throughout his work ways in which he is indebted to Heidegger. Born and raised in Freiburg of parents belonging to the Catholic intelligentsia (acquainted with Bernhard Welte, Karl Rahner, and Max Müller) and attending the university as an undergraduate, Borgmann listened to lecture series by Heidegger, and he wrote his dissertation at the University of Munich under Heidegger's student, Max Müller (Borgmann 1993, 157). The device paradigm itself, sometimes called the framework of technology, Borgmann considers a more useful and tightly developed specification of Heidegger's essence of technology: *Gestell* or framework. And just as Borgmann's critique of technology is at its base inspired by Heidegger, so too is Borgmann's proposed reform of technology, which entails, in part, recognizing this framework of technology. Another part of Heidegger's reform proposal entails a puzzling element in Heidegger's thought. The question of being is Heidegger's principal concern, yet in his later writings he begins to emphasize (following Rilke, Hölderlin, and other German poets) particular "things." Borgmann has highlighted and thoroughly developed this underarticulated side of Heidegger's thought, in which even sophisticated scholars of Heidegger sometimes interpret things to be almost any material object, for example a modern highway overpass (Dreyfus and Spinosa 1997). Borgmann's philosophy has also moved the locus of Heidegger's discussion of things more toward the special things of our lives, things of great importance to our well-being. As we will see, this last move comes from an American tradition of things rather than the European one.

This debt does not mean that Heidegger's influence on Borgmann is obvious for there is an enormous difference between Heidegger's difficult and idiosyncratic language and Borgmann's more developed, ordinary, clear, and precise language. Thinking of Borgmann as a Heideggerian may therefore do him a disservice. As he says in an autobiographical piece, "Heidegger had shown me the problem that needed attention. Rawls [with *A Theory of Justice*] set the standard for solving it" (Borgmann 1993, 158). The quest for a clear and more ordinary language, one accessible not only to specialists but to any literate reader, springs from Borgmann's aspiration to kindle a public conversation about technology's threat to the significant things in our lives. In *Crossing the Postmodern Divide* (which we will call *"CPD"* in this chapter), as Borgmann assumes the role increasingly of a public intellectual, he strives for an even wider audience by writing in a more straightforward, less technical style for a trade book audience.

Borgmann, especially with his work on focal things and practices, is every inch an American philosopher in a tradition he characterizes as concerned with "the reflective care of the good life." This tradition bears philosophical roots in the works of Emerson, Thoreau, and Melville and is carried forward in the work of Borgmann's former colleague at the University of Montana, Henry Bugbee, author of *The Inward Morning,* a Thoreau-inspired title. It is also a tradition of practitioners and writers (at times inspired by Native American traditions) who title their books, stories, poems, and essays after ponds, whales, bears, creeks, rivers, ridges, capes, islands, refuges, turtles, willows, delicate arches, farms, horses, landscapes, city squares, parks, streets and neighborhoods, villages and towns, and even suppers and motorcycles—what Borgmann calls focal things. Alternatively, these writers speak of activities, such as walking, running, woodworking, sailing, throwing pots, weaving, cooking, or "love medicine" and ceremonies—these are Borgmann's focal practices. It is out of this tradition, for example, that Aldo Leopold can begin his *Sand County Almanac* with an appeal to things and his connection to them without feeling the need to explain to his readers what he means by them. "There are some who can live without wild things, and some who cannot. These essays are the delights and dilemmas of one who cannot" (Leopold 1968, vii). It is really this tradition that Borgmann has increasingly aligned himself with since coming to America in the 1960s.

The Device Paradigm

To meet the task of philosophy in *TCCL,* Borgmann developed his philosophy of technology as the theory of the device paradigm. This theory remains basic although mostly implicit, for *CPD* and his recent *Holding On to Reality* as well. Beginning with the observation that contemporary life in technologically advanced countries exhibits a repeated pattern, Borgmann tries to provide a language of reflection, a theory, within which we can comprehend this exhibited pattern and its consequences. Ultimately, since the theory of the device paradigm yields a cautionary tale of disappointing and debilitating consequences of technology, it claims to show how to challenge and reform technology in a way that goes to the root of the problem of its affect on our well-being. Here we will present an overview of the device paradigm.

The device paradigm helps us to understand why people expect so much from technology. Certainly part of the story involves the benefits wrought by modern science and technology. After all, it was not people's prayers and rituals that brought the threat of smallpox to an end; it was modern medicine. People believe that technology has removed and can remove much, if not all, of the misery and toil that have plagued the human condition. Technology can reduce or eliminate darkness, cold, heat, hunger, confinement, and so on by bringing these harsh conditions of nature under control. Freedom from these conditions thus entails the conquest of nature.

Borgmann advances beyond this common understanding of "the promise of technology" by connecting this liberating aim of the promise with another, less well articulated, aim: enriching life. He argues we have come to embrace a vision of the good life that is inextricably bound to the technologies that shape our everyday lives. Technology in the common understanding promises not only to disburden us of our everyday hardships but also and more importantly to make us happy. It is with this twofold aim, liberation and prosperity, that the domination of nature, culture, time, and place was first undertaken and, often in a more subtle fashion, is continuing to be carried out (Borgmann 1984, 35–48).

In important respects Borgmann reflectively agrees with this promising aspect of technology. On his view technology can be used appropriately in a liberating role, in the service of centering things. In that supporting role, it can help to provide the time, space, and security necessary to pursue focal practices such as sewing, running, hiking, cooking, musicianship.

The flaw in the promise of technology shows up mostly, though not exclusively, in its enrichment role. The conventional view is that technology frees us for other, more enlivening pursuits; Borgmann argues, however, that we typically are not freed up for other centering things but only more passive consumption. Technology frees us up for more technology. Our received understanding of the good life, in popular terms measured by our standard of living, is nothing but a vision of this technologically "enriched" life. But why is it such a mistake to believe that technology can fulfill our lives? Borgmann finds that the fundamental problem lies in the details. Technology promises to liberate and enrich us through devices. A close and careful examination of devices, however, shows that when devices fill our lives, we are reduced to disengaged consumers of the commodities these devices provide. So the basic question is really: What is distinctive about the device that can be deceptively alluring?

The device paradigm shows that from roughly the Industrial Revolution on there has been a transformation of our material world from one pervaded by "things" to one dominated by devices. The best-known of Borgmann's illustrations is the shift from hearths and wood-burning stoves (things) to the central heating system (a device). A thing, in Borgmann's sense,

> is inseparable from its context, namely its world, and from our commerce with the thing and its world, namely, engagement. The experience of a thing is always and also a bodily and social engagement with the thing's world. . . . Thus a stove used to furnish more than mere warmth. It was a *focus,* a hearth, a place that gathered the work and leisure of a family and gave the house a center. Its coldness marked the morning, and the spreading of its warmth marked the beginning of the day. It assigned to various family members tasks that defined their place in the household. . . . It provided the entire family a regular and bodily engagement with the rhythm of the seasons that was woven together with the threat of cold and the solace of warmth, the smell of wood smoke, the exertion of sawing and carrying, the teaching of skills, and the fidelity to daily tasks. . . . Physical engagement is not simply physical contact but the experience of the world through the manifold sensibility of the body. That sensibility is sharpened and strengthened in skill. Skill is intensive and refined world engagement. (Borgmann 1984, 42)

The engagement with the world of things, as found in the case of the wheelwright, the blacksmith, or the musician molds the character of a person and helps to provide one with a fuller sense of self. Thus, for Borgmann, the earlier world of things defined a strong framework in which a person's particular practices defined one's character. The variety of practices created diverse communal relationships.

What has replaced the thing is the device. The device provides what Borgmann calls a *commodity,* one aspect of the original thing (for example, in the case of the wood-burning stove, warmth alone), and disburdens people of all the elements making up the world, or context, and engaging character of the thing. This world of the thing, that is, its ties to nature, culture, the household setting, a network of social relations, mental and bodily engagement, is taken over by the *machinery* (the central heating plant itself) of the device. All of these multifarious relationships are eliminated in the process.

> The machinery makes no demands on our skill, strength, or attention, and it is less demanding the less it makes its presence felt. In the progress of technology, the machinery of the device has therefore a tendency to become concealed or to shrink. Of all the physical properties of a device, those alone are crucial and prominent which constitute the commodity that the device procures. (Borgmann 1984, 42)

Devices, therefore, hide the complicated mechanisms by which the commodity is produced, and consequently result in a sharp division between the foreground (commodity) and background (machinery). Owing to this submerged character and to its variability (from coal to gas to oil to electricity in the central heating unit), the machinery becomes necessarily unfamiliar. It is very important, however, not to think of the device as only machinery. The device makes available a particular commodity—warmth. This commodity is that for which the device is intended. Just the opposite of the machinery, the commodity fills the foreground (warmth becomes ubiquitous in the house), remains relatively fixed as the means change (from coal to electricity) and becomes increasingly familiar. It follows that there exists a wide division between what a device provides in the foreground, the commodity, and how it provides its commodity in the background, the machinery. Hence, and this is Borgmann's central insight, devices *split* means and ends into *mere means* and *mere ends*.

Things, in contrast, richly interweave means and ends, so that practices are experienced as good in their own right and useful too. In the Zen tradition, for instance, one can become enlightened through giving oneself to simple tasks, like chopping wood. Of course, we can be proud of the skills we learn, or enjoy the character we develop as we acquire the sets of skill and virtues necessary to engage in furniture making, cooking, or working with a string of packhorses. Or one can enjoy the social good of all working together in the household to get things done. However, the world of the thing and the engagement it calls for are felt at times as a burden or hassle. The device frees people of all these problems.

Concerning this disburdenment, Borgmann sees in the postmodern era a trajectory to the device paradigm (in addition to the change from a thing to a device) with the further refinement of devices. The more refined a device the more it lifts these burdens from us. For that reason, refined devices disengage us even further, eliciting private passive consumption. "To consume is to use up an isolated entity without preparation, resonance, or consequence" (Borgmann 1984, 53). As consumers, we become disengaged from things and each other—our social life becomes mediated through a commodity culture.

This thing-to-device illustration in the case of the wood-burning stove is representative of the larger cultural pattern. Extensively yet unobtrusively, this technological approach to the world, Borgmann argues, pervades and informs what people think, say, and do. Organizations, institutions, the ways nature and culture are arranged and made accessible all become modeled on the device. "Technology is the rule today in constituting the inconspicuous pattern by which we normally orient ourselves" (Borgmann 1984, 105). It becomes extremely important then to consider not just the appropriateness of this or that device in a specific context, but to consider people's typical use of them and the overall consequences of the use of devices by most people in the developed world. This then is the device paradigm: Borgmann's term for the transformation from things to devices and the technological universe created by that transformation.

Now it is easy to see how we can begin to move from this description of the device to a diagnosis of the problem:[4] By destroying most or all the relationships we once had in the world of things, devices completely change our lives. Borgmann does not argue that we should all return to wood-burning stoves or the like, rather he challenges the limitless and unreflective employment of devices. If we are spellbound by the promise of technological enrichment—of a world that happily demands less and less of us in terms of skill, effort, patience, or any kind of risk—the logic of the device results in a disburdened and disengaged way of life. Television, for example, claims over fifty percent of our free time. In the force of its attraction, it exemplifies the perfect fulfillment of the promise of technology: a quick, easy, safe, ubiquitously available window on the world. So seen it is exactly what people have hoped for from technological enrichment and exactly the kind of enrichment—amusement—that devices can capture. It is ironic, then, that people do not take much pride in television—as the label couch potato indicates—and are often left dissatisfied spending so much time in front it. In general, Borgmann argues, the promise of technology

4. Borgmann argues that description, diagnosis, and prescription cannot be separated from each other ultimately. See Borgmann 1984, 68–78; Strong, chapter 17 in this volume, 317, 320–29.

pursued in this unreflective and limitless way turns ironic in the same way: technological enrichment, the life of consumption, leads to disburdenment, disengagement, diversion, distraction, and loneliness. Similarly, virtual reality turns out to be disposable, discontinuous, and merely glamorous and ephemeral in appeal. The computer will yield similar ironic results *if* pursued in this unreflective fashion.

> It has already begun to transform the social fabric, our commerce with reality, and the sense we have of our place in the world. At length it will lead to a disconnected, disembodied, and disoriented sort of life. The human substance will be diminished through a simultaneous diffusion and individuation of the person. Hyperintelligence allows us to diffuse our attention and action over ever more voluminous spaces. At the same time, we are shrinking to a source of instructions and finally to a point of arbitrary desires. (Borgmann 1992, 108)

These and other such unexpected consequences of device procurement are what Borgmann calls the "irony of technology." The good life that devices obtain disappoints our deeper aspirations. The promise of technology, pursued limitlessly, is simultaneously alluring and disengaging.

If Borgmann's task is to help us develop a language of reflection within which we can come to grips with technology, only half of that task is completed by understanding the promise and irony of technology and the roles devices and consumption play in this ironic turn. Borgmann himself notes that if his cautionary tale about technology were the limit of his prescription it would be tantamount to advising us to turn off the television without providing a genuine alternative to it. If we send the gopher down one hole, chances are it will show up at another. To get to the bottom of the matter—to actually put into practice a response to the device paradigm—we must become respectful, or "mindful," of things. That is, we must refocus our lives by turning to *focal* things and practices. The prescription follows from the diagnosis.

Things of the past were both focal (e.g., the wagon itself) and more peripheral (e.g., the chisel that the wheelwright used to build the wagon). Borgmann looks to the more central kinds of things for a foothold to get the reform of technology underway, but with two qualifications. Unlike the things that center an entire culture in the past, such as temples and cathedrals, for us these focal things are central for individuals, families, and communities but not entire cultures. Second, although many focal things today are remnants of yesterday's world of things, such as classical musical instruments, these focal things have changed in form, and, importantly, many other focal things are of more recent origin. A favorite walk, canyon, stream, slope, one's garden, or a part of the musical or theater scene of a town or city can be (or become) focal things for individuals and communities.

A thing is focal if it is what we give our time to and what we build our lives around. Like the fireplace, focal things richly interweave means and ends, point to the larger context of their setting in nature, the community, and culture, call for attention, effort, skill, and fidelity to regular practice, and invigorate individual and community life. Genuinely focal things stand over us as a commanding presence. Under the rule of the device paradigm, commodities provided by devices and consumption are what most of us spend our time on and build our lives around. Unlike focal things, our interest in any particular commodity is short-lived; the thrills of consumption are necessarily disconnected from each other, and the result is fragmentation. Commodities also are discontinuous with their larger natural, communal, and cultural settings (often blinding us to social injustice and ecological damage). They are disposable, demanding little of us. We merely turn them on and off. In sum, whereas focal things unify and gather, devices divide and scatter.

Focal things guide reform but they also require commitment. To reorient ourselves we must engage regularly in "focal practices." The "culture of the table" can be a focal practice.

> In the preparation of a meal we have enjoyed the simple tasks of washing leaves and cutting bread; we have felt the force and generosity of being served a good wine and homemade bread. Such experiences have been particularly vivid when we come upon them after much sitting and watching indoors, after a surfeit of readily available snacks and drinks. To encounter a few simple things was liberating and invigorating. (Borgmann 1984, 200)

Engagement with focal things and practices alerts us to the forces opposing them and the flawed use of devices—the irony of technology. We destroy the engagement we enjoyed with them when we try to enrich our lives through consumption. Thus, the culture of the table can be and for many of us mostly has been displaced and destroyed by fast food. We see this problem when wilderness and the natural world are destroyed as their resources are needed for increasing levels of consumption, or as nature becomes packaged in a subdivision. Walking and biking attractive city streets and parks has been displaced by automobiles, freeways, shopping at indoor malls, and private forms of entertainment. Participation in sports is displaced by spectator sports. Childhood hours spent among things outdoors are now spent in front of the television, video games, and the Internet. Temporally, spatially, socially, and bodily, centering things have been crowded out of contemporary life. We can counter these forces, Borgmann insists, but only by guarding a focal thing with a regular practice. We can "clear a central space amid the clutter and distraction" even through small steps like committing ourselves to meals with beginnings, middles, and ends, breaking through the "superficiality of convenience food" (Borgmann 1984, 204). Engaging in such focal practices, therefore, requires "resoluteness," "either an explicit resolution where one vows regularly to engage in a focal activity from this day on or in a more implicit resolve that is nurtured by a focal thing in favorable circumstances and matures into a settled custom" (Borgmann 1984, 210).

Finally, Borgmann sees that we need "deictic discourse," languages of reflection (which often turn out to be from literature) that remind us of the greater importance of these centering things and practices and help to provide the resolve to engage in them. Only then can we begin to make wise basic choices that roll back the universalization of devices. "In a finite world, devotion to one thing will curb indulgence in another" (Borgmann 1992, 116). Since consumption and devices have displaced things, the key to reform is now to displace them, not completely, but in a way that knits an unprecedented relationship between things and devices. Borgmann, as we will see, looks forward to their harmonious coexistence.

The basic issue then is something like this: Through recollection, actual practice, and the disclosures of literature and public conversation, we become conscious of the importance of focal things; from understanding the device paradigm, we become conscious of technology's threat to these same things. Having located this pivot and having been motivated to rescue things, the next step is to work out what it will take to make room for and encourage engagement with these things within a setting of technology. What will it take to overcome the conflict between focal things and technology? How, for instance, can we expect ourselves, our families, or others to live anything but the life of consumption if factors in towns, cities, and the economy virtually force such a life upon us? Overcoming this tension serves as the directive from which the theory becomes wide-ranging in prescriptions for reform. Borgmann's general answer to these questions draws in two ways. First, focal things and practices need to be expanded and complemented by public things and communal practices. Second, we need to recognize how the economy presently is in service of consumption and needs to be reoriented.

Just as individuals are faced with choices between clearing a space for focal things, so too are communities faced with letting their downtowns and other traditional centers die while developing a shopping district closer to freeway exits. Borgmann believes that communities need to become conscious of the fundamental material communal choices they face, so that rather than feeling forced to consume, members of a community feel encouraged to live lives consistent with focal practices. Fundamental material choices of this nature reinforce passive, private consumption when public places and public goods in general become increasingly instrumental in character, replacing places of celebration and encountering others in their bodily presence. This happens when downtowns are abandoned or even when trees are cut down for developments, leaving a town less attractive for walking and less livable. Reform then has to do with protecting, maintaining, and enhancing communal centers, public focal things, that many times already exist in a community in some fashion. We learn better from what we are already doing right. So communities near wilderness need to protect wilderness, while a community near the prairies may want not only to protect but restore some of the prairie. One community may put its efforts toward preserving and enhancing a different musical tradition than another community. From this standpoint, Borgmann prescribes, among a host of other things, protection of open spaces, concert halls, more preservation of historical treasures, more farmer's markets, enhancing street life, allowing nature more of a say in our built environments, more walking trails between towns and their outskirts, and more paths for running and biking. Some of these public things will favor engagement in the daily life of the community while others will favor its festive life. Given the interest of a particular town, creation of a ballpark can be centering for a community. "A thoughtful and graceful ballpark tunes people to the same harmonies" (Borgmann 1992, 135). From such shared experience, public conversation can continue to grow and address further reforms.

Both *TCCL* and *CPD* call for deep reforms of the economy. In *TCCL*, Borgmann thinks of the present economy as the machinery component of the device, like the central heating plant itself. What this device—the economy—provides is the goods of consumption: the latest, the widest assortment and the most commodities. Since this commodious good life is taken for granted, it never really comes up for examination; rather most of our discussion, whether in politics or the news media, focuses on the economy alone and how well it is performing in providing these commodities. In *CPD,* Borgmann finds this same kind of focus and hidden assumption.

> To rein in consumption has been a standard ingredient of the med-
> ication prescribed by mainstream economists. But the implication
> invariably has been that the curtailment of consumption now is

justified solely for the sake of greater consumption in the future. The present criticism of commodity consumption thus turns out to be an unreserved affirmation of commodity consumption as such and hence of commodious individualism. (Borgmann 1992, 80)

Generally, Borgmann sees that the economy is now in the service of affluence, his name for the goal of producing more, more varied, and more refined commodities. His reform tactic prescribes dislodging the economy from this service to consumption and reorienting it toward serving things and the kind of life of engagement they sponsor. The economy ought to serve "wealth" at the expense of affluence. In contrast to affluence and to poverty (where one lacks the wherewithal in terms of time, means, health care, to pursue focal practices in a secure manner), wealth is the setting within which focal things, both private and public, can flourish. It consists of the leisure, space, books, instruction, equipment, physical health, and economic security necessary to become equal to a "thing that has beckoned to us from afar" (Borgmann 1984, 223).

How wealth can be advanced politically and economically is too complex to go into in detail. Suffice it to say that Borgmann believes there is a kind "half-knowing and half-hearted going along" with present technological development. This conflicted attitude in people provides an opening for change. As we grow conscious of what our basic choices are through improved public conversations and through actual contact with focal and public things—when for instance events like communal celebrations are secured by community—we will begin to see better what other steps need to be taken. Through these kinds of steps, we may eventually come to a collective affirmation to enhance the quality of lives rather than increasing our standard of living. Accordingly, if such affirmation occurs, in *TCCL* Borgmann prescribes a two-sector economic system: one sector that is local and labor intensive; another sector, subordinate to the first, that is more automated and centralized for mass producing devices and other products and services necessary not for affluence but for wealth. With the second sector, Borgmann has in mind infrastructure, certain kinds of goods and services, and research and development. With the first sector, he has in mind industries having to do with food, furniture, clothing, health care, education, and instruction in music, the arts, and sports. He argues that the local sector should be favored through tax and credit measures to the point where its goods and services can prevail in the market.

In other words, we need a new maturity, a maturity where we are no longer spellbound by technology. Freed from the threat of their universalization, devices in this setting will no longer be in service to consumption but will instead serve centering things and the life of engagement they sponsor.

Borgmann returns to this notion of maturity in *CPD*. There he connects his critique to postmodernism, casting modernity as undergirded by the project of dominating nature through technology. What is distinctive about postmodernism is its acceptance of consumerism. So while many postmodern critiques of the modern project are correct and effective at certain levels, none goes to the root of the matter, critiquing technology as a way of life in pursuit of technological prosperity. Borgmann formulates an alternative postmodernism designed, again, to outgrow technology as a way of life and to center individual, family, and communal life around eloquent and "focal reality," a new name for focal things. He argues that not just as individuals but more so as a culture the choice of whether or not to embrace focal reality is the most important one we face as we cross the "postmodern divide."

Apart from this core theory of technology, Borgmann plants a rich garden of concepts in order to advance the general task of philosophy. He argues, for instance, for a three-way classification of knowledge. Scientific knowledge—*apodeictic* knowledge—gains power through precision and reduction, and has largely devalued and displaced testimonial forms of knowledge over the last two centuries. Art, music, writing, and other forms of *deictic* or testimonial knowledge move people to act and to understand what it is that gives context and focus to one's experiences. Deictic knowledge articulates a thing or event in its uniqueness. Apodeictic and deictic knowledge, alone or in combination, are insufficient, however, to comprehend the technological world; scientific knowledge by itself fails to select significant strands of social reality, and deictic knowledge grants the significance of particular things as individuals but is less adequate for reflecting upon repeated ontological and normative patterns found in social reality. Borgmann proposes a third explanatory possibility: *paradeictic* knowledge. It is more concrete and specific than apodeictic knowledge, and more abstract than deictic knowledge: "A pattern, then, is an array of crucial features, abstract and simple enough to serve as a handy device, concrete and detailed enough to pick out a certain kind of object effectively" (Borgmann 1984, 73). The device paradigm as the primary pattern of contemporary life is a paradeictic explanation. The theory of the device paradigm is a paradeictic explanation that reveals the technological character of the underlying repeated pattern of contemporary life.

In sum, Borgmann has accomplished the kind of philosophy of technology that we believe is worth doing. While it is ultimately concerned with the task of philosophy, with doing public philosophy about things that matter, its public character is continuous with disciplinary rigor and depth. Without the substance of the former, it would not be worth the effort and care to study and critique it. Without the disciplinary rigor, philosophers may facilely dismiss it before seeing how new, comprehensive, and profound is this philosophy of technology.

REFERENCES

Borgmann, Albert. 1984. *Technology and the Character of Contemporary Life: A Philosophical Inquiry.* Chicago: University of Chicago Press.

———. 1992. *Crossing the Postmodern Divide.* Chicago: University of Chicago Press.

———. 1993. "Finding Philosophy." In *Falling in Love with Wisdom.* Ed. David D. Karnos and Robert G. Shoemaker. New York: Oxford University Press.

———. 1999. *Holding On to Reality: The Nature of Information at the Turn of the Millennium.* Chicago: University of Chicago Press.

Dreyfus, Hubert L., and Charles Spinosa. 1997. "Highway Bridges and Feasts: Heidegger and Borgmann on How to Affirm Technology." *Man and World* 30:159–77.

Leopold, Aldo. [1949] 1968. *A Sand County Almanac and Sketches Here and There.* New York: Oxford University Press.

Mitcham, Carl. 1994. *Thinking through Technology: The Path between Engineering and Philosophy.* Chicago: University of Chicago Press.

Noble, David. 1997. *Progress without People: In Defense of Luddism.* Toronto: Between the Lines.

Philosophy of Technology:
Retrospective and Prospective Views

Paul T. Durbin

Philosophers have become interested in technology and technological problems only recently—though Karl Marx in the nineteenth century as well as Plato and Aristotle in the classical period had paid some attention to either technical work or its social implications. Within recent decades, among North American philosophers paying significant attention to technology, Albert Borgmann holds special place because of the originality of his call to citizens of technological society, urging them to rethink the way they live. What I want to argue in these brief historical remarks is that Borgmann's work might appear to be at least partially misguided—at least it might appear so to philosophers like myself who are primarily concerned with technosocial problems—unless it is interpreted in a special way.

A RETROSPECTIVE

The perspective I bring to these brief historical remarks reflects my practical (or "praxical" would be better) bent. In that, I differ with others who have recently summarized the history of philosophy of technology in the United States (Mitcham 1994; Ihde 1993). For me, the primary concerns about technology that gave rise to philosophy of technology were practical—even political. Philosophers and social commentators were worried about negative impacts of nuclear weapons systems, chemical production systems, the mass media and other (dis)information systems (among others) on contemporary life in the Western world—including negative impacts on the environment and on democratic institutions. And typically they wanted to do something—preferably politically—about the situation.

Among the first broadly philosophical works to say to those early philosophers of technology (myself included) that this might be a difficult struggle was the 1964 English translation of Jacques Ellul's *The Technological Society*. There Ellul spelled out what he called the "essentials" of a "sociological study of the problem of technology." (The word he actually uses is *Technique*—a hypostatized term for the sum of all techniques, all means to unquestioned ends.)

According to Ellul, Technique is the "new milieu" of contemporary society, replacing the old milieu, nature; all social phenomena today are situated within it rather than the other way around; all the beliefs and myths of contemporary society have been altered to the core by Technique; individual techniques are ambivalent, intended to have good consequences but contributing at the same time to the ensemble of Technique; so that, for instance, psychological or administrative techniques are part of the larger Technique, and no particular utilization of them can compensate for the bad effects of the whole.

All of this leads to Ellul's overall characterization: there can be no brake on the forward movement of the artificial milieu, on Technique as a whole; values cannot change it, nor can the state; means supplant ends; Technique develops *autonomously*.

This was the Ellul most of us knew in the 1960s, when we first started reflecting philosophically on technology. More knowledgeable students of Ellul, however, saw this as merely Ellul's warning—a warning about what Technique (technology?) demands *if* we do not heed his warning and act decisively. But how can we act, given Ellul's pessimistic conclusions? What these Ellulians say we missed was the *dialectical* nature of Ellul's thinking. Every *sociological* warning was matched by a *theological* promise; more particularly, *The Technological Society* was intended (they say) to be read in tandem with *The Ethics of Freedom* (1976). According to one of these scholars:

> Ellul's intention is to attempt to make . . . [the absolute] freedom [of Christian revelation] present to the technological world in which we live. In so doing, he hopes to introduce a breach in the technical system. It is Ellul's view that in this way alone are we able to live out our freedom in the deterministic technological world that we have created for ourselves. (Wenneman 1990, 188)

This reading of Ellul seems to have been, at that time, limited almost exclusively to a group of Ellul's fellow conservative Christians (see Ellul 1972)—a group already influenced by some of Ellul's sources in Kierkegaard and so-called existential theology (Garrigou-Lagrange 1982).

Some of these same religious critics of technology were influenced, at the same time, by translations of works of Martin Heidegger into English. But in the 1960s this did not, to any great extent, reflect Heidegger's concerns about technological society.

At the opposite end of the political spectrum, we were influenced, in the late 1960s and early 1970s, by the writings of Herbert Marcuse, especially *One-Dimensional Man* (1964)—the widely acclaimed "guru of the New Left." Where Marcuse's neo-Marxism seemed to differ from the dire warnings of Ellul's pessimism about technology was in its offering of a possible solution to technosocial problems.

Marcuse and other neo-Marxists were, in some ways, as pessimistic as Ellul. No amount of liberal democratic politics, they said, could get at the roots of technosocial problems. But there was a way out: to challenge the technoeconomic system as a whole. (Marcuse was explicit that this meant challenging, not only the capitalist technoeconomic system of the West, but also its imitator, the "bureaucratic socialist" technoeconomic system of the Soviet Union and its satellites.) Only a wholesale revolutionary challenge to the political power of technocapitalists and quasi-capitalistic bureaucratic socialists could do the trick; it was (he thought) *possible* to deal with technosocial problems, but all at once and not one at a time. The means was revolutionary consciousness-raising—and, at least for a time, Marcuse (1972) saw the vehicle for it in the worldwide student uprisings in the late 1960s. (After the New Left faded, Marcuse found hope in the radical feminist movement—but in the end he seems to have lost all hope, matching Ellul's pessimism of the right with a deep pessimism of the left; see Marcuse 1978.)

Between these extremes—in our philosophical consciousness at the time—loomed a liberal-centrist hope. Daniel Bell, a sociologist (others would say a social commentator) rather than a philosopher, had already announced the end of ideology (1962)—presumably it was the end of ideologies of the right as well as the left. Now he came forward to announce the coming of postindustrial society (1973)—a society in which experts, including technical experts, offered the hope of solving technosocial problems.

Bell was not, however, an unalloyed optimist. As much as he believed that nonideological technocratic expertise could solve at least our major problems, just as much did he also worry about the "rampant individualism" of our culture. One of his best-known books (Bell 1976)—which also influenced those of us trying to fashion a philosophical response to technosocial problems at that time—was an exhaustive documentation of the anarchy of cultural modernism in the twentieth century. Bell did not, like Ellul, counsel a return to traditional religion as an anchor for a world adrift, but he did maintain that technological managerialism could not save us if there were no cultural standards—if thinkers in the late twentieth century could not solve our "spiritual crisis."

So the first philosophers of technology in the United States, in the late

1960s and early 1970s, had a variety of approaches to turn to in the search for solutions to such technosocial problems as nuclear war and environmental destruction—technophilosophies of the right, left, and center.

In the next decade—from the late 1970s until the mid-1980s—the picture became more complex, but a political spectrum remained a useful lens through which to view the fledgling philosophy of technology scene.

Langdon Winner's influential Ellul-inspired book, *Autonomous Technology* (1977), might suggest the contrary. Early in the book Winner says: "Ideological presuppositions in radical, conservative, and liberal thought have tended to prevent discussion of . . . technics and politics." Accordingly, about liberals, Winner says:

> [The] new breed of [liberal] public-interest scientists, engineers, lawyers, and white-collar activists [represent] a therapy that treats only the symptoms [and] leaves the roots of the problem untouched. (Winner 1977, 107)

On what later came to be called neoconservatism, he has this to say:

> The solution [Don K.] Price offers the new polity is essentially a balancing mechanism, which contains those enfranchised at a high level of knowledgeability and forces them to cooperate with each other . . . [as] a virtuous elite . . . in the new chambers of power. (Winner 1977, 171)

And about Marxist radicals of the time (before the fall of the Soviet Union):

> The Marxist faith in the beneficence of unlimited technological development is betrayed. . . . To the horror of its partisans, it is forced slavishly to obey [technocapitalist] imperatives left by a system supposedly killed and buried. (Winner 1977, 276–77)

And Winner concludes: "It can be said that those who best serve the progress of [an unexamined] technological politics are those who espouse more traditional political ideologies but are no longer able to make them work."

But this is not the whole of Winner's story. He makes these points, in fact, in a book devoted to a different sort of technological politics—an "epistemological Luddism" that would set out, explicitly, to examine the goals of large technological enterprises in advance and hold them to lofty democratic standards. In subsequent books (1986, 1992), Winner has been even more explicit about this, and—though he is still generally viewed as a technological radical—he has come, more and more, to espouse participatory-democracy movements as the solution to particular technosocial problems.

More devoted Ellulians of this period were not explicitly political, but their religious philosophies were most compatible with a theological conservatism (see Hanks 1984; Lovekin 1991; and Vanderburg 1981).

At the opposite end of the political spectrum from these conservative Christians, other neo-Marxists carried on Marcuse's critique of technology even after the decline of the New Left. Philosopher Bernard Gendron's *Technology and the Human Condition* and historian David Noble's *America by Design: Science, Technology, and the Rise of Corporate Capitalism* both appeared in 1977. Both echoed aspects of Marcuse's critique even when they did not explicitly cite him. It would be over a decade before an explicitly neo-Marcusean philosophy of technology would appear, Andrew Feenberg's *Critical Theory of Technology* (1991). It makes explicit the arguments that continued to predominate in neo-Marxist critiques of technology in the late 1970s and 1980s—right up to the demise of Soviet Communism (see Gould 1988; Feenberg's book actually appeared after the official disavowal of Communism in Russia).

It was at this stage that Heideggerianism entered the philosophy of technology debate in the United States (see Heidegger 1977). I will not deal with that influence here except in terms of three avowed neo-Heideggerians.

Hans Jonas was, at the time, the best known of the three. His magnum opus, *The Imperative of Responsibility: In Search of an Ethics for the Techno-logical Age* was not translated from the German in which he composed it (though he had been a professor at the New School for over twenty years) until 1984. But he had already published an influential essay, "Toward a Philosophy of Technology," in the *Hastings Center Report* in 1979. And he was already well known in the 1970s for his "heuristics of fear" in the face of such technological developments as bioengineering: "Moral philosophy," he said, "must consult our fears prior to our wishes to learn what we really cherish" in an age of unbridled technological possibilities.

Don Ihde (beginning with *Technics and Praxis* [1979] and *Existential Technics* [1983]), with his downplaying of some Heideggerian influences in favor of a Husserlian phenomenology, may seem to be an exception to my political reading of this decade in philosophy of technology. But in later works—especially *Technology and the Lifeworld* (1990)—Ihde has espoused an environmental activism that could only be implemented politically.

At this point, while mentioning Ihde's later environmentalism, I want to digress for a moment. During the second decade of the development of philosophy of technology in the United States, there developed a parallel tradition of reflection on technology. What I have in mind is environmental ethics, since a significant portion of the literature in that field touches on negative impacts on the environment of particular technological developments: the nuclear industry and electric power companies; the chemical industry; agriculture using pesticides, herbicides, and chemical fertilizers; the automobile; and so on. Without going into these issues—and making no claims about natural affinities between philosophy of technology and environmental ethics—it seems fair here to point out how strong the political dimension is in environmental ethics. And I am not just thinking of radical environmentalism, ecofeminism, or similar approaches; almost *all* of environmental ethics, it seems to me, is and ought to be political.

Finally, we come to Albert Borgmann—one focus of this volume— and his 1984 neo-Heideggerian book, *Technology and the Character of Contemporary Life*. I have argued elsewhere (and will not repeat those arguments here; see Durbin 1988, 1992) that Borgmann's proposals for the reform of our technological culture—his appeal to "focal things and practices"—is an implicit appeal to expand focal communities. That is, it presupposes at least educational activism and probably political activism. Furthermore, the communitarian followers of Robert Bellah, who have found in Borgmann's writings an eloquent statement of goals they are striving for in our bureaucratized and technologized culture (see Bellah's comment on dust jacket of Borgmann 1992) are clearly committed to a social movement. Many view that movement as neoconservative, a charge that has also been leveled at Borgmann; but accepting that assessment is not a necessary concomitant of seeing Borgmann's work as having political implications.

In this retrospective, I have concentrated on two decades—roughly the midsixties to the mideighties—and I have made a deliberate choice to emphasize contributions to philosophy of technology that reflect a commitment to the solving of technosocial problems, typically by political means of one sort or another. There were, of course, other contributions to the development of the philosophy of technology in those years; I have myself, in fact, chronicled those other developments elsewhere (Durbin 1994) under two headings that do not emphasize the politics of technology, "The Nature of Technology in General" and "Philosophical Studies of Particular Technological Developments." However, even in many of the books I mention in that survey—books that do not seem to have a political slant—it is easy to perceive the political orientations of their authors. In any case, it is the political thrust of philosophy of technology that renders urgent the critical point I want to make in the second half of this paper.

A Prospective View: The Future of Philosophy of Technology

In *Social Responsibility in Science, Technology, and Medicine* (1992), I discuss several ways in which philosophers might follow the lead of a number of activist technical professionals who have, in recent decades, been working to achieve beneficial social change. Some of the ways I list are academic: clarifying issues or helping to move academic institutions in positive directions. Some of the ways involve working outside academia—for example, on ethics or environmental or technology assessment committees. But in addition, I join the lament of those decrying the loss of "public intellectuals" or "secular preachers"—a modern counterpart to the scholar-preachers who provided moral leadership to earlier generations of American society on issues such as slavery, child labor, or injustices against workers. The example I mention in my book—of a recent philosopher–secular preacher—is Albert Borgmann. Especially in *Crossing the Postmodern Divide,* he is explicit about playing the role of a public intellectual.

The need for vision is so great in our culture of fragmented specialized knowledge that it is time to welcome philosopher-preachers back into the mainstream. Their numbers have been exceedingly small since the death of John Dewey, but we might hope for a resurgence now.

Bringing about such a happy eventuality, however, will not be easy. And that is the thesis of this half of my paper. *Public intellectuals, visionaries, secular preachers, academic activists of any sort are going to have a very difficult time in our technological culture.*

The philosophers and social commentators I listed in my retrospective did sometimes make a public impression. Ellul was widely hailed as the first thinker to awaken American intellectuals to the dangers of technology; Marcuse's critique of technology was widely influential among student radicals and others in the New Left; and Bell served as the favorite target of abuse for those same radicals. In the next decade, Winner and Ihde were (and are) ubiquitous speakers and panelists, and both also have influenced graduate students. Ellulianism has spread slowly and continues to be influential in much the same circles as in the late 1960s. Jonas left few disciples, but his influence in biomedical circles—in particular in the Hastings Center, itself very influential—was strong.

In this volume, much attention is paid to Albert Borgmann's contributions to philosophy of technology. Whatever may be Borgmann's influence on others, whatever influence he may have that extends into the future, there remain good reasons to question the lasting influence of the other philosophers of technology that I have mentioned.

Some may think it quaint of me even to include Marcuse and Bell. Will that be the same fate, in twenty years, of Winner and Ihde and Jonas? Though an Ellulian school has persisted for twenty-five years, so far it has produced no other thinker of note.

Then there is the issue of *impact*—of solutions for key technosocial problems. No one can say that ideas of Ellul or Winner or Ihde or Jonas—or, for that matter, of neo-Marxists—have not had *some* influence on activists who have had success on particular issues. I would think, in particular, of Winner's influence on Richard Sclove, with his Loka Institute and FASTnet activist electronic mail network. But of all those mentioned, it is probably philosophers in the environmental ethics community who have had the greatest and most direct impact on particular solutions for major technosocial problems.

So if I think back to why most of us early philosophers of technology got involved, in the sixties, seventies, and eighties—and if I am right that what motivated the great majority of us were concerns over major technosocial disasters such as nuclear proliferation and widespread environmental degradation—then I believe I am not being unrealistic in saying that the field has not had the impact that I personally hoped it would. For the most part, it has not even had a great impact in academia.

What I want to talk about now is why this is so.

The key, it seems to me, is to be found in the phrase, "in our technological culture." I have always had problems with Ellul's characterizations of "technological society" in the abstract. But a description with much the same thrust—and which is both more neutral and can be tied down to specific observations in ways I find difficult with Ellul—is available in the sociological work of Peter Berger and colleagues (especially Berger and Luckmann 1966; Berger, Berger, and Kellner 1973).

Berger sometimes (Berger and Luckmann 1966) refers to his work as sociology of knowledge; at other times he describes his basic method as phenomenological (Berger, Berger, and Kellner 1973, acknowledging a special debt to the "phenomenology of everyday life" of Alfred Schutz [1962]). He is also indebted to Karl Marx (though not to doctrinaire Marxists), to George Herbert Mead (1934), and, in a special way, to Max Weber.

What Berger proposes is that we describe our culture in terms of a spectrum of degrees of "modernization," with no particular culture or society prototypically "modern." What (to Berger and colleagues) makes any particular culture "modernized" is two things: its dependence on *technological production* and its administration by means of *bureaucracy*. (Nearly all of Berger's ideas about bureaucracy seem to come from Weber— see Gerth and Mills 1958—and sociologists influenced by Weber.) Thus, the more technologized and bureaucratized a culture is, the more it makes sense to call it "modernized." And this allows comparisons both over time— historically—and cross-culturally, as between more and less modernized societies even at the present time. (Berger and colleagues do not like to refer ιο particular societies as "underdeveloped," but they think it less offensive to refer to some as "less modernized.")

With this characterization as his basis, Berger is able to identify key (he even says "essential") characteristics of workers in technological production facilities (including agriculture), as well as of citizens in a bureaucratized society—which characteristics carry over into a rigidly compartmentalized private life (for example, "modern" individuals play several roles in both work and private life; they have many anonymous social relations; they see themselves as units in very large systems; etc.) as well as into the "secondary carriers" of modern consciousness—the media in the broadest sense and mass education. These secondary carriers both prepare young people for life in such a society and reinforce the "symbolic universe" that gives it meaning—and they do so in ways decidedly different from those in nonmodernized societies. Furthermore, many people in less modernized societies envy the lifestyles of those in more modernized societies, though they often do not realize what a price—in terms of values and lifestyles—living in a modernized society exacts.

As I admitted, there are many similarities between this account and Ellul's indictment of ours as a society controlled by "Technique." (Both Berger and Ellul were influenced by Weber.) The difference, for me, lies in the attitudes of the two. Ellul views technicized society as an unmitigated disaster, inimical to human freedom. Berger simply sets out a framework to understand our society—and he remains open to various forms of resistance to modernization, in both modernized and less modernized societies (though he does not think it realistic to expect societies to return to a romanticized premodern past).

The way I see all of this impinging on the potential for philosophers of technology to have an impact on society is that they (we) must do so within Berger's secondary carriers of modernization: that is, we must exert our influence either through the media or through education. And these are, by definition, oriented toward fostering modernization, not criticizing it.

Almost all the impacts I mentioned, with respect to the philosophers of technology have been made through the media—through book publishing, magazine articles, lectures (mostly on the academic circuit), occasionally in interviews on radio or television. And we all know both the audience limitations of academic media and the ephemeral character of the impacts of the mass media. Today's hot book is tomorrow's remainder. The handful of books by academics that have had or are likely to have any lasting impact are just that, a handful—in technology-related areas, probably no more than the works of Lewis Mumford (1934, 1967, 1970) and Rachel Carson (1962). For most of us, there is little hope that our writings will have that kind of lasting impact—even if we manage to make a momentary impression in intellectual circles.

Similarly for education. Any lasting impact via mass education must come through influencing teachers and textbooks, and everyone knows by now how bureaucratized both textbook publishing and the public schools are. If we think instead of teaching the teachers, of influencing the next generation, then the impact will be by way of training graduate students; and the regimentation of graduate education is hardly conducive to producing reformers, social critics, activists who will change technological society for the better. It can happen. Some of the most critical of our current crop of philosophers of technology have survived the worst evils of contemporary graduate education in philosophy. But it is not easy, and the scholar who expects to exert a lasting impact on society via that route is almost by definition not a person who is thinking about real changes in society.

CONCLUSION

What should we conclude from this retrospective and prospective? Abstractly, it would seem there are four possibilities.

Some people will scoff. I had unrealistic hopes in the first place, they will say. Philosophy's aims should be much more limited—limited, for instance, to analyzing issues, leaving policy changes to others (to the real wielders of power whose efforts might be enlightened by the right kind of philosophical speculations); or limited to critiquing our culture (following Hegel) after its outlines clearly appear and it fades into history, imperfect like all other mere human adventures.

Others will go to the opposite extreme. I set my sights too low, they will say. We must still hold out for a total revolution. The injustices of our age, as well as its ever-increasing depredations of planet Earth, demand this. Still others are likely merely to lament the fate to which technological anticulture has doomed us; we must resign ourselves to the not-dishonorable role of being lonely prophetic voices crying out against our fate.

Then there is my own conclusion, a hope—following John Dewey (1929, 1935, 1948)—that we will actually *do* something about the technosocial evils that motivated us in the first place. That we will abandon any privileged place for philosophy, joining instead with those activists who *are* doing something about today's problems—and succeeding in limited ways in particular areas (see McCann 1986, as well as Durbin 1992).

Albert Borgmann might be read as endorsing any one of these options: limiting philosophy's scope to analyses of technology (however large scale and Hegel-like those analyses might be); or offering radical, even revolutionary alternatives to a device-dominated culture, really hoping that a revolution will come about; or merely lamenting our sad, commodity-driven fate, our culture's wasting of its true democratic heritage.

But I hope he would, with me, endorse the fourth option. We might, no matter how weak our academic base, still manage to succeed in conquering particular technosocial evils one at a time. And environmental ethicists may be showing us the best way—precisely because they do not try to succeed alone, but join with other environmental activists, fighting every inch of the way.

REFERENCES

Bell, Daniel. 1962. *The End of Ideology.* New York: Free Press.

———. 1973. *The Coming of Post-industrial Society: A Venture in Social Forecasting.* New York: Basic.

———. 1976. *The Cultural Contradictions of Capitalism.* New York: Basic.

Berger, Peter, Brigitte Berger, and Hansfried Kellner. 1973. *The Homeless Mind: Modernization and Consciousness.* New York: Random House.

Berger, Peter, and Thomas Luckmann. 1966. *The Social Construction of Reality.* New York: Doubleday.

Borgmann, Albert. 1984. *Technology and the Character of Contemporary Life: A Philosophical Inquiry.* Chicago: University of Chicago Press.

———. 1992. *Crossing the Postmodern Divide.* Chicago: University of Chicago Press.

Carson, Rachel. 1962. *Silent Spring.* Boston: Houghton-Mifflin.

Dewey, John. 1929. *The Quest for Certainty.* New York: Minton Balch.

———. 1935. *Liberalism and Social Action.* New York: Putnam.

———. 1948. *Reconstruction in Philosophy.* 2d ed. Boston: Beacon Press.

Durbin, Paul T. 1988. Review of Albert Borgmann's *Technology and the Character of Contemporary Life. Man and World* 21:231–35.

———. 1992. *Social Responsibility in Science, Technology, and Medicine.* Bethlehem, Pa.: Lehigh University Press.

———. 1994. "Philosophy of Science, Technology, and Medicine." In *The Best in Science, Technology, and Medicine.* Ed. C. Mitcham and W. Williams. Vol. 5 of *The Reader's Adviser,* 14th ed. New York: Bowker.

Ellul, Jacques. 1964. *The Technological Society.* New York: Knopf.

———. 1972. *The Politics of God and the Politics of Man.* Grand Rapids, Mich.: Eerdmans.

———. 1976. *The Ethics of Freedom.* Grand Rapids, Mich.: Eerdmans.

Feenberg, Andrew. 1991. *Critical Theory of Technology.* New York: Oxford University Press.

Garrigou-Lagrange, Madeleine. 1982. *In Season, Out of Season: An Introduction to the Thought of Jacques Ellul.* San Francisco: Harper & Row.

Gendron, Bernard. 1977. *Technology and the Human Condition.* New York: St. Martin's.

Gerth, Hans, and C. Wright Mills, eds. [1945] 1958. *From Max Weber: Essays in Sociology.* New York: Oxford University Press.

Gould, Carol. 1988. *Rethinking Democracy.* New York: Cambridge University Press.

Hanks, Joyce M. 1984. *Jacques Ellul: A Comprehensive Bibliography.* Greenwich, Conn.: JAI Press.

Heidegger, Martin. 1977. *The Question Concerning Technology and Other Essays.* New York: Harper & Row.

Ihde, Don. 1979. *Technics and Praxis*. Dordrecht: Reidel.

———. 1983. *Existential Technics*. Albany: State University of New York Press.

———. 1990. *Technology and the Lifeworld: From Garden to Earth*. Indianapolis: Indiana University Press.

———. 1993. *Philosophy of Technology: An Introduction*. New York: Paragon.

Jonas, Hans. 1979. "Toward a Philosophy of Technology." *Hastings Center Report* 9:34–43.

———. 1984. *The Imperative of Responsibility: In Search of an Ethics for the Technological Age*. Chicago: University of Chicago Press.

Lovekin, David. 1991. *Technique, Discourse, and Consciousness: An Introduction to the Philosophy of Jacques Ellul*. Bethlehem, Pa.: Lehigh University Press.

Marcuse, Herbert. 1964. *One-Dimensional Man*. Boston: Beacon Press.

———. 1972. *Counter-revolution and Revolt*. Boston: Beacon Press.

———. 1978. *The Aesthetic Dimension*. Boston: Beacon Press.

McCann, Michael. 1986. *Taking Reform Seriously: Perspectives on Public Interest Liberalism*. Ithaca: Cornell University Press.

Mead, George Herbert. 1934. *Mind, Self, and Society*. Chicago: University of Chicago Press.

Mitcham, Carl. 1994. *Thinking through Technology*. Chicago: University of Chicago Press.

Mumford, Lewis. 1934. *Technics and Civilization*. New York: Harcourt Brace.

———. 1967, 1970. *The Myth of the Machine*. 2 vols. New York: Harcourt Brace.

Noble, David. 1977. *America by Design: Science, Technology, and the Rise of Corporate Capitalism*. New York: Knopf.

Schutz, Alfred. 1962. *Collected Papers*, vol. 1. The Hague: Nijhoff.

Vanderburg, Willem H. 1981. *Perspectives on Our Age: Jacques Ellul Speaks on His Life and Work*. Toronto: Canadian Broadcasting Corporation.

Wenneman, D. J. 1990. "An Interpretation of Jacques Ellul's Dialectical Method." In *Broad and Narrow Interpretations of Philosophy of Technology*. Ed. P. Durbin. Dordrecht: Kluwer.

Winner, Langdon. 1977. *Autonomous Technology: Technics-out-of-Control as a Theme in Political Thought*. Cambridge: MIT Press.

———. 1986. *The Whale and the Reactor: The Search for Limits in an Age of High Technology*. Chicago: University of Chicago Press.

———, ed. 1992. *Democracy in a Technological Society*. Dordrecht: Kluwer.

PART TWO

Evaluating Focal Things

Borgmann's notion of focal things is central to his philosophy, and it is perhaps his most important contribution to the field. In the following five chapters, professional philosophers, with viewpoints ranging from the concrete to the political and theoretical, carefully examine and evaluate Borgmann's notions of focal things and their counterpart, devices.

What is the relationship between focal practices and focal things? On the one hand, a fair amount of attention by the American pragmatic, analytic, and Continental traditions of philosophy has been devoted to the philosophical importance of practices. Alasdair MacIntyre, notably, makes practice central to his work in ethics. On the other hand, while Heidegger calls attention to "thinging things"—focal or centering things— he neglects the focal practices within which these things are embedded. Borgmann, however, connects practices with things, and this step adds entirely new dimensions to this conversation about the importance of practices. Are focal practices, and the goods internal to them, to use MacIntyre's terminology, somehow anchored in focal things at a more fundamental level? In chapter 3, "Focal Things and Focal Practices," Lawrence Haworth examines this relationship. His meditation on things and practices, with its thoughtfully selected and well-developed examples, appeals to those who might not know Borgmann's work firsthand and deepens and clarifies, with its subtle distinctions, any sophisticated reader's understanding of Borgmann's account of this matter.

Technology's relationship with democracy is at best an uneasy one, especially given Borgmann's claim that our modern devices are not mere value-neutral instruments. Thus, technology may play a greater role in determining the character of the good life than any of the founders of the liberal democratic tradition imagined. In chapter 4, "Technology and Nostalgia," Gordon Brittan is bothered both by Borgmann's critique of liberal democracy and by his larger argument that technology fails to provide the kind of good life Borgmann believes we expect from it. After evaluating these claims, Brittan argues that a good reason for limiting technology can be found instead by an appeal to the core concern of liberal democracy: freedom. While technology provides the basis for liberal democratic freedom, it can also threaten our autonomy by making us too dependent on devices. In the interest of freedom, we should reform technology so that we regain and retain a measure of essential self-sufficiency. Thus, to reform technology we could avoid, Brittan believes, entering into the endlessly controversial questions of the good life invited by Borgmann's critique and reform of technology.

How can we evaluate what Borgmann calls the life of consumption or technology as a way of life? Brittan addresses the question of how successful Borgmann's tests are concerning the failure of technology to procure a good life. Larry Hickman in chapter 5, "Focaltechnics, Pragmatechnics, and the Reform of Technology" addresses two related questions: How can we evaluate appropriate and inappropriate technology? And how can we evaluate focal things and practices, for surely many of them are troubling? Hickman, inspired by John Dewey's work, has developed a pragmatic philosophy of technology—pragmatechnics. In this chapter he contrasts pragmatechnics with focaltechnics, his characterization of Borgmann's vision of appropriate technology. On Hickman's reading of *Technology and the Character of Contemporary Life,* Borgmann presents a rigid essentialism, splitting technology into "two ledger columns" of bad and good. The device paradigm is bad; devices as supportive of focal things are good. "Small is beautiful and big is bad." Hickman argues for a "flexible functionalism" that would counter what he perceives as a tendency by Borgmann to reduce a device to an essential property. His claim is that pragmatechnics are more flexible and better for understanding the complexities of contemporary technological life. Our technologies require ongoing evaluation carried out in context. Hickman's second disagreement with Borgmann concerns the evaluation of focal things. Borgmann maintains that focal concerns are attestable, contestable, and fallible, but they are not subject to experimental testing. Hickman argues that this move is unwise. He illustrates the difference of pragmatechnics on this matter with a discussion of traditional and nontraditional forms of the family.

In chapter 6, "Borgmann's *Unzeitgemässe Betrachtungen:* On the Prepolitical Conditions of a Politics of Place," Andrew Light develops the introduction's theme that Borgmann's reform program advocates a set of issues that any political system must address if it is to be effective in a social sphere dominated by technology. He does so by building a comparison between Nietzsche and Borgmann, an unusual pairing, into an interpretation of Borgmann's politics. Nietzsche is widely known for his unfashionable observations on contemporary culture, unfashionable because they bucked most conventions and trends at the time. However, he did not have a political theory per se, which left his work unfortunately open to adoption by any number of political persuasions (most disturbing were allegations, later shown as erroneous, about sympathies with Nazism). Instead, Nietzsche developed what Light describes as prepolitical conditions, or those "that must obtain in order for any healthy public sphere to emerge and sustain itself according to its author's diagnosis of the general cultural, social, or political conditions required for such thriving." Borgmann is similarly unfashionable, confounding both conservative and liberal critics by stopping short of presenting a full-fledged political theory. His prepolitical discussions mostly point toward the development of place as a central political referent. Light urges Borgmann to consider whether the flexibility inherent in setting forth only prepolitical conditions is worth the risk of having Borgmann's small-c conservative ideas misinterpreted and misused by those with quite different political aims.

What will it really take to challenge and reform technology in the way Borgmann outlines? Will focal things and practices serve as a sufficient pivot for this reform? In chapter 7, "On Character and Technology," Carl Mitcham considers the meaning of *character*, historically and within Borgmann's work, in order to elucidate Borgmann's theory of technology and a profound problem at the heart of the reform of technology. Mitcham finds that the reform of technology will take nothing short of a transformation of our character, and that character in turn already has been formed in a distinct way by the character of technology itself. Such a transformation of this preexisting character then poses an enormous task. In the absence of a religious teleology, Borgmann, Mitcham argues, relies on the historical promise of technology and the power of poetry for character criticism and reform. However, Mitcham questions the effectiveness of art and reflection alone to transform personal and societal character in this radical way.

THREE

Focal Things and Focal Practices

Lawrence Haworth

Albert Borgmann's *Technology and the Character of Contemporary Life* (or *TCCL*) is both critical and constructive. Prominent among the constructive ideas the book deploys are those of focal things and focal practices. The book's trenchant critique of modern technological society centers on the ways our society causes focal things and focal practices to recede. Some may dismiss Borgmann's critique as little more than an expression of nostalgia for the pretechnological (or less technological) era. To counter this reaction, it is necessary to explore a few fairly elementary issues: Do focal things have their value apart from their involvement in focal practices? Or is their "commanding presence," to use Borgmann's term, constituted by that involvement? Do focal practices have any value other than that which is derived from their utility for focal things?

It will help to have in mind some of the examples of focal practices and their associated focal things that Borgmann discusses:

cooking	the meal;
chopping wood, etc. for heating the home	the hearth;
fly fishing	trout (or, in *Crossing the Postmodern Divide* [or *CPD;* Borgmann 1992], the rod);
the arts and crafts	paintings, pots, etc.;
running	the road or the course of the run;
backpacking	wilderness;
grooming, training, riding a horse	the horse.

Two typical comments Borgmann makes about focal things and focal practices, the first from *TCCL,* the second from *CPD,* might be read as implying significantly different views of the relation between these two ideas. From *TCCL:*

It is certainly the purpose of a focal practice to guard in its undiminished depth and identity the thing that is central to the practice, to shield it against the technological diremption into means and ends. (Borgmann 1984, 209)

And from *CPD:*

Things can be focal only in the care of human practices—a wilderness in hiking; a horse in grooming, training, and riding; a rod in fishing. Focal reality is alive in the symmetry of things and practices—of nature, craft, and art entrusted to the care of humans. Human skill commensurate with the commanding presence of a thing, and human devotion corresponds to the profound coherence of the thing with the world. (Borgmann 1992, 121)

Before commenting on these passages I want to call attention to two large features of Borgmann's thought that must bear on our interpretation. First is the idea that reality, when focal, is a commanding presence, possesses depth and significance, is articulate, and that it is especially through deictic discourse that one person, who grasps this significance, is able to convey it to another. Second is the idea, suggested by the reference to "technological diremption into means and ends," of the device paradigm and of the replacement of practices by machinery that procures commodities.[1]

The Parallelism Model

If we have the device paradigm in mind, we might be tempted to read the passages cited above as asserting a thorough parallelism between focal practice/focal thing, on one hand, and machinery/commodity, on the other. Technological machinery threatens to supplant focal practices. That is, it threatens to do what otherwise it would take a practice to do. But what the machinery does is procure a commodity. So then, one might think, the "guarding" function that practices perform is that of keeping the focal thing from collapsing into or being replaced by a commodity.

1. Traditionally, Borgmann tells us, we pursued our objectives and satisfied our needs and wants by following "practices." This required that we be engaged with one another and the world. It required as well that we competently exercise the sort of skills that the practices exacted. The advent of "machinery" signifies a different manner of satisfying our needs and wants. By automating a practice, machinery brings about that satisfaction without our needing to do anything more than flip a switch or utter a command. Where once we needed to be energetically engaged with others to entertain ourselves, by playing musical instruments and singing together, say, now the entertainment is as ready to hand as the button on the TV's remote control. The concept of machinery covers actual machines, such as television sets. It also refers to social structures, such as insurance companies that similarly "procure" benefits for us that in a pretechnological era we could gain only by maintaining informal and often difficult relations with one another in a local community. The term "device" refers to what practices become after machinery has obviated the forms of human engagement that a practice entails. With this conversion of a practice into a device the outcome of the practice is converted into a "commodity."

If the parallelism between focal practice/focal thing and machinery/commodity is taken seriously, then the focal thing associated with a focal practice will always be the object of the practice, its end, aim, or outcome. This is so because commodities are procured by machinery. Moving from practices to machinery signifies achieving that object (commodity) in a significantly different way (procuring it), so that the outcome of the activity (whether practice or machinery) ceases to be a focal thing and loses its "commanding presence."

I'll call this the "parallelism" model. One reason for taking it seriously is that with each focal practice Borgmann tends to associate just one focal thing. But if for each focal practice there is to be just one focal thing then it seems most natural to let that be the thing in which the practice eventuates, which completes it, what in a sense the practice is all about. A further reason for accepting the parallelism model is that it brings great overall symmetry to the theory, by lining up the focal practice/focal thing constructs with the machinery/commodity constructs.

As well, many of Borgmann's examples of focal practices and their associated focal things fit the model. For example, fly fishing is directed toward catching trout. But the trout in its splendor, that focal thing, becomes something entirely different, a commodity, when technologically procured. It loses its splendor. That makes sense. It is because focal things are objects of ultimate concern that we worry about the transmutation of the splendid trout into a commodity. Similarly with cooking: the meal is what cooking is for. The meal would cease to be a focal thing if it were procured technologically. It would become a commodity, a Big Mac. The arts and crafts examples fit as well: pots, paintings, and the like are the objects of the practices by which they are made, and it is these pots, paintings, and the like that become commodities when the corresponding practices are replaced by machinery.

Backpacking may seem to fit too: It leads to wilderness. In that sense, wilderness is what backpacking is for. And wilderness as focal thing collapses, ceases to be an object of ultimate concern or at least to enjoy our ultimate concern, when the backpacking practice is technologically invaded to such an extent that the experience of wilderness is simply procured effortlessly. My problem here though is that backpacking doesn't lead to wilderness in anything like the way craft leads to pots and such. The obvious difference is that wilderness was there in all its splendor before the backpacker appeared on the scene, whereas the potter makes the pot.

And there are also problems with other examples. The horse grooming, training, riding/horse example does and doesn't fit the parallelism model, or at least doesn't so clearly fit. We can think of grooming and training as having the horse as their object. These practices are directed at the horse, keeping it in a valued condition, so that in a way like the arts and crafts example these practices are cases of "making." But one gets in a different relation with the horse when riding it, insofar as the riding is something more than training. It isn't clear what a technological diremption into means and end would be like in this case, but however the idea is developed we would need to identify as the resultant commodity whatever is taken to be the reason for riding. I admit to a bit of confusion here, since it isn't clear to me what the "object" of riding a horse is (unless one is thinking of the horse as a means of transportation), and probably there is a point of view from which the horse itself is that object, so that riding is a sort of worship.

In the case of fly fishing, between *TCCL* and *CPD* the associated focal thing changes from the trout to the rod. The trout works, as noted above. But the rod doesn't. We might imagine fly fishing being invaded by technology in the following way: a sonar device for detecting whether trout are in the vicinity is used; the water is seeded with a substance that causes the trout to become lethargic and move to the surface; the rod is attached to a machine that reliably casts the fly just over the spot where the trout are languishing. In these circumstances the trout is procured by the machinery and ceases to be a focal thing. It becomes a commodity, but the rod doesn't. Instead, the rod becomes part of the machinery.

Or consider running: The road or the course of the run doesn't seem to be what running is for, the object of the activity; at least it often isn't. (This is not to say of course that the qualities of the route aren't relevant to the significance of the activity and that runners don't enjoy the road they run on.) So if running were pervaded by technological devices to the extent that it ceased to be a focal practice, that wouldn't convert the road into a commodity. Instead, like the fishing rod, it would become part of the machinery, or irrelevant.

One who is impressed by the symmetry of the parallelism model might think to save it by revising the offending examples, so that the road is not identified as a focal thing in running and the trout replaces the rod as the focal thing in fly fishing. But we aren't at liberty to do this; the theory is constrained by facts. In running, the course of the run may be focal for some. And one can see the point of Borgmann's identifying the rod as focal in fly fishing. And if pots and carved wooden busts are focal for potters and sculptors, so also may be clay, wheels, burl-grained wood, and balanced knives.

It seems then that the parallelism model is too narrow. Given the number of focal practices for which the corresponding focal thing is not the object of the practice, the "menace" of technology can't be said to lie simply in its threatening to convert focal things into commodities, although that may be the danger on some occasions.

The larger problem is that there appear to be two distinct sorts of focal practices. Or what comes to the same thing, two different sorts of position focal things have in focal practices. In some practices the focal thing has a subordinate role and the practice is dominant. But in many other practices the focal thing is prominent and as it were subordinates the practice.

My strategy now will be to sketch out two models, conceived as alternatives to the parallelism model. One fits the practices that are subordinate to the associated focal thing and the other fits the practices that subordinate focal things to the associated practice. I'll call the first the "guarding" model and the second the "internal goods" model. (I don't mean to suggest that the guarding model is Borgmann's, although some might imagine that it is.) Then I'll make an effort at reconciliation by considering how these two models might be coherently joined. It won't surprise or disturb me if at the end it appears that I've done little more than arrive at Borgmann's own thoughts on these issues.

THE GUARDING MODEL

This model takes seriously Borgmann's remark, quoted above, that "the purpose of a focal practice [is] to guard in its undiminished depth and identity the thing that is central to the practice." It differs from the parallelism model only in recognizing that anything one is involved with in a practice, and not just the thing the practice ends in or makes, may be, for the practitioner, the thing that is focal. A practice doesn't just guard focal things from collapse into commodity-status. It secures everything involved in its significance; it brings depth to the tools and other "means" by which the point of the practice is achieved.

One who wanted to elaborate this model might add that at one time human life was "practicing" through and through, and just about everything people came into contact with and employed in their practices was charged with significance. The world was enchanted. This may well connect with Borgmann's allusion to religion as the unifying focal practice in pretechnological societies and to his hopeful suggestion that "there may be a hidden focus of that sort now" (1984, 218). In any case, as devices replaced practices, disenchantment set in.

Regardless of what one thinks of this "elaboration," the guarding model starts with reality and the view that there are realities that are drenched in significance. It is with reference to these that focal practices get their sense and sanction. This gives a natural reading of the practice of grooming, training, and riding a horse. It isn't quite right to say that in the practice the horse is placed on a pedestal and becomes an object of worship. But the practice is in a way devotional. In backpacking and fly fishing as well the activity centers on a significant reality that is just there and invites attention. There is a similar nearly devotional aspect. And although in a musical performance the piece, Bach's Mass in B Minor, say, needs the performance to be realized, that piece, as realized at that time and place, may stand to singers and audience alike in the same way that the horse stands to the one grooming it.

The paradigm of practices that suggest the guarding model, one may say, is religious worship. God, or whatever the worshipers take to be the sign of God's presence in the practice of worshipping Her, now literally is the devotional object. The practice may be required to make that presence focal at a particular time and place, but the significance of the focal thing is not constituted by the worshipful practice.

What I want to stress here is that from the perspective afforded by the guarding model, the thing is in the care of the practice. This involves that the thing has significance apart from the practice by which it is guarded, but is, shall we say, fragile, at least under modern conditions, and so needs the practice to preserve it. The practice is subordinate to the thing.

THE INTERNAL GOODS MODEL

I understand MacIntyre's idea of a good internal to a practice in this way: in the practice one experiences the activity one engages in as good in itself, and one experiences the end of the activity, its product or what completes it, to be similarly good (MacIntyre 1981). What may come to the same thing, one experiences the way of life that goes with being engaged in the practice as being intrinsically good. This experience of the internal good of a practice is tied to one's sensed competence in carrying on the practice, where, if one is realistic, competence is measured against the constraints imposed by the setting and by one's personal limitations. (I saw a developmentally challenged child once using a computer to do an arithmetic test. The screen presented the problem, "2 + 2 = __," along with a number of choices. He picked "4" and when the screen displayed "OK" he turned to me with a big grin and exultantly shot his fist, thumb pointing upward, into the air.)

There is evidently a close connection between the idea of a focal practice and that of a practice that has goods internal to it. Focal things, one may say, are the preeminent internal goods in those practices that have goods internal to them. The internal goods model of focal practices distinguishes itself from the guarding model, however, through its account of the source of the goodness of those internal goods. Their goodness, excellence, or significance is represented as deriving from the practice itself and the tradition behind it. In contrast to the guarding model, from the perspective afforded by the internal goods model the focal thing is subordinate to and, as it were, derives its significance from the practice.

A little concreteness will help. Parallel (if not identical) to Borgmann's "culture of the table" is the practice associated with the complex of food and wine—cooking, dining, gourmandise and connoisseurship, and food criticism—in locales that have a "cuisine." I want first to indicate the range of internal goods this practice has and how they connect with the practice itself. And then I want to indicate the ways the practice depends on tradition and the ways it transcends tradition. What I'll be getting at is the possibility of making sense of those internal goods as derived from the practice itself.

First, what makes it a practice is that there is a preferred procedure, or set of procedures, that as it were set a standard for carrying on the practice; this means that for practitioners there is something to aspire to, insofar as they identify themselves as engaged in the practice. There are a right and a wrong way, which is at the same time a better and worse way of being engaged, criteria of excellence in performance of one's part in the practice. The chef and cooks, waiters, wine steward, maitre d', so far as they really are engaged in the practice, are trying to get it right, to excel as this is defined by the practice. The practice is normative. You make a soubise this way; this is how you set the table; this is what a soufflé should taste like. These criteria, endorsed methods, etc. are the significant parts of the tradition and practice.

Note here that we really can't make an important distinction between the good things internal to the practice and the practice itself considered as an internal good—between the preferred procedure, such as "how a soufflé is made," and the dish that procedure creates. For those caught up in the practice, if well made, the soufflé is a focal thing, and the activity of preparing the soufflé is an aspect of the associated focal practice. That activity is valued and experienced as good, not because of the profits one makes but because of and in relation to the shared understanding or, better, appreciation of the right way to be engaged in it. The "good" results from measuring up to this. And it is with reference to this same standard that the soufflé and the entire meal are determined to be internal goods.

To generalize, any practice with internal goods will have associated with it a shared appreciation of a manner of carrying on the practice that sets a standard for adequate and excellent practicing and by reference to which it can be explained why those goods are experienced as good.

The guarding model seems most appropriate when the practices we have in mind center on things we need to think of as enjoying significance outside the practice, such as those cited above. The internal goods model, by contrast, seems most appropriate when we attend to practices whose focal things don't enjoy that independent significance. The clearest examples are practices that centrally involve making something. In the practice of any art or craft, the focal thing associated with the practice is preeminently the object of the activity—a pot, painting, or sculpture. But the excellence of such objects cannot be understood without having reference to the tradition out of which they are created. The tradition, of course, is not to be slavishly followed; but in going beyond it (and any good artist or craftsperson will go beyond her tradition) one as it were leans on the tradition; that is, she engages in a critique that rests on the prima facie legitimacy of the tradition. In this regard practice and science are not different; more accurately, in this regard science is a practice not importantly different from other practices.

Another way to put this is to say that the excellence of focal things in some practices is constituted by the practice and its tradition. The focal thing now presents itself a bit differently than the trout and wilderness do. The latter are most naturally seen as splendid and awesome apart from our manner of approaching them. This is not to say that the excellence of the pot or sculpture is subjective, a matter merely of the view taken by the potter, sculptor, or critic. The manner in which the artist or craftsperson works within and transcends her tradition establishes the excellence of the crafted object. That excellence is a property of the object, and one may sensibly stand in awe of it. But it has this property in virtue of satisfying standards that are historically grounded. Or so the internal goods model will claim.

It is evident that there is an epistemological issue here. The guarding model is realistic: focal things have their significance apart from the practices that guard them. The internal goods model is coherentist: focal things gain their significance from their manner of origin within a practice and its tradition.

As I have presented these two models, however, they are not alternative ways of modeling the same phenomena; rather, the domain of focal practices appears to divide down the middle and the models fit different sides of the divide. On one side are practices that centrally involve a tradition, out of which an idea of excellence in performance is derived, and have in a loose sense a product. On the other side are practices that focus on a thing that appears to have value independently of the practice and to which tradition, although present in every case, seems not to be essential. So it may be that we don't need to decide between the models, but need instead to think through how they can be coherently joined, a task to which I now turn.

The Synthetic Model

The key to this exercise in joinery may lie in the way the latter sort of practices, covered by the guarding model, depend on tradition, and the way the former sort, covered by the internal goods model, depend on realities outside the practices. Because of these dependencies the two models may be caricatures that present a distorted picture by ignoring one essential feature of all practices while overemphasizing a second: the guarding model then could be said to ignore the role of tradition while overstressing the element of independence, whereas the internal goods model could be said to ignore the role of independent realities while overemphasizing the element of tradition.

It will be convenient to begin by considering the role of independent realities. The general point will be that the goods internal to all practices are such owing to their connections with conditions external to the practices. Tradition alone does not account for these goods.

As noted, two sorts of internal goods may be involved here, the good of the practice itself considered as a procedure and the good of the goal of the practice, of that which the practice is intended to accomplish. I'll refer to these as procedural goods and end-state goods. For example, in what Borgmann calls the culture of the table the procedural goods include all the accepted ways of preparing the meal and setting the table; the end-state good is the good of the meal itself and of the individual dishes—or perhaps we should say the celebration associated with the event of sharing the meal.

Sometimes, admittedly, this distinction cannot be made. The celebration associated with the event of sharing the meal might as well be regarded as a celebration that includes all of the preparatory steps, including especially the cooking. A more explicit example: there are groups of singers who gather once a year to sing Bach's Mass in B Minor. The entirety of their practice consists in these performances. That is, they never "practice." Since the entire performance is the "end-state good"—it is what they gather to accomplish—and their practice involves nothing but the performance, there is no room for a distinction between means and end (unless that comes to the distinction between part and whole, so that singing the "Gloria," say, is a "means" to the end of performing the Mass—but even so the means would be a constituent of the end). There being no distinction between means and end implies that the performance goods here are, collectively or considered as a whole, the same as the end-state good.

The point about the relevance of independent realities can be made by reference to this practice of performing the B Minor Mass. Obviously excellence in the performance is not entirely determined by the tradition of performing it: the singers must be faithful to the music Bach wrote;

and apart from that it is not just tradition (and not just Bach's score) that determines that if they sing too loudly or too softly the singers' performance will suffer—the acoustics of the hall they are singing in and the hearing acuity of the audience are independent factors to which an excellent performance must be adapted. If the tradition of such performances is not sensitive to such factors, then the tradition is wrong and performances that slavishly follow the tradition do not achieve excellence.

The practice of gardening suggests a different kind of relation with reality. In gardening the focal thing is the garden itself. In one way the gardener stands to her garden in the same way the groom stands to the horse. Like the horse, the garden is a growing thing. Because the plants in the garden contain within themselves the meaning of their flourishing (that is, the plants have a nature with reference to which we are able to say what it means for them to flourish) in tending the garden the gardener is not making it grow but is facilitating and nurturing a process that gets its motive and direction from the individual plants and the garden ecosystem. Neither the goal of her gardening nor the day-by-day activities by which she pursues the goal are settled upon independently of reference to the realities she confronts. These realities present constraints to which she must conform—conform both her goal and the activities by which she pursues it. She can't have a flower garden if the site is in deep shade, and her schedule of watering will be more-or-less determined for her by conditions of the site and characteristics of the plants in the garden. But these same realities present opportunities. Presence of the climatic and other conditions for certain plants' flourishing brings unique opportunities for a specific sort of garden.

One may think that gardening is a special case in that most practices don't centrally involve the care of living things. But even when the "material" with which one works in a practice is nonliving a similar dependence on characteristics of the material is inevitable. A sculptor, for example, is guided by the characteristics of the marble or wood she is sculpting. These characteristics may not control but certainly are influential in determining the precise details of the form the sculptor works to bring into being.

The gardening and sculpting examples bring out the point that in any practice the practitioners are well-advised to be guided, both in their manner of carrying on the practice and in the end they are striving to achieve, by opportunities and constraints presented by the materials with which they are engaged in their practice.

The Japanese tea ceremony presents a perhaps more telling example, since here the practice is steeped in tradition. Tradition specifies that one should receive the proffered bowl of tea with the right hand and place the palm of one's left hand under the bowl; that before drinking the tea one is to turn the bowl to the left and in such a way that the most beautiful side of the bowl is turned away; that one is to sip noisily; that after sipping one is to wipe the area where the lips have touched the bowl with the thumb and forefinger of the left hand; that after eating the tea cake one is to wipe the bowl again with the napkin that accompanied the tea cake; that one is to again rotate the bowl to the left so that one may finally admire the beautiful scene on the bowl. And so on into more minute details of the practitioners' movements during the ceremony. These details have accumulated over the course of some five centuries. Concerning much of this, if asked why the ceremony is done in that way, it may seem that no answer can be given beyond that it has been done in that way for centuries and possibly that it was introduced by a particular practitioner at a particular time.

And yet behind this seemingly blind adherence to tradition, the connection with reality is made at two points. First, even though a particular movement in the ceremony may appear arbitrary and to have nothing other to recommend it than that it has always been done that way, one can discern that the required movement is arbitrary only with reference to a limited class of movements. Given, for example, that the overall object is to create an island of calm conducive to enlightenment and composure, a wide range of movements are simply ruled out as counterproductive—for example, hurriedly pouring the tea into the saucer and then slurping it out of the saucer. In the same way, a law that sets a speed limit is arbitrary in that the limit might just as well have been set a little higher or a little lower; and yet good reasons can be given for having a speed limit and for its being within some range, between say fifty-five and seventy miles per hour.

Second, many of the movements are not arbitrary even in this restricted sense, but arguably represent the most sensible way of engaging in the practice in light of the overall objective. One might imagine introduction of the "noisy slurp" as an inspired innovation centuries ago to signify appreciation of the tea and as a sort of counterpoint to the otherwise studied discipline of the ceremony. Similarly, one tea garden was laid out so that when at the end of the ceremony the practitioners walked to the end of the garden to cleanse their hands in a pool of water, on rising from the pool they would behold for the first time a beautiful view—a delightful surprise saved for the last. Our appreciation of such innovations testifies to their nonarbitrary character; they seem wholly suitable and even inspired.

An implication of this last observation is that tradition is subject to a "critique." Having perfected the ritual establishes one's credentials as adept in the practice; but a master sees the possibilities for improving on the tradition. This may involve nothing more than achieving greater coherence within the tradition, rooting out elements that are counterproductive. But as in the innovations cited it may also involve changes that respond creatively to realities that lie outside the tradition.

I assume that these comments regarding the tea ceremony apply to all practices. In no case are the goods internal to the practice wholly founded on, derived from, or constituted by the history of the practice. Rather, in all cases an account of those goods will require going outside the tradition by referring to constraints or opportunities presented by independent realities.

The bearing of these observations on the two seemingly opposed models results from the connection between "goods internal to a practice" and focal things. As indicated, a focal thing is a good internal to a practice; to some degree, every good internal to a practice is a focal thing, although as we shall see the term "thing" is not obviously applicable in every such case.

A complementary point may be made regarding the guarding model. The tie to external realities is never so strong that tradition and history play no role in determining the shape of the goods internal to a practice. This follows from the very idea of a practice. Unless tradition and history are influential in defining the goods pursued within the practice, it will not be a practice at all but a one-time activity.

We have numerous examples of such one-time activities that have internal goods and center on a focal thing: the "oceanic" experiences discussed by Freud, Goethe's "Faustian moment," the "peak experiences" of 1960s pop psychologists, James's varieties of religious experience. But take an example closer to home. Say you are working as a night watchman at a summer resort in Vermont, using the long stretch of time between dusk and dawn as an opportunity to study for your prelims. At dawn one day you walk down to Lake Champlain to watch ducks barely visible through the mist on the lake. But unexpectedly for you they aren't just ducks and it isn't just a lake but a scene of ineffable significance. Whatever specific terms it occurs to you to use in describing it, the description must point to the sense of identity of yourself with the reality laid out before you and to the sensed depth and transcendence of that reality—a transforming experience. The "internal good" met or found on such (rare, at least now) occasions may be of the same order as that had by the choir performing Bach's Mass. But because the visit to the lake is not a practice and there is no normative procedure involved, its internal good is as it were a gift. By contrast, the choir know how to bring their internal good into being.

It would seem then that the guarding and internal goods models are both distortions and that a more apt model of focal practices will incorporate the insights of each, suitably revised to accommodate the other. I would like to call the result of this exercise in joinery the "Borgmann model"; at least I don't see that it importantly departs from anything Borgmann has written. But that might be premature, so instead I'll call it the "synthetic model."

I noted earlier that the guarding model appears to fit one set of practices and the internal goods model a different set. The synthetic model brings these two sets under one scheme. But the differences among the practices aren't explained away; instead what we see are practices arrayed along a continuous scale. At one pole on this scale are practices with focal things that appear from another realm outside the practice; at the other, practices that don't obviously involve independently significant focal things at all and whose goods may seem entirely internal to the practice.

Religious worship is the paradigmatic example of the former sort of practice. From the believer's perspective, the practice doesn't invent the Thing that is focal to it. The practice is instead a response to a commanding antecedent presence and is called forth by that. What the synthetic model must claim in this connection though is that the specific characteristics of the focal thing in religious worship reflect particular traditions, by which God is determined to be He or She, wrathful or loving, savior or dispassionate judge. The synthetic model is strongly ecumenical.

At the other end of the scale are focal practices of the sort described in Studs Terkel's *Working* (Terkel 1974). Working people in very ordinary occupations—waitresses, firemen, carpenters—often strive to find something in their work that runs against the grain of its contemporary structure. The work isn't socially structured as a craft, an art, or a profession; outwardly it is just a job and something one does "to earn a living." Nevertheless, from whatever motive, the worker may invest it with internal goods and engage in it not merely as something one must do, but as being worth doing provided only that it is done right. So the waitress will take care that the plate is placed before the customer *just so* and take pride in doing this. The fireman will internalize the thought that putting out a fire creates a public good and be motivated by that realization. The carpenter will find layers of meaning in striking the head of a nail dead on so that the nail flies into the wood. As a result the waitress, fireman, and carpenter, as I have put it elsewhere, professionalize their work, even though to the world at large (emphatically) none of these jobs is a profession (Haworth 1977). That is, against the odds, they make their work a focal practice. What is hopeful in Terkel's examples is the evidence they provide of the resources ordinary people have retained or found in a world where the dominant paradigm is as described in Borgmann's book.

I won't spell out all the steps by which I come to the following conclusions—all of which are tenets of the synthetic model—but for the most part the basis for them is found in the preceding remarks:

1. A focal practice need not have just one focal thing associated with it. There may well be numerous focal things, and some may find one thing focal in a practice and others find some other thing focal. For some it may be the trout, for others the rod, and for others both the trout and the rod (there are limits of course; not everything one is engaged with in a focal practice can serve as a focal thing). For many Canadian participants in the Midnight Sun Marathon, a run five hundred miles north of the Arctic Circle, the focal thing is an abstraction, "the far north," central to Canadian identity. American participants of that marathon are perhaps more likely to find Nanisivik's desertlike landscape, not colored by the abstraction, focal.

2. In some cases the focal thing associated with a practice may not be distinguishable from the practice itself. I have cited the practices of annually performing Bach's Mass and the tea ceremony; it would seem that the entire performed Mass is the focal thing in the first practice, and that the entire ceremony is the focal thing in the second.

3. Devices are not the only threats to focal practices. Any change that leads people to engage in their practices for entirely instrumental reasons, so that the practice has no internal goods, is a threat. A cook or film director who lacks integrity, that is, who does not cook or direct with an eye to doing the work well but instead is entirely motivated by a desire to sell or to please, finds all of the goods associated with cooking or filmmaking to be external to those practices. In that case the practices are not focal and they have no internal goods. What is more important in this connection, perhaps, is not their personal motivation but that they are engaged in institutional structures that leave little room for integrity. If the director doesn't titillate the viewers she will have trouble getting another film to direct. In this way the institutional structure introduces a selection process that weeds out people who care and who take care. The waitress, fireman, and carpenter, by contrast, may find their work to be focal practices even though many of their coworkers are merely putting in time. We regard them as heroes because they are engaged in institutional structures that often don't encourage focal practice.

4. From this point of view, devices pose a threat to focal things or practices with internal goods because they remove the occasion for being engaged with integrity. When the device procures the "good" (now become a commodity) then engagement of the sort that could exhibit integrity is foreclosed; since there is no occasion for being engaged at all, there is no occasion for being engaged with integrity.

5. From the side of technology, devices pose the distinctive threat to focal practices and the characteristic contemporary threat. But in our world equally virulent threats come from the related impulses to commercialize

and commodify, including the sort of commodification that results from our giving in to the impulse to consult "experts" for advice on management of our everyday and mundane affairs.

6. Although there certainly is an autonomous fascination with devices in our culture (Borgmann's "complicity"), the proliferation of devices in ways that threaten focal practices is also motivated by commodification. The two impulses are brought together owing to the circumstance that (as Marx saw) in the search for markets, producers of commodities find it a profitable strategy to add value to non-devicelike tools, thereby converting them into devices or enhancing their devicelike character.

References

Borgmann, Albert. 1984. *Technology and the Character of Contemporary Life: A Philosophical Inquiry.* Chicago: University of Chicago Press.

———. 1992. *Crossing the Postmodern Divide.* Chicago: University of Chicago Press.

Haworth, Lawrence. 1977. *Decadence and Objectivity.* Toronto: University of Toronto Press.

MacIntyre, Alasdair. 1981. *After Virtue.* Notre Dame: University of Notre Dame Press.

Terkel, Studs. 1974. *Working.* New York: Pantheon Books.

FOUR

Technology and Nostalgia

Gordon G. Brittan Jr.

In a revelatory and eloquent way, Albert Borgmann urges us to make room at the center of our technological lives for what he calls "focal things and practices," largely traditional objects and activities such as handcrafted furniture and baking from scratch. In this way his argument seems typical of much contemporary criticism of technology, characterized by longing for a simpler, quieter, and more textured life, ostensibly of the sort people used to enjoy. This familiar longing can be identified as nostalgic. It leads many people to dismiss the criticism as romantic, the fanciful product of the affluence and free time that technology itself has made possible and to downplay the importance of work like Borgmann's. In this essay I want to make a case for the importance of the work, although in large part by way of taking seriously and criticizing one of its main themes. I also want to make a place for something like "focal things and practices," but on rather different grounds from his and with somewhat different results.

The examples of focal things and practices that Borgmann offers certainly suggest nostalgia for a pretechnological past, before the ascendancy of what he calls the "device paradigm."[1] For although at least some of them, like baking from scratch, make use of technology, the end achieved depends more on human skill and effort than it does on the autonomous

---

1. There is no human past that does not know the use of tools, and in this sense "technology." "Devices" differ from other sorts of tools in that they are defined entirely in functional terms (a device is anything that serves a certain function, which at present largely involves the procurement of a commodity), they make few if any demands on skill in their operation, and the way in which they carry out their function is quite literally (as in the case of modern computers) hidden from view. Thus in Borgmann's phrase, devices (as against more traditional tools) do not simply "disburden" but also "disengage" us. We can date the ascendancy of the "device paradigm," I think, to the 1890s, when Henry Adams first noted the ways in which the dynamo introduced an "abysmal fracture" into our understanding of what it is to be a "tool." At the same time, and as should be clear in what follows, the distinction between "devices" and other sorts of "tools" is not sharp, and when I speak of the "technological" as against the "pretechnological" and so on, it is always a matter of degree.

and anonymous functioning of pieces of machinery. Among these focal things and practices are such ancient activities as walking in the wilderness and preparing a ceremonial meal. Indeed, the discussion brims with words like "loss" and "recovery" that evoke a fall from a pretechnological state of grace. "Devices . . . dissolve the coherent and engaging character of the pretechnological world of things" (Borgmann 1984, 47) is an entirely typical passage. So it comes as something of a surprise when Borgmann denies that his motive is nostalgic, in opposition to Heidegger, whose yearning for an earlier Teutonic culture and a less cluttered European landscape is palpable. Rather, he claims that focal things and practices, however "pretechnological" they might prove to be, have force and depth only within a technological context, which for this reason should not be dismantled. In fact, it cannot be dismantled, not so much because it is already so well entrenched as because it affords the opportunities, the affluence and free time, to engage with and in these things and practices.

Despite the denial, there is a sense in which Borgmann's motive is nostalgic.[2] For the etymology of the word has to do with homesickness, and it is his contention that our lives generally have lost their center and focus, symbolized by the family hearth, and can be made whole and well only by being reoriented in such a way that we again feel ourselves at home in the world. But in the more usual present-day understanding of the word, his motive is not nostalgic. He does not seek escape, or even withdrawal, from the technological culture, still less a withdrawal into the past. Borgmann, unlike Heidegger, looks to the future as a source of value; as he has explained to me in conversation, "what is so damaging about our culture is not what it once was but what it could be." Moreover, the "engagement" that he recommends requires, unlike nostalgia, more than a particular attitude, a state of soul, or the attempt to distance oneself psychologically from the present. Finally, the objects and activities he identifies as "focal" are not necessarily the ones we made or participated in when young. Indeed, it is difficult to know how anyone these days can be "nostalgic" for a pretechnological culture, even in Borgmann's precise sense of the term, when none of us has ever lived more than momentarily in one.

All of these considerations are important in dispelling the romantic aura that might otherwise be seen to hover around Borgmann's discussion. Surely what he calls the turn to "things" has meaning only in the context of a life already structured by devices, along the same lines that a rural way of life requires an urban culture to be understood, appreciated, and, these days, made possible. But why in the midst of devices do we *need* to turn to things

2. The various senses of the word are canvassed in Davis 1979.

in the first place? Borgmann makes his case by illuminating some of the most salient features of contemporary culture.

He begins with the fact that our culture embodies, in theory if not always in practice, a liberal democratic conception of society. Before all else, the democratic conception involves principles that, to put it very broadly, ensure that though the race for life's rewards goes to the swiftest, everyone will have an equal place at the starting line and an unhindered run. These principles in part define, and if implemented make possible, what we mean by "social justice." Among them are liberty, equality, and, to make these principles "substantive," something like the opportunity to develop or realize oneself, one's abilities and interests, as completely as possible. But the democratic conception leaves open questions concerning the character of these rewards and the content of this opportunity. This in fact is of the essence of a democracy, that although no one's opportunities are to be prejudiced by circumstances—gender, race, and parents' income, say—over which one has no control, everyone is free to run the race as she sees fit, to determine where his own best interests lie. On this conception, then, we have a social or public notion of justice and an individual or private notion of goodness.

But according to Borgmann there are two large problems with this conception. First, it harbors a contradiction. On the one hand, a democratic society is to provide opportunities for all, on something like an equal footing. On the other hand, it is not to specify the character of these opportunities. But the only way in which such opportunities can be made generally available, in fact if not also in theory, is through the application of modern technology. For without technology we are hostages to the reality of scarcity and the requirements of survival, and hence are not free to choose how we are to lead our lives. The problem is that in providing us with opportunities, modern technology at the same time determines their character. That is, modern technology, applied at every level of the social contexts in which we find ourselves, closes the very questions concerning the formulation and pursuit of happiness that democratic theory is at pains to leave open. It enforces a particular set of values.[3]

We can take as an instance the development of so-called "expert systems" for medical diagnosis.[4] Such systems will allow patients in even isolated parts of the country access to the very same specialist consultants formerly available only in university hospitals, so long as their doctor has a modem

3. And, it has been maintained, a particular way of justifying them. Modern philosophy, in its universalizing and argumentative way, is of a piece with technology in attempting to dominate.
4. Following Dennett 1986.

and a cellular telephone. A doctor who opted not to make use of these specialists would thus be violating the requirements of justice. But at the same time, the new technology will transform the rural doctors' lives, making them less exciting, less interesting, and consequently less happy. From their point of view, "they will begin to sink into the role of mere go-betweens, living interfaces between patient and system, who consolidate their direct observations into machine-readable symptomatology, and execute the therapeutic directives of the system" (Dennett 1986, 139). It is in this sort of way that technology in making a democratic society possible at the same time subverts its conception.

There are, in fact, two different points at stake here. One is that technology is not, as the Enlightenment authors of the standard conception of democratic society naively thought, merely instrumental and value neutral.[5] It does not leave questions concerning how we are to lead our lives open. For it defines the range of (employment) opportunities open to us, characterizes in a very narrow way the (consumptive) rewards to which they lead, even imposes a certain (mechanical) style on the ways in which they are to be taken up and enjoyed. In short, technology employed as instrument inevitably becomes a way of life as object.

But it is not simply that technology closes questions otherwise left open. It also closes them in the wrong way. That is, the technological way of life is, according to Borgmann, "rigid," "narrow," "superficial," "distracting," "debilitating," "vacuous," and "vapid." Technology prejudices in fact the ends that democracy in theory wants to leave open, and more importantly the values it enforces are bad. At the very least, they need to be supplemented by others. Borgmann's account is as much critique as it is description of the technological culture.

The second problem with the democratic conception is that, despite its original intention, it should not leave questions concerning ends open. In part this is because to leave the ends open is to allow them to be usurped by technology in ways that lead finally to "mindless" and "unsatisfying" lives. But in larger part it is that, as Borgmann puts it in what is the key passage of *Technology and the Character of Contemporary Life* (1984),

> just as the constitutional definition of society remains incomplete and corruptible without a statement of substantive justice, so the just society remains incomplete and is easily dispirited without a fairly explicit and definite vision of the good life. (1984, 91)[6]

5. Although it is also the case that in the first phase of the Industrial Revolution the "device paradigm" was no more than latent.

6. "Substantive justice" as noted earlier involves adding a material principle of self-development to the more formal principles of liberty and equality.

Given its importance, this passage deserves careful examination before we proceed to the rest of the argument.

It is clear why Borgmann needs an explicit and definite vision of the good life. For it is only in terms of such a vision that we are in a position to criticize the lives that a technological culture leaves in its wake. It provides a standard, something to measure against, without which there is no possibility of evaluation.

It is less clear why a liberal democratic conception of society requires an explicit and definite vision of the good life. Presumably a just society requires a fair distribution of goods, a state of affairs to be ensured by implementation of the principles of liberty, equality, and self-development. A democratic society, to repeat the fundamental point once again, does not further characterize these "goods." To say that it is therefore "incomplete" is to beg the very question at issue.

Furthermore, there is something problematic about contending that without this vision, a just society is easily "dispirited." This is at least in part a factual claim about the actual state of citizens' minds. In fact, Borgmann thinks that the members of our society *are* dispirited; for one thing, they participate less and less in the political process. But it is also a claim about what is "dispiriting." As such, it is open to challenge. For one thing, it could be contended that the ideal of social justice is itself inspiring. Far as we are from having established a just society, it continues to animate the most important social movements of our time. For another thing, on the standard democratic conception it is individuals who develop visions of the good life, as explicit and definite as they might be, and are variously inspired or dispirited as they are frustrated or not in realizing them or disappointed or not with their achievement.[7]

But as a closer reading of the text reveals, it is not really the citizens of democratic societies who are "dispirited"; it is the impartial but by no means dispassionate observer of the lives they lead, in this case the author himself. For it is the author's view that justice unchecked by goodness results, in

7. In *Crossing the Postmodern Divide,* Borgmann presses the point that "the individual" to which democratic theory here makes appeal is a figment of the theorists' imagination. Within the larger technological culture in which modern democracy developed, "the individual . . . was in reality little more than an accomplice to a gigantic enterprise that, though resting on the consent of most people, was given a shape and momentum of its own" (1992, 79), although he also has come to think that even the democratic theory of the individual ("ideological individualism") is pernicious (as against a "postmodern" communitarianism). But my case does not depend on possibly naive assumptions about the role and status of individuals in theory or practice. Rather, it has to do with difficulties in combining justice and goodness in a single, coherent conception of society or, more accurately, of providing a usable definition of "goodness" that is not simply a corollary of the principles of justice.

fact if not also in theory, in mediocrity and superficiality, a dispiriting and paradoxical fact when set beside the democratic emphasis on the realization of individual human potential.

Let us grant for the sake of argument that an adequate democratic conception must make room for communal standards of goodness as well as for principles of social justice, if for no other reason than to criticize technology and keep it in its place, and in consequence to stem the tide of mediocrity and superficiality.[8] How is this goodness to be understood? Borgmann falls back on the two great concepts of ancient moral philosophy, excellence *(arete)* and happiness *(eudaimonia)*, although he construes each in what I take to be a nonmoral way.

Let us begin with excellence.

> We can measure the worth of typical technological leisure by the traditional standard of excellence in two complementary ways. We can ask what degree of excellence people have in fact achieved; and we can ask how much of their free time people devote to the pursuit of excellence. (Borgmann 1984, 127)

We can make the first question both more precise and more concrete by asking:

> (1) How well educated and literate are people? How well do they understand the structure of the world? How active and informed is their participation in politics? (2) What typically is the condition of people's physical vigor and skill? (3) How well acquainted are people with the arts and how proficient are they in making music and in other artistic practices? (4) How compassionate are people privately and as citizens? How devoted are they to helping others who suffer deprivation and hardships? (Borgmann 1984, 127)

But our contemporary democratic society fails all of these tests of excellence: in some sense, "standards" are not being kept up. As for the second question, concerning how we spend our leisure time, it is enough to point out that "all the activities which have been taken as suggesting a dedication to excellence constitute when taken together less than a quarter of the time spent on watching television" (Borgmann 1984, 99).[9]

---

8. I take it that Borgmann thinks that an explicit and definite vision of the good life will prevent technology from determining and debasing the character of our opportunities while allowing it to make them possible. That is, an explicit vision of the good life will help restore technology to its properly instrumental role, and perhaps also restrict it to more benign uses.

9. As James Allard has noted, "watching television" is somewhat vague. My wife irons, cleans the house, even cooks while "watching television," and as a family we talk constantly when it is on.

As profound as that of Heidegger for the peasant and provincial culture of Germany, a sense of the loss of Greek ideals and of the high culture that has for so long honored them pervades this litany. I share the same sense of loss for a set of norms that seems to have disappeared as quickly and quietly as Latin from the American high school curriculum, classical music from network television, and amateurs from the Olympics. Indeed, I would not deny that the items named *are* "excellences," or that many of the activities in which we engage in a technological culture are not. These activities, warming up Pop-Tarts in the toaster as an extreme example, in fact preclude excellence, for they don't allow being done well or badly. And whatever else may be said about it, the notion of excellence functions as a powerful ideal. Still, further reflection begins to dampen if it does not also undermine my own, to some degree very possibly nostalgic, commitments.

To begin with, the documentation of decline is not conclusive. How well educated and literate are people? Do we fall back on standardized examinations to decide, when lower scores on at least some of these examinations seem to be more than anything else a function of much larger numbers of people, many of them not assimilated in the cultural mainstream, taking them? Is the fall in voter turnout really a product of apathy and ignorance, or does it rather have to do with a realization that the most important decisions taken by our society are no longer political, but judicial and economic? The answers to these and other questions that could be asked about the evidence mustered are not easy. But it is well to remember that every age sees ample evidence of a decline that in retrospect is not always so apparent.

Moreover, from what point in time do we first see signs of decline? Surely not from the introduction of democratic forms of society in the late eighteenth century, or of an industrial economy in the early nineteenth. For what these developments ushered in was a vast increase in literacy, in political participation, in the creation of symphony orchestras, and in the spread of competitive sports. But if not then, when? Even to date it from the advent of a television culture, and the growth of couch potatoes, is problematic. For the introduction of printed books in the fifteenth century, and their rapid dissemination in the sixteenth and seventeenth, must, in the solitary and homebound character of reading, have seemed to cut people off from the type of communal and vigorous conversation that is itself the highest form of distinctively human excellence.

Indeed, it has been argued that, far from leading to decline, a fully technological civilization requires, among other "excellences," a more highly educated work force. Thorstein Veblen, in particular, rejected Marx's claim that technology "dumbed down" work of all kinds so that it became possible for machines to do it. It was self-evident for him that technology demands skilled workers—the more advanced the technology, the greater the skills demanded.[10] The opposing point of view, Veblen went on, is rooted in primitive or animistic ways of thinking about human beings, which identify thought with "creativity," rather than with what might be called an "engineering mentality," the orderly finding of solutions to problems. As David Riesman once put it in elaborating Veblen's position, "those who sighed over the passing of the peasant and the artisan of earlier times were gulled by a feudalistic aesthetic," hierarchical in character and patronizing in fact (Riesman 1960, 85). At the very least, it seems difficult to deny, against Veblen, that the advent of twentieth-century technology has brought with it a new demand for more educated workers and that the enormous expansion in college and university education witnessed is simply a response to this demand, particularly since most of the expansion has been in technical areas.

But the final and, I believe, most important point about excellence has to do with luck and circumstance. To appreciate it, we first have to make some sort of distinction between moral and nonmoral excellence. The distinction is not easy to draw, and the Latin word *virtue,* used to translate *arete,* blurs it. Undoubtedly Borgmann would say that to make the distinction is already to sanction the compartmentalization of our lives so encouraged by technology and thereby to undermine their coherence. Nonetheless there is something to it. Moral excellences or virtues like courage or temperance have to do with character; nonmoral excellences like health or handsomeness, Aristotle's examples, do not. Now the sorts of excellence Borgmann has in mind are, exception made for compassion, of the nonmoral type. In part because he is worried about begging crucial questions at this point, he thinks it difficult to mount a trenchant critique of technology on moral grounds.[11] A trenchant critique has more to do with mediocrity than morality. The difficulty is that nonmoral excellence is in part dependent upon luck and circumstance.[12] Health and handsomeness are obvious enough. But many of the other "virtues" listed in book four of the *Nicomachean Ethics* are

10. See, for example, Veblen 1964.

11. See, for example, Borgmann 1984, 144 ("I believe that most critics of the moral defects of technology sense the weakness of their approach") and 174 (vis-à-vis technology, "moral discourse is not cogent"). Compassion is a suspect entry on any list of *virtues,* as Nietzsche pointed out, although I very much doubt that by any common measure our own society is less compassionate that its predecessors.

12. Borgmann has chosen running and the culture of the table as "excellent" activities within everyone's reach, society permitting (in the same way that Socrates restricted happiness to mental pleasure so as not to be especially susceptible to the vagaries of luck). But even running, if not also the culture of the table, seems to be dependent on factors over

similarly dependent upon factors over which one has little control. One cannot be *magnificent*, for example, or even liberal, if he is not also wealthy, and on Aristotle's reading, "virtue" of the traditional kind is impossible if one is a woman. Even the items on Borgmann's list, much more enlightened from a contemporary point of view, depend in practice on such factors as age and genetic endowment. "What typically is the condition of people's vigor and skill?" Understandably, it is much greater for my thirty-one-year-old son than my ninety-four-year-old father. "How proficient are [people] in making music and other artistic practices?" I am not at all proficient, born with a tin ear and an underdeveloped left brain. If we proceed to identify goodness with excellence in Borgmann's way, then there is an apparent conflict with justice, which requires that a person be judged only with respect to those things over which she has control and for which she can be responsible, ultimately her character. When Talleyrand said that no one born after 1789 could know how sweet life could be he meant that no one born a Talleyrand, with his powers of observation and discrimination, and the wealth and leisure to develop and enjoy them, could know how sweet life could be.

Borgmann is attracted to the idea of excellence. Product of the same sort of upbringing and education that he is, so am I. But he realizes that it involves a set of ideals produced by a culture that was not "device-ridden" in the same way as ours and thus might be taken as an alien and unfair standard by which to judge contemporary technological life. So he moves from an identification of goodness with excellence to an identification of goodness with happiness, another nonmoral notion.

Happiness has two distinct advantages over excellence as our measure in this context: it is intrinsic to technology's own ideal, part of technology's promise,[13] and it has, at least in intention, no patronizing traces of elitism. If not everyone can excel, then anyone can be happy. The reason why the technological way of life is bad, to put it baldly, is that those who live it are unhappy. In Thoreau's famous words, they lead lives of "quiet desperation."

There are three propositions any two of which are inconsistent with the third. One is the equation of goodness with happiness. A second is

---

which one has little control, and a measure of excellence is available to only a very limited number of people.

13. As Don Mellon has noted, the reification of technology, *its* making of promises as apart from the people (and the capitalist system which they represented) who promoted its application, is problematic. But Borgmann insists on this reification. It is technology (in the "device" sense of the word) itself, as a way of life, rather than the use (or misuse) of it, that is debilitating.

that technology is bad.[14] The third is that technology leads to happiness. It follows that at least one of these propositions is false.

Borgmann thinks that the third proposition is false. It is the core of his critique that technology has not brought happiness, and is on this ground to be condemned (although not, for the reasons given earlier, to be rejected or eliminated). "It turns out that avowed happiness appears to decline as technological affluence rises" (Borgmann 1984, 124). And again, "The promise of technology was really one of happiness though that was not always explicit, [but] professed happiness [has] declined rather steadily and significantly" (Borgmann 1984, 130).

I am not at all sure how to assess this line of argument. Measurements of happiness, professed or otherwise, must be regarded with some skepticism, very possibly like measurements of intelligence, the artifacts of method and mathematics. If, in fact, there has been a decline in professed happiness,[15] then it is also true that over the period in which the measurements were carried out, there has also been a decline in real earning power per capita and other significant economic variables. There is no particular reason to isolate the supposed sterility of technological culture as the cause when it might, ironically, very well be the case that frustration over an inability to participate more fully (principally as a consumer) in that culture, because of structural shifts in the economy, is at stake.

Doubt about an alleged decline in happiness is underlined when one considers two other facts. One is that if people within a technological culture are unhappy, they are (for the most part) making no effort to change or destroy it. Difficult as it is to measure such things, my own view is that our culture is more thoroughly "technological" than it was fifteen years ago, when *Technology and the Character of Contemporary Life* was published. Technophobe though I am, we now have voice mail at home, e-mail at work, and a VCR in the bedroom. The other fact is that those who do not have technology clamor for it. In one third-world country after another, 30 to 40 percent of the population have television sets, even when their nutritional levels are inadequate.

There are, in turn, two ways in which to undermine these facts. They involve the classic excusing conditions, ignorance and impotence. The

14. And not, we have seen, the people who promote it or the economic system with which it is aligned.

15. The most comprehensive work on the subject with which I'm familiar is Veenhoven 1984, itself a survey of 245 empirical studies on happiness. Veenhoven concludes that the alleged decline in happiness, professed or not, in modern Western societies is nothing but a "myth," that people in developed countries are happier than people in undeveloped countries, and on and on.

ignorance line is simply that those who clamor for technology do not yet know that it does not deliver on its promises. They have been duped by the vast and subtle claim that technology, in this case as in so many others aided and abetted by Western capitalism, generates in its own behalf; that once they have technology—computers, CD players, industrial machinery, and all the rest—they will at the very least restore it to its properly instrumental place. The difficulty with this is that in Western countries technology has long since arrived and, once again, people continue to embrace it, whatever their professions to the contrary. One could argue that in the industrialized countries people are equally in thrall to technology's propaganda, and that although they continue to embrace it, they are not *really* happy. But Borgmann does not give much credit to this line of argument. Technology eventually "reveals itself" over time as it matures, despite its best efforts at concealment, and to distinguish between real and apparent happiness, as Thoreau sometimes seems to do, would be to beg a crucial question. No, Borgmann rests his case on "professed" happiness. But this is difficult to reconcile with the way in which people, in fact, allocate their dollars and their time.

The other classic excusing condition is impotence. This is the line that although most people clamor for and ultimately embrace technology, they do not *choose* to do so. Technology imposes itself so completely that although one may resist it in theory, to do so in practice is impossible. This is the thesis of technological determinism. But this thesis is false. If it were true, then it would seem to follow, among other things, that similar technologies would bring similar cultures into being. But similar technologies have in fact been shaped and adapted, so far as their deployment and effects are concerned, by the cultures in which they first arose, witness the advent of movable type in China and the West. And certainly Borgmann cannot hold that it is true, since the possibility of (re)centering our lives around focal practices and restoring a just balance between things and devices in our lives presupposes an ability to do so.

So one cannot allege ignorance or impotence when trying to explain why people clamor for and embrace technology. Nor does Borgmann try to do so. It is just that it is otherwise difficult to reconcile this clamor and embrace with a supposed profession of unhappiness on the part of those who have come to see the limitations of a technological way of life.

But if we do not reject the proposition that technology leads to unhappiness, then which of the other two—that goodness is tantamount to happiness and that technology is bad—has to go? Those who formulated the democratic conception of society in the eighteenth century thought that technology (by which they had rather simple, accessible "labor-saving" devices of a non-Borgmannian kind in mind) leads to happiness. They also thought that it was good, relieving drudgery, although their views, coming well before the advent of the factory system and the development of a consumer and mass-produced culture, were undoubtedly naive. But much more important, some of them rejected the equation between goodness and happiness.[16] We need to consider the grounds of this rejection. They show, to begin with, that one cannot argue from the unhappiness that technology supposedly brings to its badness. But more fundamentally, these grounds show why those who first formulated the democratic conception of society were so concerned to leave questions concerning the nature of happiness, and thence of goodness, open.

The first consideration is that the concept of happiness is indeterminate, it has no definite content. Indeed, according to Kant, "the concept of happiness is so indeterminate that although the human being wishes to attain it, he can never really say what it is that he really wishes and wills" (Kant 1785, 417–18). As such, it can never be the particular object of any activity, or the reason in virtue of which we label that activity good.

The second consideration is aligned with the first. It is that while happiness is to be understood, in some very general sense, in terms of satisfaction, and eventually of pleasure, such satisfaction is ultimately subjective in character. That is, what provides pleasure varies from person to person, and even then is heavily dependent on context. Great wines are rarely enjoyed before breakfast and meals become memorable simply as the result of being shared with witty and beautiful companions. But the claim that an object or activity is good is objective in character and hence universal. Even when we relativize happiness to the individual and to particular contexts, it is inconstant, satisfaction ebbing and flowing in ways that we cannot very well explain. But again, when we say that something is good the implication is that its being good does not depend upon the vagaries of mood and feeling.

So happiness, plausibly described in terms of satisfaction or pleasure, is indeterminate, subjective, and inconstant. If we were to identify it with goodness, we would then have to ascribe the same properties to this latter. But this would rob the concept of goodness of its normative power and render it virtually useless as a way of evaluating objects and activities.

The third consideration is more subtle. It is to the effect that we are so constituted by nature as never to be content. For as soon as an object is within our grasp, we desire another, and happiness continues to recede

16. The following remarks are inspired by Kant's criticisms of construing moral goodness in terms of happiness, and they owe a great deal to Alan Wood's unpublished paper, "Kant vs. Eudaimonism."

before our eyes, just slightly out of reach. As with Goethe's Faust, the moment never comes when we can say "stay still, thou art fair."

This point has special application with respect to the nostalgia that many people feel for "pretechnological" objects and activities. For it might be argued that whatever dissatisfaction is felt with the technological culture is a function of its having been achieved at the present moment and, never content with what is at hand (in fact, rather bored with what is at hand), we look to the past (more real than the future) for a happiness that, not now within our grasp, seems for that very reason more perfect.[17]

Two further corollaries might be drawn from these considerations. One is that goodness cannot be identified with happiness, but with moral worth. This alone is constant, objective, universal. The other corollary is that what Kant calls the moral law, respect for which is the ground of moral worth, is the only thing capable of restraining our (indefinite and ultimately frustrated) quest for happiness. We must leave questions concerning happiness open, leaving each to formulate and pursue her own conception, up to the point where we breach the moral law, i.e., up to the point where we begin to interfere with the formulations and pursuits of other people. It is the concern of a just society to make the pursuit of happiness possible for each, not to impose a particular conception of it on all.

I have argued that within a liberal democratic conception of society neither excellence nor happiness provides us with an adequate criterion of goodness. But no plausible and clearly articulated alternative is available. From which it follows that the liberal democratic conception must leave the question of goodness open.

One might reply, of course, so much the worse for the liberal democratic conception. This appears to have been Heidegger's reply, echoing Nietzsche. In its irresolution, the democratic conception both permits and invites the debasement of human life, forgoing excellence and happiness in the process.

But this is not quite Borgmann's reply. As we have already seen, in his view liberal democracy in practice answers questions concerning the good life along technological lines. Much more generally, with respect to *any* way in which the community might be organized, "[t]he question of [the

---

17. Don Mellon has reminded me in this connection of Marx's claim that we are constituted not so much by nature as by the capitalist system to want more than we at present have, and thus to keep consuming at higher and higher levels in a vain attempt to finally satisfy our desires. The worker in a capitalist economy, Marx asserts, will always remain relatively poor (frustrated and unhappy) with respect to the rising tide of expectations. See the *Economic and Philosophical Manuscripts*, First Manuscript, "Wages of Labour," IX. But the idea that "it is not in our nature to stop possessing and enjoying at some point and be satisfied" (Kant 1790, 430–31) goes back a very long way.

good] life cannot be left open, either individually or socially. In doing this rather than that, we inevitably make decisions and give our lives a direction" (Borgmann 1984, 92). Or as he has only recently remarked to me, taking my view "posits and defends an impossible case, a society that can and does leave the question of the good life open."

Is my view an impossible case? I believe that its possibility becomes clearer once two familiar distinctions are made. One concerns the very different status of individual and society. An individual cannot in the nature of the case leave questions concerning the good life open. Our behavior necessarily implicates answers insofar as it is consciously goal directed. But what is true of the individual is not necessarily true of the society in which she lives. There are trends, of course, and something like a statistical characterization of the good life, as the summed average of the goals individuals choose for themselves. There are also general moral commitments, embedded in the rules by which members of a democratic society choose to live. But it does not follow from these facts that the question of the good life, understood socially and in nonmoral terms, is thereby closed. I understand that this is to construe "society" in a particular way, as an aggregate of individuals, but just such a construction is at the heart of the liberal democratic conception.

The other distinction is between closing the question of the good life in practice and closing it on principled grounds. One might concede that in some sense every society has a shared conception of the good life, certain behavioral norms, but nonetheless maintain that there are no *principled* grounds on which such a conception can be defended or attacked. It is for this reason among others that those who first formulated liberal democratic theory left the question of goodness open, and it is for this reason primarily that members of our own society are by and large so wary of those who propose a social and public characterization of it, even when, individually, they might be in accord completely with the kinds of criticisms Borgmann makes.

Do we, then, simply accept the "reign of technology," abandoning ourselves more and more to "devices" and to the ways in which they structure our options and their exercise? Or are there principled grounds on which to make a case for engaging in at least some of the kinds of simpler, quieter, and more traditional activities that Borgmann urges. I think that there are, and that they have to do with a core principle of liberal democracy, freedom.

Borgmann notes a contradiction in the way in which technology consorts with the principle of liberty. Of course, he is not the first to do so. On the one hand, technology, in making opportunities available, affords us genuine choice. The implementation of the principle of liberty, in anything more than a formal and political sense (and perhaps in that sense as well),[18] depended on it, as Borgmann himself makes clear. On the other hand, it is undeniably the case that technology more and more permeates our lives and that we are becoming increasingly dependent on it, at a loss when it breaks down. But in this dependency we are abandoning our liberty, no longer able to make choices that the very nature of "devices" precludes. On this point Borgmann and I agree: technology untempered is "debilitating," perhaps particularly when it is most "liberating."

It seems to me that this contradiction, at least as I have sketched it, is resolvable. That is, we do not have to renounce or withdraw from our technological culture if at the same time we want to lessen our dependence on it, to become more dependent on ourselves and on each other. What we do have to do is keep alive, individually and socially, a range of basic skills, if not also a set of "focal practices."

The notion of a "basic skill" here is comparative, and presupposes something like a picture of technology on "levels," or a spectrum with full-fledged "devices" at one end and humble "things" at the other. When we "bake from scratch," we use a number of different devices; even my wife, who used to start making bread with a standing field of grain, employed a combine. The point is that "baking from scratch" is in a clear and straightforward way less "technological" than putting a package of Pillsbury rolls in the microwave, more dependent on our own skills and motive power, the result not necessarily "better" but surely more expressive of the time and experience we are able to invest in it. In a similar sort of way, we are less dependent on pocket calculators when we are able to carry out the sums ourselves, less dependent on motorcycles when we are able to walk, less dependent on television when we are able to entertain ourselves. It is futile, and not at all human, to forgo all use of tools. But still we can distinguish here between "more" and "less," and claim that our ability to do with "less" technology is a measure of our independence.[19]

The crucial word here is "ability." It is not that we regularly supply heat in our home with the wood we've selected, chopped, gathered, and split, but that we *can* do so. And our ability to do so is a function of doing

18. Surely the fact that the extension of the rights that safeguard liberty went hand in hand with large-scale technological development is more than coincidental.

19. This point must, as James Allard has made clear to me, be relativized to context. Thus repairing motorcycles is in some sense "less technological" than repairing computers (presumably it involves more hand tools and fewer "devices"), but I don't want to say that the person who can repair motorcycles is more independent or self-reliant than one who can repair computers.

them at least from time to time, of acquiring the requisite skills and of keeping them and us in shape. The "more basic" things and practices that Borgmann rightly praises are not so much the adjuncts to our happiness as they are the conditions of our independence, although it is a fact that while we often avail ourselves of the most advanced technologies, the occasions on which we fall back to some extent on ourselves fill us with a measure of self-esteem, and perhaps also of pride, and often with a kind of joy.

Borgmann too calls attention to the "debilitating" character of advanced technologies. But he also thinks that "independence, if it is to be the pivot and warrant of a critique and reform of technology, is in danger of drifting into a mere modality, something you can or could do and make sure you will be able to do, an ability acquired and sustained by whatever means."[20] By "mere modality," he implies, among other things, that virtually any ability qualifies, so long as it stands in the right ("more basic") relation to whatever "devices" are at hand, and that these abilities are not necessarily "focal," as integrating and central parts of our way of life. In his view, of course, only those activities count as "focal" that have an important aesthetic dimension, which can be done not simply well but gracefully, which partake of the divine, at least in the minimal sense of linking them to a larger whole, and which take place in our lives not on occasion but as practices.

I am very much drawn to Borgmann's account of the good life, and in my own way have, very much in company with my family, tried to lead it. But as explained, I know of no way in which to make more than a "hortatory" case for it, and can only hope that our own "engagement," in part made possible by luck and circumstance, will be suggestive to others. At the same time, I think that however "modal" it might be, the declaration of independence made here provides the grounds for a critique and reform of technology. Let me conclude with two points in this connection.

The first point is that the "reform" of technology should involve something more than *making a place* for "focal things and practices" and leading some sort of balanced life. It should also involve changing the character of the "devices" themselves.

Return to the case of the rural doctor whose "engagement" in the practice of medicine is threatened by the use of expert diagnostic systems reduce her role to that of mere go-between. This sort of case is very much to Borgmann's purposes, for it illustrates the way in which technology can drive a wedge between justice (providing the very best diagnostic procedures to all) and goodness (the immense adventure in trying to figure out what is wrong

20. Private communication.

with one's patients). But how does it help to refer at *this* point to walking in the wilderness or preparing a ceremonial meal, however much at other points it clarifies and invigorates? The doctor is no Luddite; she is not hostile to technology (there is no fear of losing a job or taking a cut in salary), indeed she would have to think that if it improved the chances of peoples' lives being saved its use would be required on moral grounds. So far as I can see, the only real option to the diminishing of her life is to develop expert systems (computer programs) that work, or work best, only if the doctor making use of them is a resourceful practitioner.[21] The design of these systems would be difficult; there is no reason to think that it would be impossible.

More generally, if dependency is the issue,[22] then independence demands that our technology be less "devicelike," more open and accessible, easier to fix, in many cases smaller and quieter, and, as in the case of the proposed new expert diagnostic systems, more involving and challenging. Indeed, it is easy to imagine the possibility that a rural doctor, understanding how expert systems work and gradually refining her own diagnostic skills by carefully comparing her results with the computer program's, comes on occasion to overrule the system, perhaps in light of facts that don't show up on the initial questionnaire. Since for Borgmann the "device paradigm" is constitutive of technology, and of a technological way of life, this is difficult for him to contemplate, much less address. But I see nothing in the logic of technological development, whatever its past history might suggest, that *forces* us in the direction of ever-greater complexity, opacity, and "disengagement," and a great deal that signals the advent of a more "appropriate" technology.

The second point is more Thoreau-like. If we are independent, and in this sense "free," to the extent that we *can* get along with less sophisticated devices and more of our own skill and motive power, then it follows naturally if not logically that we should be at our "freest" when we are engaged in the "most basic" kinds of activities. These activities do not

---

21. As Dennett puts it (1986, 143), "Compare expert systems to musical instruments: today's expert systems are similar to autoharps, designed so that anyone can learn to play them, and with an easily reached plateau of skill. We should aim instead to develop systems more like violins and pianos—instruments that indefinitely extend and challenge the powers of the individual." Dennett himself and his colleagues at the Tufts Curricular Software Studio are in the process of trying to develop such systems.

22. As I think it is in the case of the rural doctor also. The case was sketched originally in terms of a conflict between the requirements of justice and *her* happiness. But the deeper fact is that the use of expert systems very much limits her sphere of action and judgment and, unless tempered, more and more makes her dependent on them.

forgo all tools, but have more or less directly to do with survival. What I have in mind are such things as hunting and fishing, gardening, and raising and training animals.

I suggested earlier that technology liberates us from the contingencies of survival, and to this extent makes it possible for us to shape our own destinies. So there would be a paradox in urging a survivalist way of life. But this is not the point. The point is that keeping primitive skills alive, and ourselves independent of all but the simplest technologies, is compatible with generally making use (within moral limits) of whatever technology is available. Granted, neither the liberation that technology makes possible nor the freedom that comes in not having to depend upon it was part of the original liberal democratic conception of society. But both make substantive the autonomy of the individual on which that conception depends.

Moreover, hunting and fishing, gardening, and raising and training animals are elemental if not also focal, they are more communal than solitary (since the first lesson we invariably learn is that our success at even the most primitive occupations is best secured with the help of others), and, unlike some of the other activities Borgmann mentions, long-distance running for instance, they clarify for those engaged in them the nature of our need for tools and the sense in which we lose control when the tools become mere "devices." These activities are also archetypal, anchored in myth and introduced by the gods. In this way, those who engage in them partake of the divine.[23] If our society does not afford the opportunity for youth to learn the skills that they demand, or for adults at least occasionally to participate in them, then our society too requires reform.

Questions concerning the character of the good life should not be closed. Paradoxically, the most important merit of *Technology and the Character of Contemporary Life* is that Albert Borgmann incisively reopens all these questions both by bringing to consciousness the ways in which technology has insinuated itself in our lives and by challenging the values it has engendered. Such reopening should be set beside the kinds of competence I have been urging as the special hallmarks of our freedom in a "devicive" world.[24]

23. "[E]very act which has a definite meaning—hunting, fishing, agriculture; games, conflicts, sexuality,—in some way participates in the sacred" (Eliade 1959, 27–28).

24. I am grateful for their careful comments on earlier versions of this essay to James Allard, Don Mellon, David Strong, and Albert Borgmann, whose example and insight over the years have shaped not only my ideas but my practices as well.

REFERENCES

Borgmann, Albert. 1984. *Technology and the Character of Contemporary Life: A Philosophical Inquiry.* Chicago: University of Chicago Press.

———. 1992. *Crossing the Postmodern Divide.* Chicago: University of Chicago Press.

Davis, Fred. 1979. *Yearning for Yesterday.* New York: Free Press.

Dennett, Daniel C. 1986. "Information, Technology, and the Virtue of Ignorance." *Daedalus* 115:135–53.

Eliade, Mircea. 1959. *Cosmos and History.* New York: Harper Torchbooks.

Kant, Immanuel. 1775. *Groundwork of the Metaphysics of Morals.* In *Gesammelte Schriften* (Prussian Academy ed.), vol. 4. Berlin, 1902–.

———. 1790. *Critique of Judgment.* In *Gesammelte Schriften* (Prussian Academy ed.), vol. 5. Berlin, 1902–.

Riesman, David. 1960. *Thorstein Veblen: A Critical Interpretation.* New York: Charles Scribner's Sons.

Veblen, Thorstein. [c. 1914] 1964. *The Instinct of Workmanship, and the State of the Industrial Arts.* New York: A. M. Kelley.

Veenhoven, Ruut. 1984. *Conditions of Happiness.* Dordrecht: D. Reidel.

Focaltechnics, Pragmatechnics, and the Reform of Technology

Larry Hickman

There is a great deal to admire in Albert Borgmann's critique of the ways in which contemporary men and women take up with technology. His suggestions about how such interactions can be improved are both serious in tone and richly suggestive. He encourages us to go beyond what he calls "the device paradigm" in order to consider "focal things and practices," about which we are able to communicate by means of what he calls "deictic" discourse.

As I understand it, his "device paradigm" is more or less what has come to be known as the program of the domination and commodification of nature advanced by Enlightenment rationality and the crass version of means-ends relationships that Langdon Winner (1977, 228) has called "straight-line instrumentalism." "Focal things and practices," on the other hand, are matters of transcendent importance, or what Borgmann calls "ultimate concern." What is focal, he tells us, "gathers the relations of its context and radiates into its surroundings and informs them" (Borgmann 1984, 197). "Deictic" discourse is our way of talking about focal things and practices; its purpose is to express and reveal. To speak "deictically," in Borgmann's vocabulary, means "to show, to point out, to bring to light, to set before one, and then also to explain and to teach" (Borgmann 1984, 178).

What is deictic is contrasted to what is "apodeictic" or explanatory. Although deictic and apodeictic forms of communication share the trait of being fallible, apodeictic communication is more limited in its scope. It "cannot disclose to us how it gets underway, i.e., how its laws are discovered and how something emerges as worthy or in need of explanation" (Borgmann 1984, 179). It is in this sense that neither science nor technology can furnish the ends in themselves that Borgmann thinks lie outside those fields and provide human life with its ultimate meanings.

It is not hard to see what is salutary about this account. Only a few true-believer free-marketeers would want to disagree with his claim that most of us in Western industrialized countries have a tendency to get too tightly locked into patterns of consumption, and this without reflecting on the place of our behavior within the broader picture. This pattern of behavior includes activities such as buying things that we do not really need, that we only briefly desire, of which we soon tire, and with money that we do not yet have. Such behavior is frequently exhibited at the personal level, and at the social and political levels as well.

At the personal level this pattern of commodification is sometimes found even in religious practice. The attitudes advanced by fundamentalist televangelists, for example, seem based not so much on the teachings of the financially insouciant Jesus, who urged a spiritual revolution, as on the agenda of the well-heeled Euthyphro, who was sure he could find the best way of doing business with the gods. In their straight-line instrumentalist worldview, for example, even the heaven of the fundamentalist Christian becomes commodified as the equivalent of a kind of eternal Caribbean cruise: a heavily advertised and expensive commodity that must be purchased well in advance, on the testimony of celebrities, and with the stipulation that all sales are final.

At the social and political level patterns of consumption distract attention from established ecological problems such as global warming, as well as from the types of engagement that an informed citizenry would otherwise have with pressing local, regional, national, and international issues such as the growing gap between rich and poor. Once there were citizens who initiated informed debates concerning issues of public importance. Now they seem to have been replaced by consumers who buy and use prepackaged ideas. In all this, something has been lost. Some may want to call it "the larger picture," others "the aesthetic dimension of life," and still others "the ground of our Being." Borgmann calls it "focal things and practices."

So Borgmann thinks that our view of focal things and practices, or ends in themselves, has come to be obscured by the smog generated by the device paradigm. How can we dispel the smog? We don't need to tinker with the sciences, since even though they cannot tell us anything about ends or values they are at least able to provide information about the "lawful fine structures of reality." We don't need to reassess the "deictic" discourses either, since they are our best hope of diminishing the effects of the device paradigm by allowing focal things to shine.

Borgmann's solution to the problem of obscured focal things and practices is to split technology into two ledger columns. On one side is the part of technology (the device paradigm) that is bad because it involves manipulation and transformation and therefore disburdens us from intimate contact with focal things and practices. On the other side there is the part of technology that is good because it operates in the background and supports focal things and practices. At the personal and familial level, television, stereos, central air conditioners, and eating in restaurants are bad, and piano music, wood-burning fires, and preparing meals at home are good. At the public level, cathedrals are good and the space shuttle is bad.

This is a matter of crucial importance to understanding what Borgmann wants to tell us, so it deserves to be stated in his own words. In matters personal and social, private and public, the thing to remember is that technology will never be reformed from *within* the device paradigm. Reform is only possible from the outside, as he puts it, by means of "*the recognition and the restraint* of the [device] paradigm" (Borgmann 1984, 220; emphasis in original). Borgmann's proposed reform of technology, then, intends "to restrict the entire [device] paradigm, both the machinery and the commodities, to the status of a means and let focal things and practices be our ends" (Borgmann 1984, 220). This plan of action would lead, in his view, to a "simplification and perfection of technology in the background of one's focal concern and to a discerning use of technological products at the center of one's practice" (Borgmann 1984, 221). In other words, small is beautiful and big is bad. Hands-on crafts and directly legible texts are good, and machine manufacture and electronic communication are bad.

*

Borgmann's program has some interesting similarities to other critiques of technology, past and present. Like Lewis Mumford, Borgmann is concerned that the organic tends to get mangled by the machine. Like Jürgen Habermas, he is concerned that technology has begun to colonize the life world of communicative action. Like Langdon Winner, he is sharply critical of the idea that ends of production and consumption tend to determine and justify their means. Like E. F. Schumacher and Hazel Henderson, he thinks that small is usually beautiful and that big is usually ugly. And like Amory Lovins, he favors a technology that is decentralized and self-sufficient. As important as these connections are, however, it is in the work of Martin Heidegger that we find Borgmann's spiritual taproot. He follows Heidegger in complaining that contemporary technological practice (the device paradigm) distracts us from the "great embodiments of meaning" (Borgmann 1984, 198). He also follows Heidegger in claiming that technology (the device paradigm) has been responsible for a kind of diaspora of focal things and practices. For both Heidegger and Borgmann technology provides the ground for a kind of negative hope. The vacuity of technology (again the device paradigm) serves as an opening or clearing in which focal things can once more be engaged with clarity and purpose.

To Borgmann's credit, however, there are also crucial points on which he seems to part company with Heidegger. First, whereas Heidegger seems to want to return to pre-technological enclaves as a part of his romanticized search for poetic meaning, Borgmann recognizes the futility of such thinly veiled luddism. He tells us that he wants instead to go forward toward a reformation of the device paradigm from the outside in a way that will result in leaner, more appropriate forms of technology. He recognizes that we can't live entirely without devices, such as pianos and wood-burning stoves, but he just wants us to live without the big, complex, distracting ones, such as televisions, computers, and space shuttles. In other words, whereas Heidegger apparently wanted to go all the way back to stone bridges, Borgmann says that he wants to go forward by going only part of the way back, to acoustical instruments and home cooking.

Second, whereas the social dimension of focal things seems to drop out of Heidegger's work, especially after his disastrous affiliation with the Nazis, Borgmann wants to emphasize the political and social contexts of such focal things and to have them play their part in helping us develop more sympathy and tolerance for one another. If we can just strip our devices down to the bare minimum so that we can focus more intently on matters of ultimate concern, this way of thinking can begin to permeate our social and political lives.

In all this, then, Borgmann is clearly advancing one of the best neo-Heideggerian critiques of technology now available. He supplants the romantic Luddism of Heidegger's later period with a kinder, gentler form of romanticism that attempts to give technology—at least in some of its more limited forms—its due. What's more, he attempts to introduce an agenda of social and political reform into his analysis of technology in a way that almost makes us forget the disastrous consequences of Heidegger's own maladroit program in that regard.

In sum, Borgmann thinks that we need about the same amount of explanation but much less transformation and manipulation. We need to be less occupied with the malleability of things and we need to downsize our dependence on devices. We need more expression, more revealing, and more articulation. We need much less big technology, about the same amount of science, much more small technology, and, what he thinks comes down to pretty much the same thing, much more art.

*

Even those who are sympathetic with some of Borgmann's goals, as I myself am, might nevertheless find themselves tempted to tweak some of the details of his program. First, I believe that he has cast the net of his condemnation of the device paradigm too broadly. He tends to do this by reducing the many and varied functions of certain devices to one essential property. Television, for example, is unequivocally bad because it displaces social relations (Borgmann 1984, 141). But surely television does more than that. Granted, there is much that is stupid on television. Nevertheless, the medium sometimes informs and educates, it sometimes serves as soporific or aphrodisiac, as required, and during times of crisis it can even bring people together. It functions in lots of other ways, too. In other words, whether we want to dismiss a particular tool or artifact as contributing to what we think is bad about our technological culture really has more to do with the function of that particular tool or artifact within a specific context than with some property that is claimed to be a part of its essence. My first objection to Borgmann's program, then, is that it rests on a rigid essentialism. I believe that a flexible functionalism can take us further down the road to understanding the complexities of our technological milieu.

Second, there is the matter of his focal things and practices. The issue here is not so much whether we often discourse about matters that are "transcendent" in some sense, and of "ultimate concern" to us, but whether someone might want to give a different account of what such things are, how they arise, and how they function. Simply put, I believe that Borgmann has given too much weight to the integrity of focal things and practices. He does not seem to be interested in their origins and he does not think that they are amenable to testing. Taken together, these two objections amount to a criticism of his account of means-ends relationships.

Before I get into these matters in more detail, however, I want to take a step back in order to examine the ways that Borgmann characterizes some of the accounts of technology that compete with his own. Several years ago I wrote a review of *Technology and the Character of Contemporary Life* for the journal *Research in Philosophy and Technology* (Hickman 1992). In his generous reply to my review, Borgmann indicated that his discussion of rival theories was only a "disciplinary aspect" of his work, and subordinate to its "substantive concerns" (Borgmann 1992, 346). I think he may have been too modest in this regard, however, and that his discussions of theories that rival his own do in fact shed considerable light on some of the more substantive parts of his account.

He thinks that all theories of technology can apparently be fit into one of three boxes—or four, if you count his own. The first three of these boxes are labeled *substantivist, instrumentalist,* and *pluralist.* The substantivist view holds that "technology appears as a force in its own right, one that shapes today's societies and values from the ground up and has no serious rivals" (Borgmann 1984, 9). Jacques Ellul is cited as a proponent of this position. Borgmann thinks this view unduly pessimistic, and for the most part opaque too, since it tends to stop the quest for explanation in the face of a menacing, vague, and unalterable force. Anyone who has spent much time reading Ellul probably won't be moved to quarrel with Borgmann on this particular point.

*

The instrumentalist view, on the other hand, holds that "there is a continuous historical thread that leads from our ensemble of machines and tools as affording possibilities of which we can avail ourselves for better or worse" (Borgmann 1984, 10). Borgmann thinks that the several varieties of this position, including those he calls "anthropological instrumentalism" and "epistemological instrumentalism," have some important elements in common. First, they treat tools as value neutral. Second, they tend to treat matters of ultimate concern as something to be established by efforts that are essentially private.

The worst of the worst in Borgmann's account are these instrumentalists. In order to go about their everyday business, he suggests, they have to assume and make use of the reality delivered to us by the scientists. But this is the very reality that they fail to treat with sufficient respect. They seem only to be interested in how things can be used. They are not really interested in fundamental reality beyond what is concrete and quotidian, and they think that abstract science is full of "convenient and useful formalisms" (Borgmann 1984, 30). The instrumentalists are bad because they keep telling us that "whatever works is good." Borgmann characterizes this view as shortsighted because it ignores the fact that tools are never mere means, but are instead "always and inextricably woven into a context of ends" (Borgmann 1984, 11). If substantivist views collapse from the weight of their own totalizing ambition, then instrumentalist views suffer from their inability to see the big picture and from their lack of common sense.

The third theory of technology is advanced by those whom Borgmann calls "pluralists." This view attempts to take the complexity of technology seriously as a "web of numerous countervailing forces," but it "fails reality" (Borgmann 1984, 11) because it ignores overall patterns, pervasive social agreements, and coordinated efforts. If the substantive view is a kind of black hole that collapses in on itself on account of its own gravity, and if the instrumentalist view is little more than froth, with no discernible direction of movement apart from what works at the moment, then the pluralistic view tends to go flying into a thousand pieces because there is no force at its center capable of holding it together. As we shall see, Borgmann wants his own view to have the gravity of a good solid center, but he doesn't want that center to suck in everything around it.

As a working Pragmatist of an eclectic sort, I am obliged to suggest that Borgmann's taxonomy of theories of technology is at least one short. Several years ago I published a Pragmatic account of technology that has its own taproot in the work of John Dewey (Hickman 1990). This view, which I will call "pragmatechnics" for short, doesn't quite fit into any of Borgmann's three categories. It does overlap with some of them, however, as well as with some of the features of his own view, which I will call "focaltechnics." Pragmatechnics is not substantivist, for example, since it holds no brief for reifications or foundations of any sort, whether they be scientific or metaphysical. It doesn't treat technology as a "thing" or "force" as does Ellul. In fact, it is even less substantivist than focaltechnics, which appeals to the "lawful fine structures of reality."

Pragmatechnics does not fit into the box that Borgmann labels "pluralist." Not content with merely describing experienced complexities, it is instead a thoroughgoing program of problem solving that involves analysis, testing, and production: production of new tools, new habits, new values, new ends in view, and, to use Borgmann's phrase, even new "focal things and practices." Pragmatechnics thus takes up a matter that appears to be absent in focaltechnics, that is, how we come by focal things and practices in the first place. Like focaltechnics, pragmatechnics argues that if technology is to be responsible then it must be socially and politically engaged. But unlike focaltechnics, pragmatechnics argues that if technology is to be responsible then it must also be able to test our focal things and practices.

*

Pragmatechnics is not an instrumentalist view in the sense in which Borgmann employs the term. It holds that a genetic or historical understanding of tools and artifacts is important, and therefore that scientific discourse can in fact disclose how it gets under way. But it also holds that human beings are much more than simply tool makers and users. It holds that there are vast and important areas of human experience that do not involve conscious tool use since they do not call for deliberation. Like focaltechnics, pragmatechnics holds that focal things and practices generally have to do with aesthetic experience, sympathy, and enthusiasm. Unlike focaltechnics, however, pragmatechnics holds that we sometimes need to examine our enthusiasms, aesthetic experiences, and sympathies, to subject them to tests of relevance and fruitfulness, and then to honor the ones that serve common goals and to reject the ones that are unproductive because they are based on what is merely personal or sectarian.

Although some of the features of pragmatechnics overlap those of focaltechnics, then, there are important differences as well. One of the most important differences is that pragmatechnics holds that value determination, including assessment of our most cherished "focal things," is an activity of intelligence and that intelligence is not located outside of human technological activity. For pragmatechnics, the tools and artifacts of our culture require ongoing evaluation, and such evaluation must be done in context. We cannot say a priori, or even on the authority of some end in itself, that small-scale devices are more appropriate than large-scale ones. We cannot say up front that learning to play the piano is more appropriate or meaningful than learning to play an electric guitar or learning to appreciate recorded music. Pragmatechnics just doesn't admit this type of reduction: it holds that intelligence demands that what is *techne* be subjected to a *logos,* whether the techne in question involves basic activities such as using wood-burning stoves or more complex ones such as building a space station. For pragmatechnics, the *logos* of *techne* is technology.

So focaltechnics seems to want to characterize device technology reductively as an addiction to the disburdenment from attending to focal things and practices and then to work for its reform from the outside, using science and deictic discourse to achieve a small-is-beautiful "appropriate technology" alternative in which such disburdenments are reversed. Pragmatechnics, on the other hand, characterizes technology more broadly as the invention, development, and deliberate use of tools and other artifacts to solve human problems. It does not distinguish between large- and small-scale devices a priori, or even on the authority of some end in itself, but only in the context of problems and issues as they are critically articulated. It holds that technical failures are usually due to a failure of intelligence, and that most devices, especially complex ones, exhibit a whole range of values and functions from which it is the job of intelligence to select the best and most meaningful. Appropriate technology is thus for pragmatechnics not a question of essence or scale but of function and context. Pragmatechnics argues that when we encounter a problem we can only start where we are, and not where we are not. And where we are is on the "inside" of technology in the sense that our culture uses a wide range of devices, both large and small, both complex and noncomplex, some of which are used in ways that enrich human life and some of which are used in ways that are not. This is a distinction of enormous importance. I hope that it will become clear during the course of the next few pages that it is a distinction that makes a real difference.

*

Borgmann has written that he thinks there are two big differences between our two views (Borgmann 1992). The first difference involves the question of whether a strong reform of technology is needed. The second and related difference concerns whether matters of ultimate concern are testable.

As regards the strong reform of technology, Borgmann is mistaken when he identifies the type of liberalism that Dewey advocated, and that I advocate, with the type of weak or feckless reform program that ignores excellence, as he puts it, because it is content to settle for progress in the areas of justice and prosperity. Dewey also argued against that type of liberalism. In his book *Liberalism and Social Action* (Dewey LW 11)[1], for example, he identified that particular type of liberalism as outdated and called for its replacement by a more robust type that would treat individual excellence as a social goal, and not as something that occurs haphazardly or as the effect of an "invisible hand." But the point of that book cannot be properly understood without remembering a point that philosophers often tend to forget, namely that Dewey was deeply involved in educational experimentation.

In *Democracy and Education* (Dewey MW 9), as well as in *The School and Society* (Dewey EW 1) and *The Child and the Curriculum* (Dewey EW 2), excellence was precisely what Dewey was after. It is true that he thought that the pursuit of such excellence is facilitated when certain conditions are satisfied, and that these include social justice and a decent level of material well-being. Even though some of our current political leaders seem to want us to ignore the fact, it is difficult to start a school day on an empty stomach. But social justice and a decent level of material well-being do not suffice to produce excellence. The sufficient causes of excellence are many and varied, so we cannot say in advance what they are. But education, both in the schools and in a lifelong curriculum, remains one of the best means of determining such causes on a case-by-case basis.

1. Standard references to John Dewey's work are to the critical (print) edition, *The Collected Works of John Dewey, 1882–1953,* ed. Jo Ann Boydston (Carbondale: Southern Illinois University Press, 1969–91), and published in three series as *The Early Works* (EW), *The Middle Works* (MW), and *The Later Works* (LW). These designations are followed by volume and, where appropriate, page number.

In order to ensure uniform citations of the critical edition, the pagination of the print edition has been preserved in *The Collected Works of John Dewey, 1882–1953: The Electronic Edition,* ed. Larry A. Hickman (Charlottesville, Va.: InteLex Corp., 1996).

Borgmann thinks that the type of liberalism that Dewey and I propose is faulty because it leaves the pursuit of excellence to the private sphere. This may be true of some varieties of neopragmatism, but it is not true of the view I am defending here. Pragmatechnics treats learning as a public activity that engages its wider context. Dewey, for example, did not write about the school *or* society, but the school *and* society. And whereas much of current educational theory focuses extraordinary attention on either the child *or* the curriculum (at the expense of the other), Dewey emphasized the interrelatedness of the child *and* the curriculum.

*

Borgmann also criticizes the type of Pragmatism that Dewey and I propose on the grounds that its program for reform is weak because it is piecemeal. Although he sees some merit in such an approach, because it is sometimes the only type of reform available to us, he is nevertheless afraid that it will lead to a "featureless landscape wherein piecemeal meliorism is the only kind of reform that remains" (Borgmann 1992, 346). Borgmann contrasts this view with his own, which he says aims at "knowing and revealing, as distinct from making and transforming" (Borgmann 1992, 346). It is difficult to know precisely what to make of this claim, since even the small technology that Borgmann places on the good side of the ledger requires some degree of making and transforming. As near as I can determine, it seems to involve a covert dualism in which ends are separated from means. For a pragmatechnics, knowing and revealing are not separable from making and transforming, since making and transforming are the means by which knowing and revealing are brought to fruition, and it is by treating knowing and revealing as ends in view that making and transforming are made meaningful. In other words, the two types of activities are related as means and ends.

This leads directly to the second big difference that Borgmann sees between his own view and mine. It involves the question of whether and to what extent matters of ultimate concern are testable. He thinks that they are not testable, but that they are contestable and attestable. I think that in many or most of the cases in which ultimate concerns come into conflict, which is to say when they become problematic, they are also testable.

Far from being mysterious or ineffable, then, matters of ultimate concern manifest themselves in terms of whether they contribute to the enrichment of the individual and the community. The problem is that what some call matters of ultimate concern are sometimes little more than idols of the tribe or the marketplace. Were matters of ultimate concern not testable, then there could be no systematic reform of any sort, and therefore no progress.

Even though Borgmann denies that matters of ultimate concern, or final commitments, are testable, he does allow that they are "contestable, attestable, and, alas, fallible." "If my ultimate concern is impoverished or oppressive," he suggests, "you are to contest it by attesting in your speaking and acting to one that is richer or more generous" (Borgmann 1992, 347). Focaltechnics thus privileges speaking and acting over experimental testing, and this places it at odds with pragmatechnics, which treats ends as ends in view, or artifactual and provisional, and thus as subject to experimental tests. But whereas focaltechnics places speaking and acting over against experimentation, pragmatechnics holds that experimentation includes speaking and acting and much more as well.

There is more than a verbal difference in describing something as testable on the one side and contestable and attestable on the other. It is true that there are times and circumstances when adequate tests are not available, and when all we can therefore do is attest or contest. It is also true that there are circumstances under which there is a subtle gradation in which testing on the one side and attesting and contesting on the other shade into one another. But it seems to me that if a strong reformer of technology has any obligation at all, it is to seek to develop such tests wherever there are differences of opinion about ultimate concerns. The strong reformer of technology cannot be satisfied with merely attesting and contesting. To fail to take the next step beyond attesting and contesting runs the risk of endless discussion, endless claims and counterclaims, with little hope of reform, either weak or strong. Attesting and contesting, as I understand the terms, have to do with *doing,* which may or may not be productive, whereas testing has to do with *making,* or the production of new consequences.

At one point Borgmann mounts a parody of the idea of testing final commitments: "For me to test [a profound mutual commitment] the way the Consumer Union tests cars would be to jeopardize and perhaps to destroy it" (Borgmann 1992, 347). In raising this issue, he has alluded to a matter that is of high importance to the Pragmatist: tests are appropriate, and indeed possible, only when there is a perceived problem. Deliberation is required, and is possible in any meaningful sense, only when there is an experienced difficulty. Further, means and methods will vary according to the nature and context of a doubtful situation. We do not test scientific hypotheses in the same way that we test works of art, and we do not test cars in the same way we test ultimate concerns.

*

I must confess, then, that I have some serious questions about the way that Borgmann treats the matter of ultimate concerns. As I have indicated, he tells us that they are not antiscientific. "Focal practices," he writes, "are at ease with the natural sciences. Since focal things are concrete and tangible, they are at home in the possibility space that the sciences circumscribe." Moreover, "the reform of technology would rest on a treacherous foundation if focal things and practices violated or resented the bounds of science" (Borgmann 1984, 219).

But he also tells us that focal things and practices are "unprocurable and finally beyond our control" (Borgmann 1984, 219). A focal practice is "the resolute and regular dedication to a focal thing. It sponsors discipline and skill which are exercised in a unity of achievement and enjoyment, of mind, body, and the world, of myself and others, and in a social union" (Borgmann 1984, 219).

Technology, on the other hand, at least in the sense of what I regard as his overly inclusive "device paradigm," does seem to Borgmann to be hostile to focal things and practices. His device paradigm is overly inclusive because it is concerned with things in their malleability, and especially as they become increasingly malleable as a result of our increased scientific understanding of them. I am afraid that I find more than just a hint of a kinder, gentler version of Platonism lurking in the background of this vision: what is transformable and malleable is put on one side as inferior, and what is an end in itself, "unprocurable and finally beyond our control," is put on the other as superior. The problem, then, lies not so much in his criticism of his device paradigm, since pragmatechnics also criticizes reliance on faulty means-ends relationships, but in the fact that he has made his device paradigm include too much. Consequently, focaltechnics seems to be anchored in what is unprocurable and finally beyond our control, rather than in what is amenable to tests and evaluation.

It might be objected that by arguing that ultimate concerns are testable the Pragmatist is left with nothing to ground her focal things, that is, her most cherished values. Such an objection would be both correct and incorrect. If asked to ground one of her ultimate values such as her faith in democracy as a method of association in "the lawful fine structures of reality," the Pragmatist would simply deny that such grounding is possible. She is, after all, a robust antifoundationalist. What she holds most dear is not grounded in this way.

But that what she thinks valuable is not grounded in this way does not mean that it is arbitrary or without substance. With Dewey, she would say that what is valuable is constructed, but that it is not constructed out of nothing. It is constructed out of the raw materials and intermediate stock parts that we get from our histories, from our cultural interchanges, and from our personal interactions. It is constructed by common political or social action to solve common problems. And it has been subjected to the tests of long series of experiments that have culled out a good many forms of social and political organization that did not work. In fact, it is subject to ongoing tests.

*

What is of ultimate concern to the Pragmatist may change over time as new ideas and ideals are generated, and as new methods are found to bring about what is most cherished. Moreover, what one generation counts as ultimate concern may be of little account to the next. There are abundant examples of this phenomenon, from the Crusades and Inquisition to the institutions that attempted to justify slavery. This is why a Deweyan Pragmatist would argue that democracy is equivalent neither to a set of institutions nor to a set of desired outcomes, but is instead a set of provisional methods (self-correcting as long as they are actually applied and as long as they continue to be tested) for finding solutions to common problems. As Dewey put it in an address in 1939, "democracy is belief in the ability of human experience to generate the aims and methods by which further experience will grow in ordered richness. Every other form of moral and social faith rests upon the idea that experience must be subjected at some point or other to some form of external control; to some 'authority' alleged to exist outside the processes of experience" (Dewey LW 14:229).

Now there is a way of reading Borgmann's program that saves it, at least from the perspective of the Pragmatist. On this reading, the device paradigm would be identified as just and only those aspects of technology that most informed critics, upon ongoing reflection and experimentation, find to be counterproductive or undesirable. Other aspects of technology—including big-ticket items such as most medical research and most of the space program and small-ticket items such as research into sustainable agriculture in developing countries—that have led to and supported what progress we have been able to make would then be absorbed into or counted as a part of what is outside of the "device paradigm." Such items would thus take their place with goals or ends in view and science on the good side of the ledger, and only the crass straight-line instrumentalism of the device paradigm would be left on the bad side of the ledger.

But I think that Borgmann would object to being read in this way, since what seems to be his preferred dividing line between device-technology, including most of electronic technology, on the one side, and small direct-access technology, science, and focal concerns, on the other, would have been substantially redefined. I think that he would object to this model because it would have the effect of placing what he takes to be questionable, namely manipulation and transformation, on the good side of the ledger where he thinks they do not belong. What appears to be his deep distrust of instrumentalism and his profound devotion to ends in themselves seems to militate against this way of looking at matters.

I conclude with a brief example. One of the ultimate concerns that Borgmann turns to again and again involves the family. This type of discussion is, and should be, a part of any discussion of the reform of technology for several reasons. For one thing, the family is or should be a primary place of education. For another, our social and political institutions, including our ideas and practices regarding what families are and how they should be supported, are themselves constructed artifacts. Discussions of the nature and function of the family are heard today in almost every quarter, and almost all of the parties to these discussions claim to hold the integrity of the family as a matter of ultimate concern. How, then, can there be so much disagreement about what a family is and should be? And more important, how can these profound disagreements be resolved?

Borgmann tells us that we need to demonstrate, to show, to reveal, what a family can be by our practice. If we do so, in his view, we will go beyond any type of technological treatment of the subject and attest to our ultimate values in ways that will move others to action. I believe that there is a great deal of truth in this suggestion. But we must go further. This is only one strategy among many for restoring the family to its proper place as a locus of social intelligence. Other strategies involve demographic studies, longitudinal psychological studies, and other types of experimental tests that can help us determine whether our intuitions about what is worthy of ultimate concern in these matters are warranted.

It is hardly a secret, for example, that many gays and lesbians want to be accorded the benefits that accrue to legitimized family relationships. They want to be able to adopt children, to make decisions about an ill or deceased partner, and to be eligible for the survivor's benefits normally provided by life insurance policies and retirement programs. In short, they want to be recognized and respected as families in the same way that heterosexual families are. For individuals who are a part of such relationships, these are focal things. They are matters of ultimate concern.

But there are some people whose ultimate concerns run directly counter to such aspirations. Such people tell us that their ultimate concerns demand that they fight such recognition, legitimization, and respect. They see in the ultimate concerns of gays and lesbians the seeds of moral decay, transgression against the will of God, and the corruption of the young. In states such as Colorado and Oregon they have mounted ballot initiatives designed to roll back even the civil rights that gays and lesbians currently enjoy. How are such fundamental conflicts over ultimate concerns to be addressed?

I believe that Borgmann is correct when he says that attesting and contesting constitute a part of the solution to this crucial and urgent social problem. Many gay and lesbian political activists would agree. They attest to their ultimate concern by refusing to conceal their sexual orientation and by bearing the scorn of their neighbors in a public fashion. Since they do not have access to public legitimization for their domestic unions, they attest to their love and commitment to one another in private religious ceremonies. In debates, in discussions, and in the courts they contest the customs, institutions, and statutes that are arrayed against them. They contest what they take to be unfair practices by challenging existing laws, retirement programs, and adoption policies. Sometimes they even engage in civil disobedience and go to jail. It is right that they should do these things, and it is certainly the case that their attesting and contesting constitute a step toward the reform of the social pressures that often serve to stress their family relationships and render them more fragile than they would otherwise be.

Attesting and contesting in these ways is an important step toward the solution of this pressing social problem. By itself, however, it is not enough. The fact is that ultimate concerns such as those associated with family life are testable. Some of the tests involve quantifiable data. It is possible, for example, to quantify the benefits to health and psychological well-being that accrue to individuals living in stable, committed, monogamous relationships. It is also possible to test the effects on children of growing up in a same-sex household. Such studies have been undertaken, and they continue to be undertaken. To any fair and open-minded person, their results are unambiguous. Such tests reveal that in this case, where two widely diverse sets of ultimate concerns are in conflict, one is well founded, promotes health and harmony, and is salvific. The other is uninformed and moved by fear of what is not known or understood.

Such experimental results may fail to convince those whose ultimate concerns render them incapable of accepting objective evidence. This was certainly the case during the civil rights struggles in the south during the 1960s, and it is still the case during the civil rights struggles of the current decade. But such results do matter to fair-minded people. They do matter in terms of the official positions of professional health organizations. And they do matter when conflicts enter the legal system.

*

This is only one example of what the Pragmatist means by testing ultimate concerns. I could have discussed any number of equally important matters, such as disputes concerning the direction that our form of democracy should take, or whether wilderness areas should be preserved from development. These matters also involve ultimate concerns, and they are also hotly contested.

I believe that the type of appeals I have just discussed, though they may appear too "instrumental" to some purists, will turn out to have greater positive long-term effects than any appeal to ultimate concerns as ends in themselves. This claim, too, is at least potentially testable.

My intuition is that Borgmann recognizes that there is this danger in talking about "final structures," "ultimate concerns," and "things in their own right," and that he tries to temper his treatment of these matters by appeals to science, or "the lawful fine structures of reality." He does so because he is also a democrat, a pluralist, and a person who believes deeply in the possibility of reform. I suspect that he also knows that public policy decisions are best made on the basis of experimentally informed discussion and open-minded debate, rather than on the basis of appeals to ultimate concerns. This is because what is accepted as ultimate is hardly ever also universal.

As I have tried to indicate, I am in general sympathetic with some of Borgmann's goals for the reform of technology. We need to move beyond narrow consumption and use models for living, and we need a new commitment to social intelligence. Further, I think that his emphasis on ultimate concerns will be especially attractive to those whose lives are influenced by the claims and interests of liberal theology and those who already feel strongly about environmental issues.

Nevertheless, I wonder if Borgmann's suggestions will enjoy wide appeal. Some of his readers, especially those who live in urban areas, will probably be uncomfortable with his suggestion that when we get beyond the simplest of devices such as acoustical musical instruments and wood-burning stoves we have allowed our ultimate concerns to become clouded. Some of his readers, especially those who are struggling with inherited religious and other cultural values that don't seem to be applicable to their everyday lives, will probably be uncomfortable with his view that ultimate concerns are unprocurable and finally beyond our control. And some of his readers, especially those who view electronic communication as one of the antidotes to provincialism, may reject his argument that our culture has too much technology and that technology is the source of our current political and social ills. I myself am uncomfortable with these ideas because my conception of technology is pluralist and functionalist. My Pragmatism leads me to think that where technology fails us, it is not technology that is the problem. It is ourselves. It is our lack of interest, our lack of insight, and our lack of devotion to the solution of pressing problems. And above all, it is our lack of ability to invent new tools and to criticize our own highly cherished values.

REFERENCES

Borgmann, Albert. 1984. *Technology and the Character of Contemporary Life: A Philosophical Inquiry.* Chicago: University of Chicago Press.

———. 1992. "Reply to Larry Hickman." *Research in Philosophy and Technology* 12:345–47.

Dewey, John. *The Collected Works of John Dewey, 1882–1953.* Ed. Jo Ann Boydston. Carbondale: Southern Illinois University Press, 1969–91.

Hickman, Larry. 1990. *John Dewey's Pragmatic Technology.* Bloomington: Indiana University Press.

———. 1992. "Technology, Final Structures, and Focal Things." Review of *Technology and the Character of Contemporary Life,* by Albert Borgmann. *Research in Philosophy and Technology* 12:337–45.

Winner, Langdon. 1977. *Autonomous Technology.* Cambridge: MIT Press.

SIX

Borgmann's *Unzeitgemässe Betrachtungen:*
On the Prepolitical Conditions of
a Politics of Place

Andrew Light

[E]verything that is truly productive is offensive.
—Friedrich Nietzsche, *Unfashionable Observations*

A PERSONAL INTRODUCTION

I don't know why I am drawn to Albert Borgmann's work for inspiration in my own philosophy of technology. In this volume alone, Andrew Feenberg's and Doug Kellner's politics are closer to my own; Carl Mitcham's style and attention to the contributions of analytic philosophy to our discipline is closer to my training; Larry Hickman's and Paul Durbin's pragmatism is more informative to my own methodological views; and Paul Thompson's interests in environmental issues are more pertinent to my central area of research.

But I have a kind of answer to offer here. I think I know now why I have stayed with Borgmann even though I've made it a point on several occasions to publicly demonstrate my disagreement with some of his central claims (Light 1995, 1997). The answer is simple, and struck me as pretty funny when I first realized it: Borgmann's work reminds me of my first philosophical love—Nietzsche. Before I turned to analytic and pragmatic approaches to environmental ethics and political philosophy, I was one of those precocious undergrads who found it all in Nietzsche. Fortunately I got over it (thanks to an assignment by Bernd Magnus to write on Nietzsche's political theory, which, contra the views of several Nietzsche scholars, I don't believe actually exists—a point I will return to later). But when you've read Nietzsche seriously, and with some openness and charity, he stays with you long after you think you've left him. Even though you may abandon any sense of philosophical reliance on Nietzsche's work, something remains. Perhaps what remains is a will, not to power hopefully, but to want to be different, to occasionally step outside your disciplinary boundaries either in style or substance, and to be contrary, troublesome, or just plain ornery.

Of course, you can inherit the same tendency from many other thinkers (Socrates, Diogenes of Sinope, Marx, and Wittgenstein spring to mind to varying degrees on different topics), but Nietzsche more than most pulls you in and dares you to be "unfashionable"—in a particular sense of that term that I will explain below—or else lose all self-respect as a critical thinker.

I have no idea whether Nietzsche had much of a social conscience. I doubt that he did. But what he did have was a deflationary cultural conscience: a critical take on the cultural biases of his day. And this desire to poke the ribs of certain received views is something Borgmann, and just about everyone else in philosophy of technology, shares today. I believe that philosophy of technology in general, and Borgmann's philosophy in particular, is as much about advancing unpopular cultural opinions as anything else. Based on that intuition I hope to show here how comparing the structure of Borgmann's and Nietzsche's work helps to reveal how philosophers of technology share something of the legacy of the Socratic torpedo fish as an important part of the role of their subfield in the discipline of philosophy in general.

But suggesting that we philosophers of technology have a particular cultural philosophical niche isn't an attempt to straitjacket the subfield into one or another role, but more an attempt to continue a conversation about why we are doing this sort of work at all, and in response to the nagging sensation that philosophers of technology ought to be more concerned with the question of their larger social role than many of our colleagues in other philosophical subfields. Because much of philosophy of technology is almost exclusively focused on the social effects of an increasingly technological culture, our sense of ourselves as philosophers will have an undeniable public dimension. This dimension may temper our cultural criticism through the realization that while our work may not be the sort of thing that will necessarily have a social impact, it should nonetheless be gauged such that it can contribute to a broader conversation in the public sphere about larger and more immediate issues than some other philosophers are engaged in. The upshot of this suggestion is that the political message generated by philosophies of technology needs to be taken seriously by us as a scholarly community. If anything, the Unabomber episode proved that the ideas of philosophers of technology—no matter how half-baked—can still resonate with some part of the population. Of course, whether the contributions of philosophers of technology to broader discussions of social problems and cultural controversies will ever be given a hearing by a more sober audience is quite another matter.

In this chapter, using Borgmann as an example, I'll first explain how at some moments the general drift of philosophy of technology takes on the flavor of what Nietzsche praised as *unzeitgemäss,* or "unfashionable." In accomplishing this task I'll briefly discuss the issue of what *unzeitgemäss* meant for Nietzsche—a question that still plagues the pages of Nietzsche scholarship but has recently taken a very interesting turn. But this foray into the Nietzsche literature will be used only to serve a larger goal. By arguing that Borgmann's way of doing philosophy of technology is unfashionable in Nietzsche's sense, I will show how this stance helps to make Borgmann's cultural conservatism more attractive, and not necessarily amenable to a politically conservative or reactionary agenda, as some have suggested. Many sympathetic critics of Borgmann, including me, have had a tendency to imagine a politically unsavory potential in the cultural commentary at the heart of Borgmann's work. Such a move, however, may not be fair. A careful look at the political component of Borgmann's work will show that his political theory is in some respects as absent as Nietzsche's arguably is. That is not to say that politics itself is absent from Borgmann's work. It is there, but, as in Nietzsche's, the political implications of Borgmann's work are what I will call "prepolitical conditions" rather than the components of a political theory per se. The question is whether Borgmann's cultural critique invites a specific political response to fulfill those prepolitical conditions, or whether those conditions are malleable enough to resonate with different kinds of political views. I'll finally argue that some of the most interesting prepolitical conditions that can be derived from Borgmann's philosophy of technology involve a commentary about the politics of "place." And this notion of place can be unpacked in several different political contexts, which while possibly unfashionable, need not necessarily have the romantic conservative implications often attributed to Borgmann's work. I will situate Borgmann's cultural conservatism as something more amenable to a positive view of the general role of philosophy of technology in cultural critique than it might at first appear. It is Borgmann's unfashionableness that is part of the reason I am drawn to his work. It is Borgmann's willingness, like Nietzsche's, to sometimes say things I don't feel comfortable saying, but in a way that I can embrace even given our other differences.

BORGMANN AND NIETZSCHE

I take my cue for a comparison of the argumentative structure of Borgmann's and Nietzsche's cultural and political views (rather than a comparison of their substantive philosophies) from Nietzsche's *Unzeitgemässe Betrachtungen,* or, as this work has recently been translated, *Unfashionable Observations.* Why this text and why this new name for it? Richard Gray, author of the new translation,[1] gives a nice reason why Walter Kaufmann's formerly accepted translation of *Unzeitgemässe Betrachtungen* as *Untimely Meditations* ought to be rejected. Starting with *Betrachtungen,* Gray argues that there was actually no good reason to suggest, as Kaufmann argued, that the figurative meaning of "meditation" or "reflection" ought to be preferred to the more straightforward "observation." Kaufmann had argued that in some respects Nietzsche's text was parallel to and a potential comment on Descartes's *Meditations.* But this attempt by Kaufmann to fit Nietzsche more neatly into the philosophical canon (a frequent move of Kaufmann's) gave a false impression of what Nietzsche was up to in these essays (Gray 1995, 395).

Nietzsche's essays in this book ("David Strauss the Confessor and the Writer," "On the Utility and Liability of History for Life," "Schopenhauer as Educator," and "Richard Wagner in Bayreuth"), as Gray points out, are not directed inward as a meditation (and largely not on the themes taken up by Descartes), but instead outward as observations about the culture of his day (Gray 1995, 396). Kaufmann's gift to philosophy, in the form of a more palatable Nietzsche connected more closely to other figures in the canon, also represented in some sense a loss to social and cultural philosophers and critics. Nietzsche's essays are also not "untimely" in any sense as they are quite concerned, again, with contemporary events. Says Gray: "This expression *[unzeitgemässe Betrachtungen]* succinctly spells out the common impulse linking these in many respects extremely divergent essays: Nietzsche's inimical attitude toward his 'time,' understood broadly as all those mainstream and popular movements that constituted contemporary European, but especially German, 'culture.' " (Gray 1995, 395). Nietzsche's goal in these essays is not, however, to simply appeal to the excesses of his contemporaries and answer with a call for us to return to the virtues of a classical age (which is arguably his project, at least for the development of an aesthetic sense, in *The Birth of Tragedy*), but more generally to take a stand against the growing cultural tide and fashions of his day. Nietzsche's attention here is to the critical stance that must be taken up against the very idea of fashionableness itself. This stance, says Gray, is central to Nietzsche's critique of contemporary life, "a critique that culminates in the Schopenhauer essay in the vilification of 'the three M's, Moment, Majority Opinion, and Modishness' " (Gray 1995, 398). *Unzeitgemäss* therefore

1. Volume two of the authorized translation of the complete works of Nietzsche (based on the Colli-Montinari edition). The new series was launched by the late Ernst Behler and represents the first complete rethinking of the English translation of Nietzsche's corpus, including the unpublished notebooks. Bernd Magnus is the editor of the series.

refers more to Nietzsche's sense of belonging to an isolated critical minority of those Germans resisting the Bismarckian tide, rather than to a retreat to the justifications and standpoints of another time. Nietzsche is fully engaged with his own time here. It would be a mistake then to interpret the project of these essays as part of a retreat to some form of conservative romanticism. For those familiar with the rest of Nietzsche's mature work, such a conclusion should not be a surprise.

Explaining this change in perception of the focus of Nietzsche's *unzeit-gemässe* essays in this new translation is not as much of a digression as it may appear. Reflection on this new appreciation of Nietzsche as a cultural critic may help us to decide how to interpret Borgmann's work in the same light. Of course, with Borgmann, there is less confusion over the focus of his work. But while there is no direct translation problem here from one language to another, there is a similar worry concerning the perception of the role of philosophy of technology in relation to philosophy in general, as was found with the interpretation of the meaning of Nietzsche's unfashionable essays. Similar to the mistaken impression that Nietzsche's work may be apolitical, or even acultural, there also exists a tendency to see philosophy of technology as an obscure topic of philosophical inquiry focusing on the ontology of artifacts. But arguably the field instead represents a potential bridge between the narrow terrain of philosophical criticism and the broader practice of cultural commentary. As the essays in the present volume should demonstrate, the field in general has a strong undercurrent of cultural criticism insofar as technology is one of the most important media of culture.[2]

Borgmann's work represents an exemplary instance of this overall direction in the field: providing his own version of cultural commentary is increasingly the focus of his work. The very form of Borgmann's *Crossing the Postmodern Divide* (or *CPD*) (1992) reveals this tendency: *CPD* is arguably an attempt by Borgmann to communicate the central theses of his more philosophically nuanced *Technology and the Character of Contemporary Life* (or *TCCL*) (1984) in a more prosaic form, amenable to a wider audience. And what is the tone of Borgmann's message to his public audience? Largely that contemporary culture has come to be dominated by its more recent technological innovations, combined with a moral and social critique stemming from this cultural observation. Borgmann identifies this problem thorough his figure in *TCCL* of the "device paradigm." Overwhelmingly,

___

2. In some respects, one could even argue that the field has become too exclusively focused on socially normative cultural criticism of technology. David Roberts and I argue for this point in more detail in "Toward New Foundations in Philosophy of Technology: Mitcham and Wittgenstein on Descriptions" (2000).

this approach makes Borgmann's work *unzeitgemäß,* unfashionable. Consider, for example, Borgmann's running criticisms in *CPD* of environments of virtual and hyperreality. Borgmann's critiques of such technologies, now embedded and embraced through popular culture, are unfashionable in the same vein as Nietzsche's *Unzeitgemässe Betrachtungen,* in particular Nietzsche's essay on David Strauss. Just as Nietzsche sought in that essay to deflate the growing excitement over the technical-military organization of the German state, Borgmann goes after the heady American love of the new and the rapture over the supposed new freedoms unleashed through mass technologies.

The unfashionable moments in Borgmann's work that are most important for understanding his politics emerge in his cultural conservatism, which seems inextricably part of his work. At the very least, Borgmann has raised such fears of romantic (or even reactionary) conservatism by his interlocutors. Early on, Andrew Feenberg voiced such worries, suggesting that Borgmann's apparent yearning for a premodern set of technological relations evidenced an inherent romanticism (Feenberg 1992; for a discussion of the relation between Borgmann and Feenberg see Light 1997). In Larry Hickman's contribution to this volume, one can see an argument for the evidence of a rightward ideological bent in Borgmann's corpus. Additionally, in a review of *CPD* by Andrew Cutrofello, Borgmann's views are summarized as "uncritical, undialectical nostalgia for premodern values" that are "dangerously conservative" (Cutrofello 1993, 96). Certainly, judging the accuracy of such claims is difficult. It is clear, however, that the political implications of many of Borgmann's views about focal practices, technologies, and human social relationships imply at least a form of small-*c* conservatism. Let me make it clear, however, that I think there is nothing a priori wrong with this conclusion. The same is probably also true of most of Borgmann's critics as well. Many (if not most) of Borgmann's interlocutors often preface their remarks with the claim that they agree with the direction, intent, or even conclusions that Borgmann is making. Even as strident a critic as Cutrofello admits that Borgmann's project of attempting to discern the divisive impacts of modernity on social life is worthwhile. But then, as do many others, Cutrofello goes on to voice objections that include a worry about the range and possibly regressive extent of Borgmann's cultural conservatism.

But if there is anything like a time that Borgmann's work resonates with, then it is certainly not the "morning in America" that has dominated the right-wing agenda in the United States since the Reagan years. Cutrofello admits that Borgmann "strongly supports many liberal political positions," while conservative critics like David Hartman find Borgmann completely unsatisfactory (Cutrofello 1993, 96; Hartman 1994). So even though Borgmann's *unzeitgemässe Betrachtungen* may sometimes appear untimely—appearing to pine for an older, simpler order of things—they are really more properly unfashionable, even to the right.

Welcome, in this context, is the fact that Borgmann brings us along in his observations without getting "preachy," as we used to say back home in Georgia. Those worried about the possible romantic implications of Borgmann's critique of popular culture could contrast it with another once popular attempt at unfashionable criticism: Alan Bloom's *Closing of the American Mind.* Bloom's critique also includes an implicit criticism of the rapture over technology prominent in American culture, and is clearly in service of a conservative, some would say reactionary, agenda. In one passage Bloom typifies his view of mass culture with the image of a thirteen-year-old boy doing his math homework while "wearing his Walkman headphones or watching MTV." "A pubescent child whose body throbs with orgasmic rhythms; whose feelings are made articulate in hymns to the joys of onanism or the killing of parents; whose ambition is to win fame and wealth in imitating the drag-queen who makes the music. In short, life is made into a nonstop, commercially prepackaged masturbatory fantasy" (Bloom 1987, 75). Bloom goes on to argue that only an immersion in Western classics can save us from this hyper hell. But surely Bloom's target, at least in this example, is too easy. There is much in the picture that is uncomfortable, but nothing that is seriously appealing as a viable alternative. The lure of the medium, contemporary music technology, is not grappled with but instead left behind through a focus on the message of the musicians. In contrast, at least Borgmann gives the new technologies a close look and provides a careful explanation for why they are attractive to most people. Perhaps Borgmann does not reduce the medium to the message, but he does evocatively show how the two go hand in hand in the device paradigm. In terms of his descriptions of everyday life then, Borgmann is arguably more timely than self-avowed conservative critics who eschew analysis of the objects of their critique in favor of inflammatory rhetoric.

But even if a definitive argument could be made that Borgmann's own views are not themselves regressive, the worry could be that the approach to cultural and political theory that he champions is itself amenable to a troublesome agenda appropriable by others. Let us therefore delve deeper. What about the structure of Borgmann's argument? Other than the conservative cultural implications of Borgmann's arguably untimely romantic lapses, is there anything in Borgmann's general approach to political questions that may seriously serve any particular ideological stance, even a stance that Borgmann himself does not hold? Clearly, if Borgmann's work could serve a pernicious ideological view then the wariness it sometimes generates would be justified.

NIETZSCHE AND BORGMANN'S PREPOLITICAL CONDITIONS

The strongest objection by progressives working through Borgmann's cultural criticism would have to be that his work may in fact contain inherently conservative elements that cannot be abstracted away from his generally unfashionable approach to the reform of everyday life. One version of that claim could be that Borgmann's unfashionableness is not simply reflective of the fact that he is doing philosophy of technology but that his particular neo-Heideggerean approach is itself inextricably regressive despite its pretensions to the contrary. But I do not believe that the structure of Borgmann's political work implies any necessary conservatism. What is the character of the structure of Borgmann's cultural and political logic? In answering this question let us return to Nietzsche for comparison.

Nietzsche, like Borgmann, is a cultural conservative, certainly even more so. Depending on how one reads Nietzsche, he may even be an elitist, even though that characterization is difficult to make stick. It is widely held by most readers of Nietzsche today that whatever the actual content of his political views, it was largely lost in the transformation of his thought to the service of Nazi philosophy, thanks to the dedicated work of his thoroughly fascist sister, Elisabeth Förster-Nietzsche (see Macintyre 1993). The Nietzsche-as-Nazi myth has of course been discredited, thanks largely to the work of Walter Kaufmann (see Kaufmann 1974). But the question remains how this transformation occurred in the first place. The dominant picture often represented is not so much to blame Nietzsche's unfashionable cultural criticism, or even the structure of his philosophical work, but to lay all of the blame on Elisabeth Förster-Nietzsche's inheritance of her brother's literary trust. Förster-Nietzsche had her brother's manuscripts pulled out of the waste heap (against her brother's wishes that they be thrown away), stitched them together, and published what Bernd Magnus calls the "non-book," *Der Wille zur Macht* (see Magnus 1988). This book would become the linchpin of the Nazi appropriation of Nietzsche.[3]

Without going too much further into the facts of the tragedy of Nietzsche's literary legacy, one can quickly move from this brief history to the question of the contemporary political appropriation of Nietzsche. If one looks at the literature on Nietzsche's political philosophy (of which there has been a small explosion in the last fifteen years—see for example

3. Of course, there has also been a steady stream of political commentators extrapolating a progressive view from Nietzsche's work. See Thomas (1983) and Strong (1996).

Detwiler 1990, Thiel 1990, and Warren 1988) the framework in which the portrayal of Nietzsche's views is taking place actually has not improved all that much. Today, philosophers like Bruce Detwiler are engaged in vigorous and, from my perspective, fascinating debates over an extrapolated Nietzschean political philosophy that has arguably tenuous roots in the rational reconstruction of Nietzsche's texts. Mostly what Detwiler, Warren, Thiel, and others have done is to derive a political philosophy out of the cultural criticism without a necessary reliance on any explicit Nietzschean political philosophy in the published works.[4]

Now these political theorists and philosophers are certainly not Nazis, but they may be stretching the limits of Nietzsche's work in order to get their interpretations. Much of these works involve the continuing project of explicating what it might mean to characterize Nietzsche's political thought in the same way as one would characterize the political thought of Marx, Mill, or Rawls. What would it mean, for example, to make sense of the characterization of Nietzsche's work as a form of "aristocratic radicalism"? Consistent with such projects we might ask what the necessary and sufficient conditions are for a community that would provide for the maximum flourishing of *Übermenschen*. One may leave this literature with the sense that if Nietzsche had a political theory, its reconstruction requires theoretical moves that are so difficult as to raise the issue of whether Nietzsche's political philosophy is worth paying attention to at all.

But one need not come to such a conclusion. The reason is that it is arguably the case that Nietzsche does not have a coherent, and certainly not a complete, political theory that could be interpreted one way or another. Let me simply suggest for now that what Nietzsche provides us with is at best an account of what I call the *prepolitical conditions* for some future political theory (in other words a kind of metatheoretical account), rather than a normative political theory itself. On my view, a prepolitical theory describes the conditions that must obtain in order for any healthy public sphere to emerge and sustain itself according to its author's diagnosis of the general cultural, social, or political conditions required for such thriving. In contrast, a political theory proper (or, perhaps more accurately, the theoretical justification for a political program) describes the optimal structure, role, and function of the public

---

4. The importance of relying on the published works, rather than the *Nachlass*, the unpublished notebooks is a matter of some scholarly debate. Warren (1988, xii–xiii) says that the polls of the debate have been defined by Magnus, who argues that the notebooks should not be used at all and Heidegger who argued that any of Nietzsche's writings were fair game. As such, Magnus rejects use of *The Will to Power*, culled from the *Nachlass* by Förster-Nietzsche, and Heidegger claims that the book contains the philosophy that Nietzsche really intended to write.

sphere itself. While certainly difficult to separate we could say for example that a theory concerning the necessity of private property ownership for the effective maintenance of a desired condition of personal freedom would be a prepolitical condition, while the argument for, say, a representative theory of government grounding and ensuring private property ownership would be a political theory proper that would in part respond to such prepolitical conditions. Certainly, most political philosophers produce both kinds of theories. (Also, one can imagine that some prepolitical theories produce conditions that demand a fairly narrow range of theories that can adequately respond to them.) My strong intuition (which I will not fully defend here), however, is that Nietzsche's work contains a robust account of prepolitical conditions but very little in the way of a substantial political theory, and that consequently many kinds of political theories can respond to his conditions.

Tracy Strong (1996) confirms these intuitions by arguing that while Nietzsche's work is amenable to a wide range of political interpretations, he by and large "does not write on political matters" (138). Nietzsche's political opinions "while more complex than often thought, are not of particular philosophical importance" (125). Strong also persuasively argues that Nietzsche does not reject politics as such but the politics of the modern world, which he finds particularly impoverished (138).[5] Of course, for Strong, there is a unique reason why Nietzsche does not produce a political tract, or set of political views, that is grounded in Nietzsche's rejection of the philosophical attempt to find a truth against the "reality" of the world, as, for example, it was represented in the Greek polis (Strong 1996, 141). But nonetheless, I would characterize Strong's conclusion about what we can derive from Nietzsche in political terms—that we must learn "to let uncertainty and ambiguity enter one's world, to let go the need to have the last word, to let go the need that there be a last word"—as a very good statement of Nietzsche's prepolitical conditions on any future politics (Strong 1996, 142). Importantly, for Strong, this "antipolitics" of Nietzsche's ensures that Nietzsche's work ultimately resists easy appropriation to any political theory.[6]

5. While again my intent is not to offer a substantive comparison between Borgmann and Nietzsche, one could indeed argue that Borgmann's work similarly rejects the politics of modernity (especially in *CPD*) even while he is arguably more hopeful than Nietzsche of a future constructive "postmodern" politics.

6. In earlier work, Strong (1975, 1988) also suggested something like this view. The difference in Nietzsche's political theory from that of a modernist theory is recognized by Strong in Nietzsche's commitment to a political analysis that goes beyond that of "selves meeting each other and seeking forms of mutual agreement" (Strong 1988, 162). Modern political theory can be more accurately identified according to Strong with "Hegelian or liberal politics." He continues, "For Nietzsche, the problem lies deeper: it is the having of

Let us assume then for now that I could provide a more rigorous argument that there is no proper political theory or political philosophy in Nietzsche's work but rather many opinions on different cultural and political issues of his day. Further, let us assume that using such texts as *Thus Spoke Zarathustra* one could argue that Nietzsche's political project, such as it is, is instead to provide an explanation of the prepolitical conditions for what we may loosely describe as a "postmodern political theory." Among the issues that Nietzsche's work helps us to understand is how a postmodern political theory would have to address certain normative questions regarding the possibility of articulating any system of justice, or even a common view of the good, out of antifoundationalist premises. Once this is outlined, we can imagine that many different political theories could meet the prepolitical conditions we could reconstruct out of Nietzsche's work and that there could be important philosophical differences between these theories. The same would be true of the fit between any competing political theories and the prepolitical conditions that they respond to. There are, for example, many different forms of government that could preserve rights to private property.

But if I am correct about the profile of Nietzsche's contribution to political thought, how do we reconcile the political theories attributed to him with the political views that are actually in his texts? Is Nietzsche's lack of an explicit political theory an error on his part? Should Nietzsche have had a more explicit political theory responding to his own prepolitical conditions, thus ensuring that something called "aristocratic radicalism," or more importantly some form of proto-Nazism, would or would not clearly describe his views? Yes and no. Certainly something is wrong if we expect that Nietzsche should have anticipated that the lacuna in his work would produce support both for the political implications of his views offered by his sister and for the readings offered by contemporary political philosophers. Clearly one should answer no, both if there appears to be no

---

selves at all that is first in question, then, second, and necessarily conterminous, the kind of self that is attained" (ibid.). It is at this level that I think Nietzsche is describing prepolitical conditions instead of articulating conditions for political selves. This prior level of analysis is articulated by Nietzsche before an investigation of social interaction between humans and exists at a level before that which makes a determination of the types of political institutions we may want to establish. Consequently, there should be different formulations of political agreement and institutional arrangement that can adequately respond to these prepolitical conditions. And because a variety of formulations of responses to these conditions can be voiced, such prepolitical conditions are loosely interpretable, that is to say, they are open-ended with respect to the political ideas that can be compatible with them. I first articulated my own views on the absence of Nietzsche's politics in "Zarathustra's Postmodern Politics," presented at the Northwest Conference on Philosophy held at the University of Washington, Seattle, November 1991.

reason for Nietzsche to have anticipated such problems and if one believes that there is in fact no lacuna in his theory at all, that is, if one believes that Nietzsche does have a full and complete political theory in his work and that it is securely antifascist.

But without embracing a completely open theory of textual interpretation whereby any reading of a text is equally valid, I would argue that nothing in Nietzsche's work necessarily cries out for a full political theory. This is especially true given the still open question of what sort of philosophy Nietzsche was actually engaged in. But while Nietzsche should not be held responsible for the later interpretations of the political implications of his work, perhaps his contemporary political interpreters should have their feet held to the fire. After all, if one claims that Nietzsche does have a full-fledged political theory, and that this theory is incompatible with fascism, then is it not the case that such arguments leave disputable the possibility that Nietzsche's work can be interpreted as protofascist in the end, that is, that this is a question that we can debate? If we instead defended the view that Nietzsche did not have a political theory, which could be protofascist or otherwise, then wouldn't we be in a better position to deny the claim that Nietzsche's work was protofascist? After all, if we argue that Nietzsche only has a theory of prepolitical conditions, then his political opinions (such as they are) do not necessarily serve any particular ideology or political theory even if they sometimes resemble the opinions of ideologues. Nietzsche's substantive political views, on the other hand, would serve as preconditions that a range of political theories would have to meet to be consistent with a Nietzschean philosophy. Thus, resisting the temptation to reconstruct a full political theory out of Nietzsche's work may be prudent for the continuation of Nietzsche's political rehabilitation, as well as being more philosophically defensible.[7]

For similar reasons, I am against attributing a political theory to Borgmann's work as well. Of course, nothing of the particulars of my inter-

7. Strong (1996) comes close to this sort of conclusion but does not state it as explicitly as I would like. After surveying the various political appropriations of Nietzsche's work (and poking a few holes in the more hopeful resuscitation of Nietzsche by commentators like Kaufmann) he concludes that while he (Strong) is not trying to show that there is danger that Nietzsche would have been a Nazi, he is "also not trying to exclude that possibility on the grounds that [Nietzsche's] texts 'show' us that he wasn't (or would not have been)" (131). Nonetheless, Strong also concludes that "Nietzsche is available to a wide range of political appropriations, indeed perhaps to all" (138). While we may quibble over what Strong means by "available" here, this view seems inconsistent with the overall direction of Strong's article, if it is not entirely vacuous (after all, anyone can attempt to appropriate any thinker for any reason). At the least, an appropriation of Nietzsche that claimed that his work provided a validating *theory* for some particular political system would be on completely tenuous ground if it is the case that Nietzsche does not have a theory to offer.

pretation of Nietzsche need be accepted to see this point with respect to Borgmann. Still, the comparison between the political misappropriation of Nietzsche and the misinterpretation of Borgmann is worth noting. I must confess that I came dangerously close to attributing such a full political view to Borgmann in a critique of his reality/hyperreality distinction in *CPD* (Light 1995). Even though I flirted with such a claim before, I now think more care needs to be taken in figuring out what sort of political theory Borgmann has, if indeed he has one at all.

The best political moments in *TCCL* are all prepolitical conditions, or even assessments of conditions by implication, and not a worked-out theory of politics or even the state as such. Starting with the area of political action itself, Borgmann's account is not prescriptive but almost entirely descriptive: "Political action, when it faces a crisis, finds its orientation in the device paradigm. Politics has become the metadevice of the technological society" (Borgmann 1984, 107). Unfashionable? Yes; especially to activists who think they are fighting "the system." But Borgmann does not have a political theory here that prescribes the right form of action or activism to obtain a specific ideological end. The implied critique of political action in this claim amounts to a prepolitical condition for activism—avoid a political discourse that entails the technological entanglements of the device paradigm (Borgmann 1984, 113).

Similarly, the embrace of pluralism toward the end of *TCCL* is indiscriminate with respect to the specific content of political activity. There are two kinds of pluralism open to us, "shallow pluralism afforded by the availability of many different commodities and the more profound pluralism of a diversity of focal practices" (Borgmann 1984, 228). Correspondingly, the direction for the achievement of pluralistic focal ends is a requirement for any political organization or institution seeking to break out of the device paradigm. Presumably (since Borgmann is committed here to some form of political pluralism) there are several alternatives for achieving a society that engages in or supports the pursuit of focal ends and practices. Borgmann says that the public reform of technology

> can only be achieved through a collective affirmation in a shared and public commitment to a certain kind of behavior or enterprise. It is to the body politic as a practice is to a group of persons. It allows us as a natural community to accomplish tasks that would lie beyond the capacity of individual decisions. (Borgmann 1984, 233)

Unfashionable? Again, yes; but the reason is that we know the kinds of reform of technology advocated by Borgmann are not popular to most people and not usually commensurate with the activity of most institutions. But Borgmann's account of the decision apparatus to reform technology is not part of a complete political theory. The public reform of technology is a prepolitical condition for any sort of public sphere supportive of a rejection of the device paradigm and its "shallow pluralism." Whatever public sphere adopts Borgmann's ends must do so consistent with his framework of a pubic process of social reform. Of course, Borgmann's ends make certain sorts of theories incommensurable with his prepolitical conditions. But his ends are not determinate enough to prioritize one of any number of theories that could meet his prepolitical conditions.

In *CPD,* however, Borgmann does embrace what he calls a "celebratory communitarianism," possibly indicating a preference for a particular kind of political theory, namely communitarianism. Perhaps this view does take us closer to a particular political theory. But a good case can be made that Borgmann still does not articulate a full political theory in *CPD.* The ends that Borgmann seeks to achieve through the celebratory character of this form of communitarianism makes this a communitarianism with as small a *c* as his cultural conservatism mentioned earlier. We can again imagine that different political systems (or at least different forms of communitarianism) could meet the prepolitical conditions of encouraging this celebratory stance. Borgmann has not embraced a formal Aristotelian or Hegelian communitarianism that stands against liberalism as it is formally theorized. (In fact, given Borgmann's deep admiration for Rawls it is hard to imagine him fully embracing any robust communitarian view.) Borgmann's is a stance that many systems, with varying degrees of changes to their structure, could achieve.

So far so good. If Borgmann does not have a proper political theory then we can be less concerned about the possible conservative implications of his views for similar reasons to why Nietzsche's unfashionable (sometime conservative) political views should not worry us. But we may still wish to ask, What, if any, are the criteria for deciding among the different political theories that could meet Borgmann's prepolitical conditions? In a more recent article in *Inquiry* Borgmann provides a nice answer in a brief meditation on the central requirement of any theory of the good life in the context of contemporary North American culture:

> [T]here is an urgent need to ask about the conclusiveness of social theories, because the culture at large neither leaves the gap between the necessary and sufficient conditions open, nor does it fill the gap in substantially diverse and incompatible ways. Rather there is a definite and prevailing style of life, a notion of the "good" life that most people in this country endorse and pursue. If that style of life

is essentially flawed, then any theory whose necessary conditions can be smoothly augmented to yield conditions sufficient for that style of life fails to be relevant to what actually distinguishes and troubles contemporary culture. Necessary conditions augmentable in this way need not be vacuous. They may well, if implemented, render contemporary culture more pleasant, prosperous, secure, or egalitarian. Yet in the end they may serve only to entrench a way of life not worth living. (Borgmann 1995, 148)

Returning to my beginning, one paraphrase of this claim would be that any theory of contemporary culture (or the political sphere that is home to that culture) that is not unfashionable—that is, that does not challenge what I take it that Borgmann believes to be the flawed theory of the good life entrenched in that culture—does not meet his prepolitical conditions for an adequate social or political theory. Thus, as long as the politics derivable from Borgmann's work fulfills this requirement, and anecdotally as long as the far right continues to object to Borgmann's work, then a far-right appropriation of Borgmann is highly suspect. Borgmann's *unzeitgemässe Betrachtungen* are just too unfashionable for such appropriations.

Borgmann's Metapolitics of Place

One last question remains: If Nietzsche's was a prepolitical condition for a postmodern political theory, what range of political views do Borgmann's prepolitical conditions, especially in *TCCL,* serve? The most obvious answer is that the conditions direct us to include in our political theories an attention to our relationship to technology. And just as one could claim that Nietzsche's prepolitical conditions challenge any theory that takes as uncontentious the relevance of foundationalism to normative political discourse, Borgmann's conditions maintain that any well-formulated political theory must address his concerns about technology as a necessary condition for establishing any political foundation for achieving the good life.

But I think there is something more specific going on in Borgmann's contribution to political thought that may serve a theory of the state. Borgmann's work provides us with an account of the prepolitical normative criteria for the literal ground of the public sphere. In a strong sense, Borgmann's work helps to define what a *place* should be, or more specifically, the place in which the good life can be pursued. In what remains of this chapter I want to sketch out first how an analysis of *TCCL* informs a theory of space and place and second how this theory amounts to a prepolitical condition for any prescriptive theory of the organization of the public sphere.

First, briefly recall Borgmann's ontology of technology. Borgmann divides technologies into "devices" and "focal things." Things are inseparable from the context in which they are embedded. To stand in some relation to a thing is to engage the social milieu of which it is a part. The social relations created by the wood stove in the North American frontier prairie home is Borgmann's paradigmatic example of the complex effects on human relationships of a focal thing. Each person fulfills a special function in relation to the stove, with some gathering wood, others stoking the stove, and others cooking (Borgmann 1984, 42). The stove is in this sense a medium for family life.

The device, on the other hand, is the foreground of modern mass technologies for which the world of things is the rapidly receding contrasting background. In contrast to the wood stove, a central heating plant "procures mere warmth and disburdens us of all other elements" (Borgmann 1984, 42). As artifacts that need no particular context, devices do not connect us to particular places, social relationships, or forms of agency. We can enjoy devices almost anywhere, anytime, with anyone. As a consequence, we lose whatever social identity we could have had with a comparable connection to focal things. This in turn diminishes the quality of any social relations that may have been influenced by our identification with specific practices involved with focal things. While the social relationships that are implied in the device/thing distinction are understandable and are the primary focus of many of the commentaries on Borgmann's supposed conservatism and romanticism, I think that this distinction implies larger suggestions about social organization than only those connected to the artifactual world. Specifically, Borgmann's device paradigm suggests a prepolitical view of what it means to be connected or attached to a particular place in a normatively significant sense.

In principle, Borgmann's device paradigm should be extendible to discussions of space for a couple reasons. The first reason is that one common method of making a normative argument for evaluating social spaces is to evaluate their component parts. On this account the overall characteristic of a built space is the result of the material characteristics of its component parts. This is a defensible assumption providing that an important rider is attached: while such a theory may provide a necessary component of a full description of a space, it probably will not be a sufficient description in most cases. Such arguments do not claim then that cities, for example, are solely reducible to their parts. Several writers have adequately pointed out that a full description of the city must include other elements. But an analysis of cities as collections of things and devices is at least an important part of a full account of the character of cities as built spaces (see, for example, Lynch 1960 and de Certau 1984). If a space is dominated by one type of artifact, and we have a good theory of the normative implications of that artifact type, then a legitimate claim can be made that the normative assessment of the artifact type can be extended as an important part of the description of that space.

Given the intuitive plausibility of this sort of argument, it is no surprise that many commentators on technology have also been theoreticians of social space, in particular cities and suburbs. Diverse theorists such as Murray Bookchin (1986, 1992), John McDermott (1986), and especially Lewis Mumford (1934, 1938) not only have written on both topics, but often use the same theoretical base for their criticism of certain types of artifacts and specific types of spaces. For Borgmann then, following the device paradigm, we can say that there are both *thick* spaces (constituted by focal things) and *thin* spaces (constituted by devices). We can expect to find thick and thin (or thinglike and devicelike) spaces of many types and many combinations. The trick is to unearth the terrain of complexity of a space under the weight of the device paradigm and then to carefully discern the character of the space among the things and devices that constitute it.

The second reason that the device paradigm ought to be easily extendible to questions of space is that Borgmann's theory is concerned with how technologies affect human identities and human relationships. By extension, whatever can be understood about the dominant artifacts of a space should help us to assess the probable social effects of that space. Therefore, by providing a theory for how artifacts socially influence us, Borgmann has already provided us with a way to account for spatially influenced identities and relationships. The preferred normative content of social space given the device paradigm should be clear: thick spaces that give us a sense of *place*. Borgmann's theory is in essence a theory of place and how artifacts shape our sense of place.

We are better off, on Borgmann's account, when we attach ourselves to a specific location for specific reasons. Borgmann's is not a theory of how isolated individuals relate to artifacts, but how artifacts shape us in relation to others to form places, or thick spaces, and, I would argue, consequently thick public spheres. This is not, however, to argue that an extension of Borgmann's account would result in a critique of thin spaces (interpreted as sparsely populated places). If thick spaces are inhabited by focal things, or at least share the relationships Borgmann wishes us to have with focal things, then thin spaces are those that are made up by devices or inhabited by devices or that share the relationships we associate with devices. A device-generated space can come in many forms. A typical thin space might be an overpass crowded with the same gas stations, fast-food restaurants, and chain motels that are found all along North American highways. There is nothing distinct about these areas as places, and their virtue (to their owners) is that such franchises can be found anywhere and have devoted consumer followings. A thick space, in contrast, could be either a multifaceted, wonderfully crowded urban area or a simple mom-and-pop store in the country that is marked by a distinct connection to its surroundings. As a prepolitical condition then the diagnosis of the device paradigm asks a political theory responding to it to normatively prioritize the protection, preservation, and even restoration of communities of thick spaces as the places of political interaction.

Again, there is much in *CPD* to support such an interpretation of Borgmann's work. But in general if we have a theory of place in *CPD*, any requirement of movements and institutions to work toward establishing particular kinds of places for political life is open in the sense that such spaces are part of a prepolitical condition rather than a demand for a particular kind of political system or ideology. We can imagine several types of political theories meeting Borgmann's "thick place criteria" in different ways as a priority, for example, for schemes of distribution or as part of a more robust definition of community. And, we can also imagine how such a focus on place would be unpopular for a variety of political theories. Take for example a political theory advocating some sort of vulgar corporatism. A preference for thick spaces as the place for political life would discourage systems of organization that are disconnected, e.g., those *requiring* a workforce to keep mobile, that is, to be disconnected to any particular place, or those encouraging an ethic of consumption that prefers the homogeneity of places as part of a vision of the good life. Vulgar communists would be required by Borgmann's conditions to abandon utopian, or reductionist visions of a classless society as the sole criterion of a good community. Thick spaces as places cannot be entirely satisfied through economic categories of analysis such as changing the mode or means of production or the rules of ownership. Similarly, those wishing to reduce place attachment to one exclusive set of ideological relations of blood or nation would find no fertile ground in Borgmann's mosaic of place.

Conclusions

Are such observations as these, suggesting ways in which Borgmann's work provides a check on the kinds of political use to which his theories can be put, sufficient to ensure that his prepolitical conditions cannot be co-opted into unsavory political theories? Is it enough to ensure that Borgmann's work won't come to serve the interests of fascists, or neoconservatives, for example? Is it, again to evoke the Nietzsche discussion, enough to ensure that Borgmann's work is unfashionable but not romantically untimely?

Perhaps the inherent structural limitations of Borgmann's views are comforting, but maybe they are not enough to put all these worries to rest. I have two final recommendations to help to make sure that Borgmann's work is as unfashionable as we would all want it to be, especially unfashionable enough to never be too appealing to cultural reactionaries like Bloom. First, Borgmann himself needs to think through whether the risk of misappropriation of his work is high enough to warrant a more explicit political theory to meet his own prepolitical conditions. If this risk is high given the past readings of his political intents (real or imagined), then Borgmann may wish to put an additional obstacle in the path of the rational reconstructions that will no doubt follow his work. After all, this is not something Nietzsche ever did (see Strong 1996, 139). But second, because the threat to the reading of Borgmann may be more at the level of philosophical misuse rather than popular political abuse, we, as philosophers of technology who seek to maintain the unfashionable character of our colleague's work, and in general of our discipline as a whole, need to watch for those misappropriations that may occur. We are a community of scholars. We will be a better community when we scrutinize our own scholarly practices in these untimely times. Perhaps only then can we achieve as a philosophical guild the sort of celebratory community that Borgmann hopes for the larger society.[8]

References

Bloom, Allan. 1987. *The Closing of the American Mind.* New York: Simon and Schuster.

Bookchin, Murray. 1986. *Post-scarcity Anarchism.* Montreal: Black Rose Books.

———. 1992. *Urbanization without Cities.* Montreal: Black Rose Books.

Borgmann, Albert. 1984. *Technology and the Character of Contemporary Life.* Chicago: University of Chicago Press.

———. 1992. *Crossing the Postmodern Divide.* Chicago: University of Chicago Press.

———. 1995. "Theory, Practice, Reality." *Inquiry* 38:143–56.

Cutrofello, Andrew. 1993. "Must We Say What 'We' Means? The Politics of Postmodernism." *Social Theory and Practice* 19:93–109.

de Certau, Michel. 1984. *The Practice of Everyday Life.* Berkeley and Los Angeles: University of California Press.

Detwiler, Bruce. 1990. *Nietzsche and the Politics of Aristocratic Radicalism.* Chicago: University of Chicago Press.

Ellul, Jacques. 1964. *The Technological Society.* Trans. J. Wilkinson. New York: Vintage.

Feenberg, Andrew. 1992. *Critical Theory of Technology.* Oxford: Oxford University Press.

8. My thanks to Albert Borgmann, Eric Higgs, and Ron Perrin for helpful comments on this chapter.

Gray, Richard T. 1995. Translator's afterword to *Unfashionable Observations,* by Friedrich Nietzsche. Stanford: Stanford University Press.

Hartman, David. 1994. "The Holy Spirit Reduced to the Postmodern Spirit." *New Oxford Review,* November, 28–29.

Kaufmann, Walter. 1974. *Nietzsche: Philosopher, Psychologist, Antichrist.* Princeton: Princeton University Press.

Light, Andrew. 1995. "Three Questions on Hyperreality." *Research in Philosophy and Technology* 15:211–22.

———. 1997. "Wim Wenders and the Everyday Aesthetics of Technology and Space." *Journal of Aesthetics and Art Criticism* 55:215–29.

Light, Andrew, and David Roberts. 2000. "Toward New Foundations in Philosophy of Technology: Mitcham and Wittgenstein on Descriptions." *Research in Philosophy and Technology* 19: forthcoming.

Lynch, Kevin. 1960. *The Image of the City.* Cambridge: MIT Press.

Macintyre, Ben. 1993. *Forgotten Fatherland: The Search for Elisabeth Nietzsche.* New York: HarperCollins.

Magnus, Bernd. 1988. "The Use and Abuse of *The Will to Power.*" In *Reading Nietzsche.* Ed. Robert C. Solomon and Kathleen M. Higgins. Oxford: Oxford University Press.

McDermott, John. 1986. *Streams of Experience.* Amherst: University of Massachusetts Press.

Mumford, Lewis. 1934. *Technics and Civilization.* New York: Harcourt, Brace and Company.

———. 1938. *The Culture of Cities.* New York: Harcourt, Brace and Company.

Nietzsche, Friedrich. 1995. *Unfashionable Observations.* Trans. Richard T. Gray. Stanford: Stanford University Press.

Pitt, Joe. 1995. "On the Philosophy of Technology, Past and Future." *Society for Philosophy and Technology Electronic Journal* 1: np.

Strong, Tracy B. 1975. *Friedrich Nietzsche and the Politics of Transfiguration.* Berkeley and Los Angeles: University of California Press.

———. 1988. "Nietzsche's Political Aesthetics." In *Nietzsche's New Seas.* Ed. Michael Allen Gillespie and Tracy B. Strong. Chicago: University of Chicago Press.

———. 1996. "Nietzsche's Political Misappropriation." In *The Cambridge Companion to Nietzsche.* Ed. Bernd Magnus and Kathleen M. Higgins. Cambridge: Cambridge University Press.

Thiel, Leslie Paul. 1990. *Friedrich Nietzsche and the Politics of the Soul.* Princeton: Princeton University Press.

Thomas, R. Hinton. 1983. *Nietzsche in German Politics and Society 1890–1918.* Manchester: Manchester University Press.

Warren, Mark. 1988. *Nietzsche and Political Thought.* Cambridge: MIT Press.

SEVEN

On Character and Technology

Carl Mitcham

The central word in the title of Albert Borgmann's *Technology and the Character of Contemporary Life* names a special reality. The pivotal and longest of the three parts of the book, "The Character of Technology," argues for understanding that which gives our world its distinctive historical features in terms of this same unique reality. The primacy of character is further implied by the statement of the problem in part one, which adopts a scientific realism to characterize our understanding of phenomena, and by the reform pointed toward in part three, which is a correction not so much of technology as of its dispositional context. Yet remarkably enough there is no extended discussion of character in the text itself, and the term "character" is conspicuous by its absence in the index.

What follows is thus an attempt to reflect on the character of the argument in this remarkable book by means of reflections centered around the remarkableness of character. No claim is made to have done full justice to the richness and complexity of Borgmann's own reflection. There is only an aspiration to complement the nobility and difficulty of its guiding task: disclosure of the character of focal things and practices and the life that, when properly honored, they may impart—and that, in their absence, withers.

THE WORD *CHARACTER*

The word *character* is derived from the Greek χαρακτήρ, naming initially a tool to engrave (χαράκτειν) and derivatively the mark engraved or impressed especially on coins and seals. Metaphorically, as early as the fifth century B.C.E., the word comes to include the mark impressed on persons and things, that is, their character. For Herodotus, Aeschylus, and others the term had already been adapted to what were to become the spheres of moral psychology and literary criticism. Theophrastus, as Aristotle's successor, wrote his ἠθικοὶ χαρακτῆρες, *Moral Characters*, with its thirty sketches of personal vices, as an extension of the analysis of virtue in the *Nicomachean Ethics,* where the word ἦθος, ethos, in fact has a denotation similar to χαρακτήρ.

In classical Latin, adoption of the word, not to say the idea, stumbled— although in revealing ways. For Cicero, in what is presented as a gloss on Aristotle's *Topics,* with regard to the discussion of "what something is," "additur autem descriptio, quam χαρακτῆρα Graeci vocant" (*Topica* xxii, 83) "What something is includes description, which the Greeks call χαρακτῆρ." In Roman moral discourse the terms *natura, indoles, ingenium, animus,* and *mores* are not always distinguished.

But in medieval Latin it became common to transliterate the word, which acquired special theological significance. From the time of Augustine *character* was applied as a technical term for that which, according to Christian theology, is impressed upon the soul by the sacraments, especially baptism, confirmation, and ordination. Thomas Aquinas, for instance, speaks not only of grace as the principal effect of the sacraments, but also *"de alio effectu sacramentorum, qui est character"* (*Summa theologiae,* 3a. 63, 1), that is, "of the other effect of the sacraments, which is character." Throughout the Middle Ages, however, less spiritual meanings retain their force, and from late Latin usage the word *character* found its way into English, where it exhibits a diversity of associations. Not infrequently it simply denotes a letter of the alphabet.

The character sketch was first given a modern force and a new twist in that bourgeois criticism of bourgeois culture represented by the work of Ben Jonson, Jean de La Bruyère, Joseph Addison and Richard Steele, and others—which is a manifestation of the moral criticism of human affairs that can be traced from Michel de Montaigne to Albert Camus. In La Bruyère's *Les caractères de Théophraste avec les caractères ou les moeurs de ce siècle* (1687), the character sketch becomes an essay in social criticism that may even be interpreted to point toward the revolution of a century later.

Following the French Revolution, Immanuel Kant distinguishes between an empirical or physical and intelligible or moral character. The latter is grounded in a unity of the will that likewise provides a basis for transforming the self and the world. No longer a comment on the world and its persistent foibles, character becomes that which will remake the world in its own image.[1]

---

1. For a slightly expanded historical review, see especially Eucken 1908. For more extended treatment of the historico-philosophical background on the theme of character and related ideas see Funke 1958. This extensive scholarly work, following an introduction (part A) and a foundational analysis (part B) of nature and "second nature" as well as the phenomenon of *Gewohnheit,* reviews *hexis, ethos,* and *habitus* as they occur in, among

Borgmann's own concern for character and its reform is not unrelated to some of these developments.

CHARACTER IN THE SOCIAL SCIENCES

The preceding sketch of the history of the word *character* has been in some measure guided by an entry in James Hastings's *Encyclopedia of Religion and Ethics* (1908). It could not have been so guided by Paul Edwards's *Encyclopedia of Philosophy* (1968), Donald Borchert's supplement to same (1996), or Edward Craig's *Routledge Encyclopedia of Philosophy* (1998), because none of these later reference works includes an entry on the topic—thus reflecting the modern disembedding of moral theory, a disembedding that involves the replacement of moral character with moral principle and moral calculation.[2] For most of the twentieth century the discussion of character was largely abandoned to the social sciences.

In the social sciences, especially psychology, character has been an issue of some dispute. A. A. Roback's *The Psychology of Character* (1928), for instance, defines character, under Kantian influence, as "an enduring psychophysical disposition to inhibit instinctive tendencies in accordance with regulative principles" (450).[3] But European and North American psychologists and psychoanalysts have taken different approaches to interpreting this inhibitive disposition, the former emphasizing general and determinative features, the latter largely replacing character by the notion of personality, with its more plastic and individualistic connotations. With regard to the formation of character, there are two basic views: internalist and externalist, by nature and by culture. Again, European emphasis tends toward the former, so that as a stable and invariant feature of human behavior, character may even be related to the animal phenomenon of imprinting (Konrad Lorenz) or to unconscious archetypes (Carl Jung).

There nevertheless exist character disorders that call for psychoanalytic treatment. Insofar as character includes stability, strength, and goodness,

---

others, Plato, Aristotle, Plotinus, Thomas Aquinas, Luther, Montaigne, and Pascal (part C); discusses habit or custom as a systematic idea in Berkeley, Hume, Reid, Kant, Hegel, Husserl, and others (part D); and concludes with a terminological study (part E) and analytic index (part F).

2. The lacuna at issue is only to the smallest degree remedied by Nussbaum 1992.

3. This book remains the single most comprehensive scholarly survey of theories of character ancient and modern, philosophical and psychological. See also A. A. Roback's entry on "Character" in the *Encyclopedia of the Social Sciences* (1930). Note that the *International Encyclopedia of the Social Sciences,* ed. David L. Sills (New York: Macmillan and Free Press, 1968), contains no entry on "character"; instead it has extensive entries on "personality" and "trait theory."

disorders are evidenced by emotional instability, inability to pursue long-range goals, or failure to be concerned about others. Yet character disorder is not the absence of character, since the disorders have their own stabilities and are as such difficult to alter. Wilhelm Reich, however, in appropriate North American exile, went so far as to describe character itself as a disorder; it then becomes one of the main tasks of psychotherapy to break down that "character armor" that conceals the inner self in order to release its rich fluidity. (Another presentation of this view of character as determinative prison may be seen in Dutch director Mike van Diem's 1998 movie *Character,* which received an Academy Award for best foreign film.)

In more mainstream North American psychology the preferred concern is maladaptive personality disorders. Personality is conceived as a provisional integration of diverse psychological traits, and it is these traits with their stabilities that can be mixed and matched. Under different terminology, John Dewey, in his *Human Nature and Conduct* (1922), which argues for bringing social psychological conceptions to the attention of philosophy, foreshadows this distinction. For Dewey, habit is "a certain ordering or systematization of minor elements of action," while character is "the inter-penetration of habits" (Dewey 1922, pt. 1, sec. 2 [pp. 39, 37]).[4]

For Dewey as social reformer, the primary barrier to increasing scientific intelligence in human affairs is not human nature but habit and custom, the latter defined as "widespread uniformities of habits" or "collective habits" (pt. 1, sec. 4 [p. 35]). Human nature and its impulses or instincts are, as it were, no more than a possibility space for habit and customs. (Dewey's "custom" is much the same as "culture" in the social-scientific sense.) Habits and culture formed under and adaptive to one circumstance establish a character that may be maladaptive to a changed environment—hence William Fielding Ogburn's notion of "cultural lag."[5] Progressive reform proceeds piecemeal but interactively on two fronts: within persons through education and in the environment through engineering. With regard to industrial or technological activity, the root maladaptation is "found in the separation of production from consumption—that is, actual consummation, fulfillment" (Dewey 1922, pt. III, sec. 9 [p. 185]).

---

4. Dewey references are to both the original edition (by part and section) and to the critical edition in the collected works (by page). The original division into part and sections, which begin renumbering within each part, is for some reason replaced in the critical edition by part and continuously numbered chapters.

5. Ogburn 1922, pt. 4, "Social Maladjustments," presents the "cultural lag" thesis. For more on Dewey's strategic use of the concept of habit in relation to technology, see the miscellaneous references in Hickman 1990. Interestingly enough, as Hickman notes (16), Dewey even describes habits as dynamic technologies.

Despite his absence, there are echoes of Dewey's argument in Borgmann's analysis of the machinery/commodity divide and his call for replacing commodity consumption with focal things and practices.

The Aristotelian Metaphysics of Character

According to Dewey, nature, at least human nature, is almost nothing. Character, including habit and culture, is everything. The traditional view is subtly different.

In Aristotelian metaphysics it is possible to identify three somewhat overlapping types of reality: first, those entities or beings that are by nature; second, human beings (which are by nature a very special kind of being); and third, those things that are made by humans, that is, artifacts. A fourth type of reality, which is never entitive, is that of accidents, which have no being in themselves but have being only through some entity.

The first and second of the entities possess natures or essences, although the essences of the second are in a crucial sense incomplete; the third do not, at least not in the strict sense. Within Aristotle's *theoria,* natural entities are understood as self-generating substantial unities of form and matter, act and potency. Artifacts of human making, although the external generating process imitates nature by further informing some natural reality (wood or clay, for example) treated as matter, are pseudo-entities, because they never achieve a substantial integration of form and matter. If a bed were to sprout, says Aristotle, not a bed would come up but an oak (*Physics* II, 1; 193a12–15).

For those realities that have natures, such as rocks, plants, and animals, what is important is to come to know their natures or essences, a knowing that takes place through or by means of form. For those things that are made by human beings, what is important is to come to know their uses.[6] For human beings, with their special nature—a nature that is in potency to or able to be stamped with a second nature of habits[7]—what is important is to come to know their character. Character is the special, historically received form of that unique entity-in-potency that is human nature (see *Nicomachean Ethics* II, 1; 1103a24–26). In the world of human beings, in between essence and utility, one finds character, an integration of nature treated as matter and of cultural form.

Setting aside the contrast with utility, the first treatise on character as

6. Evidence that this conclusion is not wholly dependent on an outmoded Aristotelian metaphysics is provided by the analysis of artifacts in Searle 1995, which explicitly adopts a science-based metaphysics.

7. At one point, for instance, Aristotle goes so far as to describe habit as becoming nature. See, e.g., *Problems* XXVII; 949a27–29.

such emphasizes contrast with nature. As is proclaimed in the very first sentence of his *Characters,* Theophrastus never ceases to wonder at the way in which all Greeks breathe the same air but do not all exhibit the same character. Nature does not impose character, although the humors provide differential foundations for its formation and exercise (see, e.g., Aristotle, *Problems* XXVII–XXX; 947b10–957a36). Neither is human nature simply a possibility space for character. Human nature is oriented toward, in need of, in potency to, character. Character may not be a truly substantial union of matter and form, but it is more substantial than the union found in artifacts. Although character has its being through another and is therefore to some extent accidental, it is not an accident with the same contingency as, say, the color of a triangle, which in no way helps realize the essence of triangle.

Here then is a contrast with Dewey's understanding of the plasticity of character. For Dewey, all unities, in the physical as well as in the social world, are more or less contingent or accidental. Certainly character is accidental, and as such readily subject to alteration by means of education and social engineering. The only legitimate guidelines for its transformation are "ends in view" and their effective, intelligent, scientific pursuit. For Aristotle, however, essence or substantial union of matter and form cannot be changed. Artifacts or nonsubstantial unions of matter and form may rather easily be changed. Character or the historical union of matter and form may be changed, but not easily; or differently stated: cannot be changed, except with difficulty. Moreover, as a perfection of nature, character is properly guided not simply by external ends or relations but by its reflection and manifest realization of inner depths.

In his own appeal to connections, internal as well as external, manifest through focal things and practices, especially as exhibited in a focal concern for nature and wilderness, there is surely a touch of Aristotelian sensitivity in Borgmann.

NATURE AND CHARACTER

According to Aristotle, Theophrastus's teacher, philosophy begins in wonder—and in the recognition of a distinction between *phusis* (nature) and *nomos* (law and custom or convention). Prior to philosophy, human beings noticed that all things exhibited more or less stable and distinctive patterns of behavior. Fire burns, dogs bark and wag their tails, human beings speak. There are also the ways or customs of the various human communities: the way of the Egyptians is different from the way of the Hebrews, and the way of the Greeks is different still. All such ways, nonhuman and human alike, were originally explained as having been given by a god or the gods. The first philosophers were "those who discourse on nature"—as opposed to "those who discourse on the gods" (see, e.g., *Metaphysics* XII, 6; 1071b27). What god or the gods explain for those who preceded philosophy, nature, as "the principle of motion and being at rest" (*Physics* II, 1; 192b21), explains for the first philosophers. But what nature is not able to explain, or to explain as immediately as it explains the ways of fire and dogs, are the ways of human beings. Outside of and apparently at odds with nature is custom or convention.

The first philosophers thus focused attention on nonhuman realities, especially the realities of the heavens. When Socrates brought philosophy down from the heavens to dwell in the cities (Cicero, *Tusculan Disputations* V, iv) attention necessarily shifted from what is by nature to what is by custom or law, especially justice (see *Gorgias* 382d ff.). It nevertheless remains the argument of Aristotle that some justice is not merely by custom but also by nature (*Nicomachean Ethics* V, 7; 1134b18–1135a4). The argument that all justice is a matter of custom rests on a radical split between nature and culture. Nature is unchangeable and the same everywhere (just as fire burns the same in Egypt and in Greece), whereas culture varies from one people to another. The counterargument is that culture may be grounded in or related to nature, not unlike the way speech is by nature, although the particular language spoken is from culture.[8]

The reflective analysis or *theoria* of localized or embedded customs in pursuit of human good and their relations to human nature constitutes the discipline of τὰ ἠθικά, ethics, a substantive form of the adjective ἠθικός, a cognate of the noun ἦθος, meaning an accustomed place or, in the plural, the haunts and abodes of animals and, with respect to human beings, their customs, manners, and characteristics. The word ἦθος (beginning with a smooth eta) is, Aristotle himself says (*Nicomachean Ethics* II, 1; 1103a17–19), related to ἔθος (beginning with a smooth epsilon), meaning simply custom or habit. (It is also etymologically related to ἔθνος, a group of animals or people living together.) Individual disposition both grows out of and establishes customs and laws. In the individual human being ἔθος as what is possessed is, however, called ἕξις, a stable disposition (or habit), which is in turn the foundation of ἀρετή ἠθική, moral virtue or virtuous character.

In this usage character is now to be contrasted not with the unchange-ableness of nature but with the even greater changeableness of passion and opinion. Character is that which unifies an individual, transcending the

---

8. Cf. the analysis of language as in tension between nature and convention in Borgmann 1974, esp. 19–23.

particular acts and ephemera of human affairs, so that it can account for the stabilities of moral action. It is the person of courageous character who acts or is likely to act courageously. It is one of temperate character who can be expected to act temperately. It is one of just character who will do justice. A man or woman of character exhibits true unity across the vicissitudes of fortune.

In this sense, again, character is not opposed to nature but is itself a second nature, the source of motion and rest in those things that can take on definition beyond that provided by nature—and, in a derivative sense, in those things dependent on that mind and hand embedded in such a second nature. Thus, contrary to Heidegger, human making and using in the case of modern technology does not have an essence, not even one that is "by no means anything technological," but a character. ἦθος ἀνθρώπῳ δαίμων, wrote Heraclitus (fragment 119). "For the human being, character is daimon (divinity, destiny)." Likewise with contemporary life, and with technology: ἦθος πόλι δαίμων. "For the polis, character is destiny." ἦθος τέχνη δαίμων. "For technics, character is destiny."[9]

THE INEFFABLE PRESENCE OF CHARACTER

Like destiny, character is invisible, even when it is present. We are reminded of both this invisibility and presence by our inability to pin down character despite our repeated attempts to do so, if not under the name of character then under other names—*daimon,* pattern, model, role, ideal type, paradigm. It is of the core of character to be difficult to define or debatable.

Metaphysically, character is neither essence nor accident, neither universal nor particular. Epistemologically, character is neither conceived nor perceived. Anthropologically, character is neither determination nor freedom. Character is a reality always in-between, neither universal essence nor particular accident, a limited determination and an equally limited freedom. In it lies the heart of the human condition.

Although I would without hesitation affirm the possession of a certain character, indeed that my person could not manifest itself except through those patterns and roles in which my actions are embedded, I am not able immediately or with any certainty to say what this character is. My character is not revealed to me in any straightforward way by what I know about the world or about my body, neither simply by my actions or my feelings. At

9. The influence of Leo Strauss should be apparent in the first two paragraphs of this section. For relevant discussions of the Aristotelian theory of character beyond those cited in note 1 above, see material in Annas 1993; Dunne 1993; Nussbaum 1986; Sherman 1989. For Martin Heidegger's theory of technology, see Heidegger 1954; the brief quotation is from the third paragraph of this text.

the same time we are always ready to describe the character of others in precisely such terms. She has the character of a scientist; he is a computer nerd. He is an ectomorph; she is a dumb blond. He is a Chicano; she is feminist. She is a woman; he is a man.

As soon as such characterizations are stated, they immediately become subject to critical objections. They appear to be "caricatures"—stereotypes, unfair generalizations—as much impositions from without as the naming of some pattern of behavior arising from within. When introspection fails and we ask others for insight into our own character, we readily object to what we may be told—even when silently admitting its truth. "But that is not the whole truth," we insist. "There is more to me than that."

One of the paradoxes of character: what is known through character is both more and less than that in which character rests. My nature, in Aristotelian terms a form that requires more form than it has even to become the form it is, cannot be grasped except as it takes on character. Yet what character manifests about my nature is only partial. This second nature that it takes on to become what it is, hides at the very same time that it discloses. Character is embedded in a hermeneutic paradox.

The same paradox is manifest at the societal or institutional level. For the Greeks, the state exhibits a variety of characters: tyranny, aristocracy, democracy, and so on. For Marxists, societal character orders are feudalism, capitalism, socialism. For Weberians, leadership exhibits one of three ideal types: charismatic, traditional, or bureaucratic. But such ideal types are neither simply perceived nor subject to any covering law. Thus empiricist objectors readily complain that they are no more than nominalist constructions.

When technology is analyzed in terms of character, objections are even more pronounced. In *La Technique ou l'enjeu du siècle* (1954) Jacques Ellul proposes, in opposition to Heidegger's essentialist thinking, a "characterology of technology." Borgmann (1984, 9–10), along with others, nevertheless attributes to Ellul a substantialist theory of technology, even as his own characterology is subject to similar charges. Character study is neither a microempiricist social deconstructivist examination that can document all the actors and actants involved nor a general deductive-nomological examination that can subsume technological particulars under some covering law.[10] It is sociological theory that aims to identify higher-

10. Ellul 1954. See also Ellul 1977. To reject an opposition between Ellul and Borgmann (not to mention one between Ellul and Dewey) in regard to the primacy of character is not, however, to deny a difference between the two characterologies. Borgmann's philosophical elucidation of the character of technology under the device paradigm that separates commodity from machinery in order to procure disburdened availability is more penetrating

level features that persist even amid microlevel changes—although many dismiss it as poetry.

Borgmann's attempt to think, between science and aesthetics, technology as device and its characterological reform through focal things and practices is the attempt to live honestly in the distinctly human realm, the realm where character matters.

CHARACTER MATTERS, THINGS MATTER

Character matters. Character cannot be avoided. To try to avoid character is to intentionally dismiss a central dimension of reality.

Liberalism argues that insofar as character matters it does so only at the personal level, and that what matters more is freedom, including individual liberty to choose and to form one's own character. For liberalism it is thus crucial to reject a substantialist or essentialist view of both the state and technology in favor of an instrumentalist view, and to structure both accordingly; only this leaves all substantive decisions about character formation up to users of neutral means (legal and technical). Borgmann sides with liberalism in rejecting a substantialist view of technology, but argues that in the instrumentalism of technology there is nevertheless an informing character that cannot be so easily rejected.

Antiliberalism argues that insofar as freedom matters it must do so primarily in the service of character and that the idea of individual liberty to choose and to form one's own character is at once illusory and perverse. Character formation slips up on us when we least expect it. Those kinds of reality that readily take on a second nature, precisely because they are ordered toward it, tend in the first instance to take on character unawares. People acquire character before they know what has happened, the same way they learn language before they know what language is. Only once someone has character is he or she able to think about changing it—which is more difficult than acquiring it. To refuse to recognize this pervasiveness of character is in fact to open oneself to perversions of character. Borgmann sides with the antiliberals by agreeing about the inconspicuous and pervasive presence of character, but argues for its reform.

For Borgmann human character is manifest in both technology and contemporary life. The problem is that pattern or character in contemporary life is hostage to the influence of the pattern of technology.

---

than Ellul's sociology of technology as conscious, rational efficiency constructing a world of rationality, artificiality, technical autonomism, self-augmentation, monism, universalism, and autonomy. The former can actually be used to elucidate the latter. For micro-actor/actant analysis, see, of course, Latour 1987.

Technological devices are extremely shallow. One trait or function is predominant; all others are arbitrarily exchangeable and progressively eliminated. An autographic work of art is paradigmatically deep. We cherish it (the original) and refuse to exchange it for a work to which a necessarily limited number of traits have been transferred (a copy) because the latter may bar or mislead deeper insight than we now have. We must recover [the human] as a being of absolute depth and learn to realize that if we increasingly surround [human beings] with shallow things, [they] will become shallow also. (Borgmann 1973, 35)

Absent modern technology, the human condition was characterized by hard-won splendors emerging from an enforced but multileveled engagement with the richly textured things present in economies of subsistence. Through technology the human condition is disembedded and transmuted. The harsh work of making and using artifacts is divided up and apportioned out to scarcity-reducing machinery that supplies commodities increasingly free of embodied burdens. Hunting and gathering gives way to shopping, agriculture to suburban development, citizenship to voting and polling, the crafts of making and using to mass production and consumption, meal and pilgrimage to snack and tourism.[11]

But why should we seek to reform the commodification of life and its glamorous hyperreality? What is missing that deserves to be retrieved—that could not be added as another commodity, another virtual experience?

It is things, focal things. Things matter.

Teleology and Promise

Traditional arguments for character reform are teleological. As a secondary form responding to the active receptivity of a primary form in potency to greater or more intensive reality, character may nevertheless—precisely through its historical contingencies—enter into and react upon that potency in a way that ignores or fails fully to honor the substantial imminence it realizes. The child who never learns to speak and is thus unable to sing the beauties of the world and the one who never learns to play with others and is thus unable to imbibe the pleasures of convivial solidarity have both taken on characters unworthy of human nature. Natural law theory, with

11. Judicious citations from the extensive literature on this transmutation are provided in Borgmann 1984. Complementary references alluded to in the present paragraph include Lothario dei Segni (Pope Innocent III), *De miseria humane conditionis* (late 1100s), and associated reflections; Guardini 1927; Polanyi 1944; Illich 1982; Achterhuis 1988; and Hoinacki 1996. The next paragraph alludes, as well, to critical categories that restate the device and focal thing distinction developed in Borgmann 1992.

its elaboration of the basic forms of human virtue, is the flowering of this teleological judgment of character. In articulating the transhistorical orderings toward goodness (to preserve being, to bear and nurture children, to live in community and to know truth) natural law provides a more than human reference for measuring human affairs.

Modern science, which Borgmann sees as in itself an unqualified achievement and as true, deprives him of the teleological line of argument. In the world as revealed by modern natural science there is no teleology, no potency. Potency has been replaced with possibility, immanent to everything but ordered toward nothing. All that exists are options for arrangement and rearrangement, none of which can be judged by means of the possibility precipitates of physics, chemistry, and evolution. In a way, everything becomes character of a peculiarly indeterminate type, if not accident. The typically modern response has been to make of character what we want, to undertake a value analysis, to settle for "ends-in-view" (Dewey) or an aesthetics of the technologically extended senses (postmodernism). What Borgmann argues is that there is nevertheless a possible richness, depth, and integration of experience that is not being seized upon, even though it was promised in the early modern formation of commitments to technology and then democracy. Historical promise replaces metaphysical potency as the ground for the criticism of character.

The promise of technology and the promise of democracy are one—yet vitiated by internal failure if not contradiction. The technological and democratic promise was a disburdened realization of a free encounter with reality and a more richly developed and diversified human experience. Foreshadowed in the harsh engagement with pretechnological things were freely realized goods of character that await reincarnation in metatechnological focal concerns. The irony of technology, for Borgmann, is that the technological melioration of the cruelties, harshnesses, and sufferings of the human condition has at the same time and unnecessarily abandoned its splendors. The promise of technology was a world less "nasty, poor, brutish, and short" (Thomas Hobbes) in order to be more richly human. But that life that was once deep engagement and rich diversity has become the shallow availability of fashionable commodities. The original promise calls out, through the inchoate dissatisfactions of contemporary life and in the marginal but powerful presence of focal things with their possibilities of reorienting practices, for renewal.

There is nevertheless a problematics of this divided promise. Might not the historical trajectory of the promise in fact reveal its hidden truth? For some (Niccolò Machiavelli, for instance) the promise was surely not so much an attempt to achieve in disburdened manner the splendors of character found among the ancients as it was an effort to "lower the standards" of character for the moderns (Strauss 1958, 174 ff.). Even insofar as the two sides of the promise can legitimately be held together, as Borgmann strives nobly to do, is it not possible that there is a danger of asking too much? The modern attempt to overcome technologically the fragility of goodness as experienced by the ancients may itself be more fragile than was realized by those who promised more than could be delivered. *Corruptio optimi quae est pessima* (The corruption of the best is the worst). Finally, absent teleology, how is any thing, not to mention character, to retain a focal presence? Is it not inevitable that any metatechnological focal concern will be seen within the perspective of scientific realism as no more than an "end-in-view" or aesthetic construction?

ECHOES

In Borgmann's interpretation of the promise of technology one may discern echoes not so much of Machiavelli as of Walt Whitman's interpretation of the promise of democracy. For Whitman, as for Borgmann, the promise of democracy was best captured by John Stuart Mill, who identified "two main constituents": "1st, a large variety of character—and 2nd, full play for human nature to expand itself in numberless and even conflicting directions."

"It may be claim'd," Whitman also wrote,

> that common and general worldly prosperity, and a populace well-to-do, and with all life's material comforts, is the main thing, and is enough. It may be argued that our republic is, in performance, really enacting to-day the grandest arts, poems, &c., by beating up the wilderness into fertile farms, and in her railroads, ships, machinery, &c. . . . I too hail those achievements with pride and joy: then answer that the soul of man will not with such only . . . be finally satisfied. . . . Out of such considerations, such truths, arises for treatment . . . the important question of character.

Despite "general good order, physical plenty, industry, &c., (desirable and precious advantages as they all are,) . . . society, in these States, is canker'd, crude, superstitious, and rotten."

To address the problem of the democratic character, Whitman argues that "a new founded literature . . . consistent with science . . . is what is needed." It is poetry that will redeem the promise of democracy.[12]

---

12. All quotations here are from Walt Whitman, *Democratic Vistas* (1871), paragraphs 1, 12–14, 15, and 19, respectively.

Poetic Orientation and Reform

Entities that do not invite character also do not need orientation. Rocks and trees know by nature what they are and how to behave. People do not. Character, as a stable disposition (whether viewed as second nature or as possibility precipitate), thus also constitutes an orientation. Orientation is grounded either in nature (teleologically interpreted by the senses or the gods) or in poetry. In the latter case, reform of character and its orientation will originate with poetry.

Technology and the Character of Contemporary Life is divided into three parts. Part one argues a distinction between three types of orientation: apodeictic, deictic, and paradeictic. Apodeictic orientation is provided by the deductive-nomological explanation of phenomena typical of the physical sciences. In the strongest case, given the possibilities defined by a law of nature (F = ma) and specific values for a sufficient number of variables (for example, a force of 1 Newton and a mass of 1 kilogram) other variables are determined (the acceleration is 1 meter/second). By contrast, deictic discourse selects out and calls attention to the reality of particulars, and is typically found in poetry and the arts. Thomas Cole's "The Oxbow" (1836) shows forth the grandeur of North American wilderness; Walt Whitman's *Leaves of Grass* proclaims the emergent order of pluralist democracy (in, e.g., "Song of Myself," 1855) and the grandeur of its technology (in, e.g., "To a Locomotive in Winter," 1876). Finally, paradeictic argumentation articulates order at a slightly more general level and is typical of history and the human sciences. *The Communist Manifesto* (1848) lays out the prototypical bourgeois form of production; psychology analyzes human behavior in terms of character and personality types.

Part two presents a paradeictic argument concerning modern technology. The character of modern technology is grasped neither through a covering law nor as the playfully contingent social construction of some particular technology such as bicycles, Bakelite, or fluorescent light bulbs.[13] The multiplicities of modern technologies—from automobiles and airplanes to televisions and computers—are unified by their manifestation of the device paradigm. There is a pattern or character that is the stability and strength of technology. But it is a pattern or character that we experience as

13. See Bijker 1995. As an aside: An argument might even be made that the kind of social deconstructivist studies pioneered by Bijker would be possible only if the device paradigm of machinery/commodity separation obtained. Only when the commodity is disembedded from a culture and supplied by a machinery that is inconsequential for the user could the designing of that machinery become so much an object of playful alternatives, with the commodity interpretation turned over to the interactive but ephemeral drama of advertisers and consumers.

both promise and as disappointment if not threat, and that therefore needs to be reformed.

Part three explores possibilities for the reform of this ambivalent and characteristic paradigm. The reform at issue, Borgmann argues, must be not *within* the technological framework, but *of* the technological framework. Not being within the framework means two things. First, it must not be within in a way that would constitute no more than an extension or perfection of that framework; second, it cannot be within, since the framework exhibits a kind of perfection in its own right that is not to be vouchsafed. Reform must therefore be "of" the framework in a somewhat restricted sense. That is, although the character of technology is not to be escaped, perhaps it may be delimited, put in its place, by the perspective of and commitment to deeper experience. Such paradeictic reform is ultimately possible, for Borgmann, only on the basis of deictic discourse, by turning from metaphysics to poetics, in order to manifest and serve the profound reality of focal things around which we recognize it is worthwhile to reorient our lives.

Does poetry have the power to initiate such character-transforming experience?

CHARACTER FORMATION AND TRANSFORMATION

Prior to the modern period, poetry was not thought to play a major role in the transformation of character, although it was given one in character formation and preservation (cf. Plato's *Republic*). The modern idea that poetics has the power to sponsor transformations in, for example, both technology and character may be associated with the poetic reconception of philosophy (see Rosen 1988), not to mention the privatization of religion and the abdication of liberal politics to deal with character. Nevertheless, any attempt to respond to the question of the role Borgmann would allocate to a quite different kind of poetry must consider the related problems of incontinence and temptation.

The transformation of character is different from the formation of character. Corresponding to the hermeneutic circle of knowing is a pragmatic circle of doing; as Aristotle notes at the beginning of the *Nicomachean Ethics,* the improvement of character depends on the possession of good character. At the conclusion of his *Ethics* Aristotle further observes that theory and teaching are seldom effective in forming virtue; just as seed sprouts only in well-prepared soil, so will good character be able to develop only in the prepared soul (*Nicomachean Ethics* X, 9; 1179b23–25).

The origins of good character, which then is able to become better, and which prepares the soul for perfection, are *paideia* and *nomos,* education and law, the examination of which is to be found in a study to follow the *Ethics,* namely, the *Politics.* Although one of Aristotle's lost works is the *Protrepticus* (from προτρέπειν, to urge on or persuade), classical ethics paid little attention to the issue of defective mature adults who desire to reform their character. Not so much in the philosophical as in the spiritual life was attention given to this problem, where it is related to the issue of μετάνοια, turning around or repentance. The despiritualization of metanoia or repentance may be associated with the modern rise of both technology and utopian revolutionary politics.

According to traditional teaching with regard to character, excellence of character or virtue is theoretically grasped as a mean between the two extremes of defect and excess. Indeed, Borgmann does precisely this in his own analysis of the character of technology, which he grants is in itself a perfection, although historically it has been manifest in the extremes of deficiency and overabundance: in the past there was too little of it, in the present there is too much. A new balance in contemporary life will be achieved by means of a focal thing and practice delimitation of excess, without going to the opposite extreme of deficiency.

There is no doubt, as Borgmann realizes, that this new balance will be difficult to achieve, precisely because of our present overindulgence, not to say addiction, to device and commodity technology. In the tradition of Enlightenment humanism, he argues that what is required is the exercise of a "new adulthood and maturity" (1984, 212). There are echoes here of the Kantian human being freed from a self-imposed tutelage, the socialist "new man," and even Dietrich Bonhoeffer's "humanity come of age." The problem broached by the inevitable references to such illusory appeals is only slightly addressed by his defense of John Rawls's more modest Aristotelian principle. According to this (pseudo) Aristotelian principle, "other things equal, human beings enjoy the exercise of their realized capacities (their innate or trained abilities), and this enjoyment increases the more the capacity is realized, or the greater its complexity" (Rawls 1971, 426).

While Borgmann admits that the principle has been technologically compromised by the device paradigm and its commodification of the world, he also argues that it cannot finally be subverted. The reason is that truly human engagement is ultimately found in multidimensional connections that device procurement and commodification necessarily truncate. His examples include the experience of wilderness, the culture of the table, and the Zen of running. The mountain with handrails (Sax 1980) is no longer the challenge of nature; the microwaved meal is not a sustaining celebration with friends; the runorama/skiorama is not a disciplined engagement with body and world (Borgmann 1992, 87 ff.; 1994, 39 ff.). In each case what the focal experiences draw forth and nourish, which their virtual imitations do not, is character as a second-order natural perfection that in its human stability is able to delimit technological excitements. The plurality of such focal concerns and their character-enhancing activities may threaten attenuation of such power. But in another appeal to "a new kind of maturity and adulthood" (1984, 218), Borgmann argues that focal pluralism need not be finally debilitating.

Is maturity enough? What Borgmann fails adequately to consider is the way maturity itself is also being turned into a device. Maturity increasingly means the technological-like management of a plethora of superficial and focal engagements, the effective budgeting of work and leisure, balancing business and quality time with family and friends, the therapeutic adjustment of allopathic and alternative medicine, and so on. This view of character is already hinted at by John Dewey, who describes character as a kind of tool, but it has come even more to the fore in contemporary self-help and self-management programs. How could it not? Maturity in the technological world is seldom an orientation of technology; more often is proclaimed as the fundamental orientation in and toward technology.

INCONTINENCE AND TEMPTATION

At the heart of traditional character are ἐγκάτεια, *continentia,* self-control, and ἄσκησις, *ascesis,* exercise or self-discipline, which are not so much virtues as doorways to and guardians of virtue. Obstacles to the necessary *ascesis* are, in philosophical terms, ἀκράτεια, incontinence or weakness of the will, the inability to act on knowledge of the good, and, in spiritual terms, πειρασμός, temptation, the attractive mispresentation of evil. With regard to incontinence, Borgmann's analysis of the social-scientific data that point up an unhappiness within the character of contemporary life could also be used to characterize our social state as one shot through not so much with the possibility of reform as with weakness of the will. Does poetry have a capacity to energize the will, to disclose the inherent powers of focal goods with a clarity that shames both sullenness and hyperactivity?

With regard to temptation, is it possible that not only specific commodities but even the very device paradigm might be described theologically as manifestations of deception or beguilement? According to Borgmann, the most disquieting aspect of the character of contemporary life is the suggestion that "in technology we may suffer a radical loss of humanity which remains largely concealed, however, and may well be accompanied by an apparent feeling of comfort amidst material security" (Borgmann 1982, 113).

According to the Gospel of Matthew (4:1–11), following his baptism, Jesus was led by the spirit into the desert, where he fasted for forty days. Then Satan appeared to him and said, "If you are the Son of God, command these stones to turn into bread." Next the tempter took him to Jerusalem and set him on the parapet of the Temple, and said, "If you are the Son of God, throw yourself down," and you will be protected. Finally the Evil One took him to a high mountain and displayed all the kingdoms of the world in their magnificence, promising, "All these will I bestow on you." Is it not the machinery of technology that is able to turn even stones into the commodity of food? Is not health to be made available by technological medicine for all who would throw themselves down from the Temple, with its ability to treat even such injuries as may assail them in the fall? Does the device paradigm not place into human hands in the most dazzling manner all the powers of the world?

The rite of baptism, however, in its own poetics, asks the catechumen to reject Satan: to reject "all his works," to reject "all his empty promises," and to turn aside from "the glamour of evil." Upon those who are able to affirm these liturgical rejections and are then anointed not with ideas but with water, there is conferred a new character, a character that in turn opens one up a new life.

From the spiritual perspective there are two selves struggling with each other in the soul, two selves which cannot both be satisfied. Indeed, having rejected the temptations of Satan, Jesus spoke to his followers in the following words:

> Do not lay up for yourselves treasures on earth, where moth and rust consume and where thieves break in and steal. . . . For where your treasure is, there will your heart be also. . . . No one can serve two masters; for either he will hate the one and love the other, or he will be attentive to the one and despise the other. . . . Therefore I warn you: Do not be anxious about your life, what you are to eat or drink, nor about your body. . . . Look at the birds of the air: They neither sow nor reap nor gather into barns. . . . Learn a lesson from the way the wild flowers grow: They neither labor nor spin. . . . If God can clothe in such splendor the grass of the field, which blooms today and is thrown on the fire tomorrow, will he not provide much more for you, O weak in faith! (Matthew 6:19–30)

Is such poetry able to speak to those who have been formed by the character of contemporary life, itself informed by the character of technology?

THE UNBEARABLE LIGHTNESS OF TECHNOLOGY

Traditional life and character constitute a kind of oppressiveness, a destiny or fate that manifests itself in repetition. Modern or technological character is concentrated in its liberating powers to be anything, that is, to be new, to never repeat itself. The changes of contemporary fashion are not so much accidental features floating on top of technological prowess as they are bright and glamorous expressions of its deepest currents. Modern technology conveys to modern life an unprecedented and heretofore unknown lightness. To paraphrase Karl Marx, it is now possible to do one thing today and another tomorrow, to be a mother in the morning, a corporate lawyer during the day, and a sexual playmate after dinner, without ever becoming mother, lawyer, or sexual sexist (*German Ideology* I, 1, a).

Milan Kundera's *The Unbearable Lightness of Being* (1984) is a poetic meditation on this technological transformation of the human condition that tells the erotic parable of a Czech physician, Tomas, drawn toward two women: Tereza and Sabina. Tereza comes from a small town outside Prague, through her commitment to Tomas becomes a photographic journalist, escapes but then returns to Czechoslovakia after the Soviet invasion of 1968, and longs to return even more to the countryside; associated with her is a burden of marital fidelity. Sabina is an artist, a painter engaged with change, who moves with cosmopolitan ease and pleasure from Prague to Geneva, then Paris, and finally New York; associated with her is the lightness of a sexual affair.

The first paradox of Sabina is that although she appears to step freely from all confines of being, including especially the constraints of ideological politics and kitsch art, yet it is precisely her lightness that carries Tomas's traditional medical techniques into an arena where they transform the world. That is, in her escape from the unbearable weight of traditional ways of life she is at the same time one who radically transforms the world, not just spiritually but also physically.

The second paradox of Sabina is that in the very lightness of her way of being in the world there is also an unbearableness. Sabina finds herself rootless. Correspondingly, Tomas discovers in the very weighted constraints of post-1968 Czechoslovakia, especially when the new rulers deprive him of his medical practice and he is forced to move with Tereza to a small village, a kind of freedom. But he is only able to experience this discovery of lightness in weight because of a historically enforced choice. He was never able to choose it on his own.

Indeed, this is a third paradox of the unbearable lightness of technology: its repeated attractiveness, even in its unbearableness. Unlike the unbearableness of the traditional way of being in the world, an unbearableness marked by weight and stability, unbearable lightness exhibits a marked attractiveness and constant change. One does not have to be destined to it. One chooses it, willingly, repeatedly. And it is not clear that, once chosen, it is possible again to choose otherwise. It may have become a fate—not to say, a fall.

This is one of the meanings of the myth of the Fall: it helps members of a religious community to understand and to bear the unbearable repetitive weight of pretechnological life and character. But what could help us to understand and to bear the unbearable lightness of technological life and "the tradition of the new" (Rosenburg 1959)? Is there anything beyond entertainment and postmodern irony—practices that only swell the lightness rather than nurture any *metanoia?*

Imagine Borgmann making his measured rational appeal to Sabina or to Tomas, in their separation from any religion or community, to turn from lightness to weight, from device to focal thing and practice. Could such an appeal possibly be effective? Or is it not the case that in his narrative of Tomas's inability to choose, Kundera's parable poetically calls into question any such petition to maturity as a basis for reform? Is the unbearableness described by Kundera ever sufficient or substantial enough in character to lead to other choices, for either individuals or communities?

COMMUNITY AND CHARACTER

Although the sacrament of baptism confers a new character—as does, in the Buddhist tradition, the taking of "refuge in the Buddha, the Dharma, and the Sangha" followed by the shaving of the head and the donning of the yellow robe—it is a character not to be realized except in the spiritual community with its focal things and practices. In the Christian tradition this new *polis* is the *ecclesia,* the church. The new things are bread and wine, water, incense, chrism, candle, cross, an oriented architecture of space and light.[14] It is within the confines of this new embedded and embodied order that both the motivation of character reform and its poetics have been most manifest. It is not clear that any poetic reform of character can take place outside such a community. Certainly the only consistent restrictions to modern technology have been within the confines of such communities, especially in the monastic and anabaptist traditions.

The church as community and as alternative to traditional community separated itself from and opposed itself to existing communities reflecting national, geographic, or economic bases. But in a human world that has become characterless in the precise sense that it eschews the formation of

14. For an analysis relevant to these things see Rykwert 1996, esp. 43 ff.

character—even though this itself confers its hidden character—the alternative community depends on a double movement of negation, negation of something that does not even appear to be present.

In a world in which both *polis* and *ecclesia* have themselves have become immeasurably influenced more by the device paradigm than by their traditional focal concerns, the first step to return to focal practices may be to search for small communities of friends and the most basic things, sometimes no more than a candle.[15]

LAST THINGS

The most moving things are, however, called the last or ultimate things. The traditional Christian community once opposed itself to communities of the world through its reflection on the last things: death, judgment, heaven, and hell. But to dwell upon such things is now taken as a sign of morbid character, not to say lack of mental health. Eschatology, the study of last things, has been displaced by the study of artifacts, technology.

Through technology, as well, things have been displaced. Independent of any philosophical overcoming of Aristotelian metaphysics, there is a clear sense in which the traditional understanding of artifacts is anachronistic. If an oak were genetically engineered to be a bed it would sprout not an oak but a bed. The onco-mouse is an artifact in which matter and form have been engineered to be substantially one. Across the spectrum of scientific making and using—from chemical engineering to biotechnology—technology has progressively deepened its penetration into the very being of the artifice it designs. For the scientists and engineers at the frontiers of this work, as Borgmann observes (1984, 118, 216), modern technology itself is a focal practice. The scientists and engineers among us appear to be creating new last or ultimate things, around which life—theirs as well as ours—is being reoriented.

What is the character of such a reorientation? Are we not becoming the last humans, by becoming cyborgs? Our new character is grounded in human-technology symbiosis. Prior to reflection, technology transforms character. The burden of Borgmann's reflection is to disclose that such symbiosis remakes character and to argue for its delimitation. But is such reflection alone sufficient to transform technology, especially in the absence of catastrophe?

REFERENCES

15. This, in very different ways, may be taken as the burden of complementary arguments by Stanley Hauerwas (1995), Ivan Illich (1982 and 1993), and Thomas Tierney (1993).

Achterhuis, Hans. 1988. *Het rijk van de schaarste: Van Thomas Hobbes tot Michel Foucault.* Baarn: Ambo.

Annas, Julia. 1993. *The Morality of Happiness.* New York: Oxford University Press.

Bijker, Wiebe. 1995. *Of Bicycles, Bakelite, and Bulbs: Toward a Theory of Sociotechnical Change.* Cambridge: MIT Press.

Borgmann, Albert. 1973. "Functionalism in Science and Technology." In *Proceedings of the XVth World Congress of Philosophy,* 6:31–36. Sofia, Bulgaria: Sofia Press Production Center.

———. 1974. *Philosophy of Language: Historical Foundations and Contemporary Issues.* The Hague: Martinus Nijhoff.

———. 1982. Review of *Man and Technology,* by Edward Goodwin Ballard (1978) and *Being Human in a Technological Age* (1979), ed. Donald M. Borchert and David Steward. *Man and World* 15:107–15.

———. 1984. *Technology and the Character of Contemporary Life: A Philosophical Inquiry.* Chicago: University of Chicago Press.

———. 1992. *Crossing the Postmodern Divide.* Chicago: University of Chicago Press.

———. 1994. "The Nature of Reality and the Reality of Nature." In *Reinventing Nature? Response to Postmodern Deconstruction.* Ed. Gary Lease and Michael E. Soulé, 31–45. Washington: Island Press.

Dewey, John. 1922. *Human Nature and Conduct: An Introduction to Social Psychology.* New York: Henry Holt; reprint, *John Dewey: The Middle Works, 1899–1924,* vol. 14. Carbondale: University of Southern Illinois Press, 1967.

Dunne, Joseph. 1993. *Back to the Rough Ground: "Phronesis" and "Techne" in Modern Philosophy and in Aristotle.* Notre Dame: University of Notre Dame Press.

Ellul, Jacques. 1954. *La Technique ou l'enjeu du siècle.* Paris: Armand Colin. Reprint, Paris: Economica, 1993. American edition, *The Technological Society,* trans. John Wilkinson, New York: Knopf, 1964.

———. 1977. *Le système technicien.* Paris: Flammeron. English version, *The Technological System,* trans. Joachim Neurgroschal, New York: Seabury, 1980.

Eucken, Rudolf. 1908. "Character." In *Encyclopedia of Religion and Ethics.* Ed. James Hastings, 3:364–65. New York: Scribner's.

Funke, Gerhard. 1958. *Gewohnheit.* Vol 3. of *Archiv für Begriffsgeschichte: Bausteine zu einem historischen Wörterburch der Philosophie,* ed. Erich Rothacker. Bonn: H. Bouvier.

Guardini, Romano. 1927. *Briefe vom Comer See.* Mainz: Matthias-Grünewald. English version, *Letters from Lake Como: Explorations in Technology and the Human Race,* trans. Geoffry W. Bromiley, Grand Rapids, Mich.: Eerdmanns, 1994.

Hauerwas, Stanley. 1995. *In Good Company: The Church as Polis.* Notre Dame: University of Notre Dame Press.

Heidegger, Martin. 1954. "Die Frage nach der Technik." In *Vorträge und Aufsätze,* 9–40. Pfullingen: Neske. English translation by William Lovitt, "The Question Concerning Technology," in *The Question Concerning Technology and Other Essays,* 3–35. San Francisco: Harper and Row, 1977.

Hickman, Larry A. 1990. *John Dewey's Pragmatic Technology.* Bloomington: Indiana University Press.

Hoinacki, Lee. 1996. *El Camino: Walking to Santiago de Compostela.* University Park: Penn State University Press.

Illich, Ivan. 1982. *Gender.* New York: Pantheon.

————. 1993. *In the Vineyard of the Text: A Commentary to Hugh's "Didascalicon."* Chicago: University of Chicago Press.

Kundera, Milan. 1984. *The Unbearable Lightness of Being.* Trans. Michael Henry Heim. New York: Harper and Row.

Latour, Bruno. 1987. *Science in Action: How to Follow Scientists and Engineers through Society.* Cambridge: Harvard University Press.

Nussbaum, Martha C. 1986. *The Fragility of Goodness: Luck and Ethics in Greek Tragedy and Philosophy.* New York: Cambridge University Press.

————. 1992. "Character." In *Encyclopedia of Ethics.* Ed. Lawrence C. Becker and Charlotte B. Becker, 1:131–34. New York: Garland.

Ogburn, William Fielding. 1922. *Social Change with Respect to Culture and Original Nature.* New York: Viking.

Polanyi, Karl. 1944. *The Great Transformation.* New York: Rinehart.

Rawls, John. 1971. *A Theory of Justice.* Cambridge: Harvard University Press.

Roback, A. A. 1928. *The Psychology of Character: With a Survey on Temperament.* 2d ed. New York: Harcourt Brace. (1st ed. 1927.)

————. 1930. "Character." In *Encyclopedia of the Social Sciences.* Ed. Edwin R. A. Seligman and Alvin Johnson, 3:335–37. New York: Macmillan.

Rosen, Stanley. 1988. *The Quarrel between Philosophy and Poetry: Studies in Ancient Thought.* New York: Routledge.

Rosenberg, Harold. 1959. *The Tradition of the New.* New York: Horizon Press.

Rykwert, Joseph. 1996. *The Dancing Column: On Order in Architecture.* Cambridge: MIT Press.

Sax, Joseph L. 1980. *Mountains without Handrails: Reflections on National Parks.* Ann Arbor: University of Michigan Press.

Searle, John. 1995. *The Construction of Social Reality.* New York: Cambridge University Press.

Sherman, Nancy. 1989. *The Fabric of Character.* Oxford: Clarendon Press.

Sills, David L., ed. 1968. *International Encyclopedia of the Social Sciences.* New York: Macmillan and Free Press.

Strauss, Leo. 1958. *Thoughts on Machiavelli.* Glencoe, Ill.: Free Press.

Tierney, Thomas F. 1993. *The Value of Convenience: A Genealogy of Technical Culture.* Albany: State University of New York Press.

PART THREE

Theory in the Service of Practice

In the next part Borgmann's theory of the device paradigm and his thinking about focal things and practices are applied in fresh ways to film, farming, design, and ecological restoration. While the authors of this section are indebted to Borgmann in their respective employments of his ideas, they extend Borgmann's initial ideas in new and sometimes challenging ways.

What is the relationship between art and technology? Both Heidegger and Borgmann hold that art in the past, before the advent of modern technology, was preeminent in its power to reveal and orient the world. Both philosophers believe that in the future art may (or may fail to) regain this status in our culture; for now technology has taken over this orienting role of art. Borgmann especially, as Carl Mitcham showed in chapter 7, looks to the "deictic discourse" (the disclosure of focal things in their importance) of art for its potential to challenge this dominant role of technology.

Phillip Fandozzi in chapter 8, "The Moving Image: Between Devices and Things," points out how art, in particular film, can challenge technology in Borgmann's sense. Films, such as *The Conversation,* can help us to criticize technology by demonstrating vividly its logic and irony. Other films, such as *Babette's Feast* and *Local Hero* can disclose in moving detail Borgmann's focal things and practices. Films such as these may bring us out of the cave of consumption into the light of day. However, Fandozzi first shows how film, television, and advertisement fuel the device paradigm and keep us chained in the cave, as it were. Here he distinguishes the above kinds of films from the way cinema as mass media can be reduced to a mere device, a source of entertainment and little else. In relation to this, he discusses the development of montage in advertising, television, propaganda, and film. Here "[t]he audience must sit back passively and accept the inevitable linking of associations presented on the screen" since the very meaning of images is controlled. In contrast, Fandozzi favors a tradition of realism in film. Here, he argues, materials are allowed to speak for themselves and the poetic implications of daily life and the commonplace come alive. Physical reality is redeemed.

Paul Thompson in chapter 9, "Farming as Focal Practice," argues not only that farming is an exemplary focal practice, but, moreover, that it is the most primary and comprehensive of all focal practices. For Borgmann, the material event of the shift from things to devices is the most significant of our time; Thompson finds that land should be seen as the largest of these public, focal material things because land and nearly all cultural practices are intertwined. Arguing for the fundamental importance of place over things, Thompson maintains that farming is correlative to a place, not a thing; he warns against a reading of Borgmann that would reduce place to a function of the things that occupy it, a kind of "Kantian environmentalism." This subtle difference between places and things becomes important when we remember that inhabiting a place is the real issue of reform and when we consider that "land as place is replaced by a version of the device paradigm in which land presents itself as but one of many purchased inputs in the production process."

A genuine reform of technology in Thompson's view must come to terms with production, finally embracing an appropriate sense of it. Farming as a focal practice is unlike contemporary wilderness, a place we do not inhabit, nor is it like the urban, with its separation from yet material dependence on the rural. It teaches us to be productive in a very basic way and teaches us what it means to inhabit a place. Producing being by being there, farming is productive of both farmers and farms.

If we overcome the obstacles to reform, how would reformed technology appear? It may seem as though every technology for Borgmann is a device, but a moment's reflection will show otherwise, for the wood-burning stove is a technology that is a thing and not a device. He also distinguishes running shoes and snow skiis as *instruments* because they mediate our engagement with the world rather than obscure the world and detach us as devices do. Even though devices (and the device approach) are the pervasive technologies of our time, a reform of the device paradigm may advance these instruments and other nondevice kinds of technologies. Accordingly, in chapter 10, "Design and the Reform of Technology: Venturing Out into the Open," Jesse Tatum, drawing on his studies of the home power movement, distinguishes between technologies and practices that *center* our existence and those that *enable,* in a supportive role, that center to exist and flourish. With this basic distinction in hand, he draws out reform possibilities in the design of technologies. Along the way, we learn, to use a phrase of Langdon Winner's, that artifacts have politics; Tatum finds that the redesign of technology encourages participatory research and development and participatory democracy.

Can the notion of focal things and practices be extended to a new and respectful postmodern relationship with nature? In *Crossing the Postmodern Divide,* Borgmann argues that we as a culture are now in the process of leaving modernism behind and are faced with a fundamental choice between two versions of postmodernism. Choosing "hypermodernism," he argues, will limitlessly extend the device paradigm and take it to new extremes. Alternatively, we could choose postmodern realism, a culture that puts focal realism and communal celebration at its center and limits technology on that basis. Similarly, Eric Higgs in chapter 11, "Nature by Design," shows that as we practice ecological restoration, even wild places like the Canadian Rockies force us to give up modernist assumptions about nature. However, attempting to practice ecological restoration in a postmodern manner, we face the same sort of cultural choice as Borgmann's between "technological restoration" and "focal restoration." The former practice results in the commodification of nature, e.g., Disney's "Wilderness Lodge," and the commodification of practice, e.g., when corporations restore nature through landscaping. As an instance of a focal practice, focal restoration requires multifaceted engagement and "the realization of a new kind of relationship with nature, one that enforces humility and respect." For this focal practice to become viable for many people, Higgs believes more must be done to reform the political economy than Borgmann outlines.

EIGHT

The Moving Image:
Between Devices and Things

Phillip R. Fandozzi

I'll never forget my first film by Ingmar Bergman, *The Seventh Seal.* I saw it as a university freshman. Before that time, I had always enjoyed films, but seldom considered their significance. However, this film touched me deeply and gave a shape to many feelings and thoughts that until then had been obscure, unfocused. Many years later I began to teach courses in film and came to regard them as an important source for exploration and study. And while there is much to deplore in the current state of cinema and, for that matter, art in general, there are works that hold promise and perhaps even attain greatness.

While Albert Borgmann claims in *Technology and the Character of Contemporary Life* (or *TCCL*) that art is the supreme deictic discipline, he laments the absence of significant art in contemporary life. However, I believe that art, and particularly film, still has the potential to reveal salient aspects of our world, that it can assist in the ongoing effort to recognize and resist the technological trends that engulf us.

While cinema can be reduced to the status of a device, a source of entertainment and little else, it also has the capacity to rise to an art form and reach out beyond the confines of the screen, illuminating our lives. In this chapter, I will address the way the moving images of film vacillate between a technological device and a revelation of the world of significant things.

PHILOSOPHY AND FILM

In his translation of *The Republic,* F. M. Cornford notes that "a modern Plato would compare his cave to an underground cinema, where the audience watch the play of shadows thrown by the film passing before a light at their backs. The film itself is only an image of 'real' things and events in the world outside the cinema" (Cornford 1977, 28). Here it seems we have the ultimate Borgmannian device: while its unifunctional aim is global—the "education" or should we say the indoctrination of an entire society—it fits the device paradigm in that it is contextless, becoming a substitute for the world of things. Engagement is reduced to "forced" watching of the various images projected on the screen. We can even remove Plato's sense of imposition or deprivation by projecting the images in such a way that they entertain the viewers, so that they end up enjoying their passive lives—the entire enterprise becomes commodious.

As in a device as Borgmann describes it, the machinery is concealed, isolating the function, and requiring a minimum of skill, strength, or attention on the part of the viewers. Even today many speak of film (and of course TV) as an easy medium, one that requires no prior training or education. In fact some critics would agree that the modern media do exactly what Plato's cave did—provide a surrogate reality, a hyperreality that is replacing the real world.

Another aspect of Plato's image of the cave is the inability of the spectators to communicate with each other, creating a solipsistic society, lacking social intercourse. This again is strongly suggestive of the privatized world of our present computerized-TV society. What we have in this strange Platonic vision is a civilization of cineastes who live within the ultimate device—a self-enclosed manufactured world. Of course, Plato cannot fully account for all the varied aspects of life, but for good reason; he is mainly concerned with matters of epistemology and metaphysics. As Frank McConnell notes in his book, *The Spoken Seen,* "It is a vision . . . of a socially mediated metaphysics, of a metaphysics of media" (McConnell 1975, 89). Long before we had the know-how to produce such a theater of life, Plato envisioned the possibility of a device that could explain the apparent passivity and error-prone life of his contemporaries.

However, things change: one day one man escapes from his cinematic prison and is released into the sun-drenched world above. At first blinded by the light, he soon adjusts to the brightness and is able to discern things as they really are. However, his return to the cave is fraught with danger—his wild ideas and brightness of vision are suspect in the lower world; he is unjustly accused of corrupting the darkened social order and put to death.

At this point there is a twist in the story that is important for my purposes: Plato argues that the experiential world—the everyday—*is* the cave, that it is the world that surrounds us in its thingliness that is in error, a mere copy of the real; so that the underground cinema is merely "a copy of a copy." He is denying the ultimate reality of things in their everyday appearance. For Plato, only a higher, more complete reality can explain and provide a sufficient foundation for things. This version of a conceptual realism opens the door for science and a host of epistemological questions that persist to this day.

I would suggest that Borgmann, following Heidegger, has provided another answer to Plato's quest. The "essence" of things is not dependent on a conceptual framework of a higher order, but rather is found embedded in the rich, experiential context discovered through engagement, in what he calls focal concerns and practices. In this scheme, it would seem that cinema does function as it does in Plato's myth—it takes us out of the real world of contextual experience, replacing it with a virtual realm of fleeting images and shadows. Even though film is only once removed from reality it is still deficient and reductive.

In order to begin to vindicate cinema, I will turn to Heidegger's "The Origin of the Work of Art" (Heidegger 1977, 149–87). As Borgmann notes in *TCCL,* Heidegger chooses pretechnological things—fountains, temples, jugs—to ground his analysis. In attempting to understand the "equipmental quality of equipment," their usefulness, Heidegger chooses a pair of shoes, peasant's shoes, but to "facilitate" his explanation he turns to Van Gogh's painting and here he finds the words that begin to describe the way these shoes open up a world of meaning and significance. Whether we agree or not with the details of Heidegger's description, we now have an opening for art and, in particular, film. Contrary to Plato, Heidegger reverses the order of priority by making art a necessary condition for the revelation of things. Whether it is Van Gogh's painting, Meyer's poem ("Roman Fountain"), or a Greek temple, Heidegger finds an artwork to explicate his ontology. Art can be an opening of context, a revelation of how things stand in the world.

So while art in a technological society can be reduced to its function—whether that be entertainment, therapy, or decoration—it would be premature to deny its capacity to reveal reality. Once we discard Plato's conceptual realism and adopt something akin to Heidegger's artistic or contextual realism, we can reenter Plato's cave and perhaps discover a fruitful reciprocity between the world outside the darkened space and the silver screen.

Borgmann basically agrees with Heidegger in his approach to art, giving credit to art in the past, e.g., Greek temples and medieval cathedrals, but finding little of significance in the contemporary world. However, here we must invoke the dichotomy that Borgmann applies to the discourse on focal concerns, that is, on the one hand the more individualistic, pluralistic concerns such as running or the festive meal and on the other the more inclusive, socio-politico-religious concerns. It is much more difficult to achieve consensus or even limited acceptance of a political or religious vision in our diverse culture than to inspire or even gain assent on a personal level regarding health, enjoyment, or a general sense of well-being.

While I agree that contemporary art is hard-pressed to offer the former (as are all other endeavors as Borgmann acknowledges), art and in particular film can and does explore the more individualistic focal concerns that animate our lives. In fact film is especially adept in this regard. Recognizing that the cinematic image, just as the words on a page or the brush strokes of a painting, is reductive and not a substitute for actual experience; nevertheless the suggestive power of film is able to open a world of significance.

INSERTING BORGMANN INTO THE DISCUSSION

However, by and large, this salutary outcome has not been the case. Film, often spurred on by TV, has tended to develop a style that nicely complements technology: the advertising industry has created a paradigmatic form for a technological society. Ads do refer to the outside world, but in a way that justifies and reduces it to an unlimited series of choices, restricted only by financial means. While the obvious goal is to get the viewer to become a consumer by buying this or that particular product, it also reinforces the attitude that all is available, that freedom of time and from effort are the ultimate goals. Borgmann skillfully captures this aspect of advertising in his conception of "the foreground of technology" —those means used to promote commodities and their consumption.

The genius of advertising is the way it has developed its methods over the years to make ads that are entertaining—at times dramatic, at times humorous—holding the attention and, according to some, more interesting than the actual programming. So, unlike Plato's cave, it is not necessary to force people into watching by the scruff of the neck, but rather to induce them to go out into the malls and markets to consume. The world then comes to be viewed as so many products and services; and indeed images on the screen come to dictate the things of our world.

To better understand the advertising style, we should turn to the history of film and, in particular, the creation and development of *montage:* an editing technique that juxtaposes various images, usually in fast succession, to condense or heighten a specific action in process. Fast-changing images, quick cuts, action-packed sequences that hold the eye and stimulate the imagination. It is a technique that has been used to great effect within a rich context of film narrative; but when divorced from context and used exclusively to grab and hold the attention, it is reductive and superficial.

Montage was used to promote dramatic intensity and emotional emphasis. Inspired by D. W. Griffith's films such as *The Birth of a Nation,* Soviet filmmakers developed the theory of montage, also drawing upon the associative psychology of Pavlov. Louis Giannetti in his book *Understanding Movies* explains the theory:

> [I]deas in cinema are created by linking together fragmenting details to produce a unified action. These details can be totally unrelated in real life. . . . The emotion is produced not by the actor's performance, but by associations brought about by the juxtapositions. In a sense, the *viewer* creates the emotional meanings, once the appropriate objects have been linked together by the filmmaker. (Giannetti 1990, 132)

Of course there were critics of this technique who agreed that it was too manipulative and deprived the viewer of any choice. The audience must sit back passively and accept the inevitable linking of associations presented on the screen. Significantly, political considerations were also involved, for the Soviets tended to link film with propaganda. Sergei Eisenstein went farther and developed a montage of abrupt changes, harsh linkages—sharp, jolting, even violent—"shots of contrasting volumes, shapes, designs and lighting intensities" (Giannetti 1990, 134).

Outside of any narrative or explicit ideological context, this technique can be seen to set the pattern for advertising and rock videos, and has influenced TV programming as well as film. It certainly makes for easy viewing, requiring little skill, strength, or attention. The constantly changing images mesmerize the eye and feed the imagination. (A current beer ad on TV nicely illustrates this mind set: a mundane, i.e., boring, wedding ceremony is suddenly invaded by a throbbing, montage-filled sequence that transforms the wedding into an exciting sports spectacle.)

However, there is another tradition in film that emphasizes the more realistic aspects and relies on mise en scène—the arrangement or framing of a single shot or scene. Andre Bazin is recognized as one of the founders of this school, which sees films as a presentation of the physical world. He believed that formalist techniques distort reality, making film too "egocentric and manipulative" (Giannetti 1990, 144). He was one of the first to point out that certain filmmakers preserved the ambiguities of reality by minimizing editing. He condemns the editor who makes the choices for the audience, who cuts and pastes the film to direct and manipulate attention (Giannetti 1990, 145).

Bazin is not a naive realist, believing in newsreel objectivity, but he seeks a balance between selectivity, editing, and the objective nature of the medium. "The materials should be allowed to speak for themselves. . . . The director must reveal the poetic implications of ordinary people, places, and events. By poeticizing the common place, the cinema is neither a totally objective recording of the physical world nor a symbolic abstraction of it. Rather, the cinema occupies a unique middle position between the sprawl of raw life and the artificially recreated worlds of the traditional arts" (Giannetti 1990, 147).

Another important theorist who champions the realism of film is Siegfried Kracauer, whose book *Theory of Film* is subtitled *The Redemption of Physical Reality*. According to the author, science and technology have influenced our habits of thought, leading to a penchant for abstraction in our everyday life. "The technician cares about means and functions rather than ends and modes of being. This cast of mind is likely to blunt his sensitivity to the issues, values, and objects he encounters in the process of living; he will be inclined that is, to conceive them in an abstract way, a way more appropriate to the techniques and instruments of his concern" (Kracauer 1971, 292). Agreeing with a host of other critics, he describes the way in which the medium often overwhelms the matter transmitted so that its abstract qualities (e.g., color, sound, vividness) are prized over and above the specific contents. Through this habit of abstract thinking, physical reality becomes elusive; "so things begin to recede" (Kracauer 1971, 380).

However, rather than turning to pretechnological art forms as does Heidegger or to poetic descriptions as does Borgmann, Kracauer argues that "film renders visible what we did not, or perhaps could not, see before its advent. It effectively assists us in discovering the material world. Its imagery permits us . . . to take away with us the objects and occurrences that comprise the flow of everyday life" (Kracauer 1971, 300). Of course the author is going to emphasize those filmic features that strengthen this world-revealing capacity—long takes, deep-focus shots, close-ups—and those films that excel in them. Quoting Lewis Mumford, Kracauer underlines this capacity: film "enables us to think about that world with a greater degree of concreteness" (Kracauer 1971, 259). Furthermore, this richly detailed type of film always offers more than the director intends; there is a "fringe of indeterminate meanings" surrounding the explicit narrative. "A street, a face open up a dimension much wider than that of the plot which they sustain" (Kracauer 1971, 303). That is why good films reward repeated viewings, sharpening the often narrow practicality of everyday activities. According to Gabriel Marcel, cinema helps the viewer break out of the habitual ways of seeing, deepening and rendering more intimate "our relation to this Earth . . . this power peculiar to cinema seems to be literally redeeming" (Kracauer 1971, 304). Therefore, if technology is "a way of taking up with the world," making available endless opportunities, "scattering our attention and cluttering our surroundings," which in turn disengages us from the world, then those activities and pursuits, including film, that would counter this tendency should be allies in the effort to curb technology and its insidious effects.

Borgmann warns us that "the peril of technology lies not in this or that manifestation, but in the pervasiveness and consistency of its pattern" (Borgmann 1984, 208). Here we face the problem that film (and TV even more so) is understood as a *medium,* so that particular products simply become part of the pattern and lose their distinctiveness and capacity to stand apart. Can we go too far in following the McLuhan dictum of "the medium is the message"?

> But where danger is, grows
> The saving power also. (Heidegger 1977, 316)

Quoting Hölderlin, Heidegger concludes his essay "The Question Concerning Technology" with an appeal to consider art as a source of hope even in the contemporary world. (We could add to this appeal Borgmann's statement that "the turn to things cannot be a setting aside and even less an escape from technology but a kind of affirmation of it" [Borgmann 1984, 200].) We have to make a distinction, therefore, between film as "mass medium," as part and parcel of the entertainment industry, and film as a prospective redeeming power.

In connection with this point, I would like to invoke another salient observation of Borgmann's analysis: the difficulty of staying true to one's focal concerns and practices in a technological society. Perhaps similar to the way in which Plato's escape is besieged by the cave dwellers, so the devotee of nature or running is often overwhelmed by the demands and preconceptions of our society; he or she may be unable to stay true to the experience of a particular focal practice. Furthermore, Borgmann cites examples of the technological reduction of things even when in direct contact with them, e.g., the wood products worker who sees trees in terms of board feet. In other words, the *attitude* that technology promotes becomes central—the inability to let anything stand on its own, to be seen in its own right.

I propose that a similar problem besets the cinema. Films that offer a critique of technology or that attest to focal experience are lost in the proliferation of films that saturate our society. Those films that are deeply revealing and/or morally transformative are unable to achieve a voice of their own and tend to become a part of the general cacophony. I believe that the reason film often seems to have a more difficult time being taken seriously compared with books or the other arts is that it is seen as "easy," directly accessible and open to relativistic interpretation. Of course some of these same problems beset the other arts, but film is particularly vulnerable because it is so popular and so many poor films are made and distributed in the most commercial manner. Another problem is the negligible effort that is made by our educational institutions to teach film literacy. Too often the prejudice against film (and the visual arts in general) is ingrained in our curriculum and in our efforts to protect the literary arts against the mass intrusion of the visual.

Before I discuss some noteworthy films that mitigate Borgmann's suspicion about the whole medium, I want to question his emphasis on bodily engagement in his analysis of focal concerns. Certainly physical activity within a natural setting is an essential part of a full and vigorous life. However, there are other engagements with persons and things that are not directly expressed through bodily activity and yet they also are worthy of focal concern. Our human relationships are so complex, extending from extreme fragility to ultimate commitment; and although almost impossible to describe outside of an artistic context, they are often at the core of our lives. Even our relationships with natural things are at times expressed through silent meditation or perhaps daydreaming; they may not take an obvious physical form, but they too can be focal. I find that one of the great strengths of film is its capacity to subtly address issues such as these.

EXAMPLES

First, a film that offers a critique of technology and clarifies how it undermines focal concerns. *The Conversation,* made in 1974 by Francis Ford Coppola, tells the story of a wiretapper, a surveillance man who defines his life in terms of a technological practice: Harry Caul is "the ear" who can listen in on any conversation for a price. He prides himself on his ability to figure out a way to invade the most private of circumstances; he is known as "the best bugger on the West Coast." He also prides himself on his "objectivity"—he only listens to "sounds," never allowing himself to acknowledge their meaning.

Yet through the film we see that he has perfected his skill at the expense of all other aspects of his life: he lives in a perpetual state of paranoia, obsessively guarding his own privacy with an intricate set of locks on his apartment door, an unlisted telephone, and a social life strictly limited to one pathetic relationship that he can never really acknowledge or affirm. His most revealing statement comes when his landlady unlocks his apartment one day to give him a birthday present: he is horrified to find out that not only does she know his birthday, but she has a key for emergencies such as a fire. He vehemently tells her that none of his possessions are important to him—only his keys!

Later we discover that in spite of himself Harry is bothered by one of his previous assignments back east that led to the death of three people. He had tried to rationalize his part in it in terms of his professionalism, but his most recent assignment—to record the private conversation of a young couple in a crowded plaza in San Francisco—has reminded him of the possibility of another murder. He begins to reconstruct the conversation that he has now composed through his clandestine techniques. Slowly but surely, he hears ominous overtones of murder and intrigue. He finally becomes convinced that the couple he had been bugging in the plaza is in danger from his client and that he must destroy the tapes. When the tapes are stolen from him, first he desperately tries and fails to get them back, then attempts to warn the couple by going to the hotel where he believes the murders will take place. He bugs their room and takes the room next door. As he listens to what sounds like a brutal murder, he is only able to listen, cowering in his bed, overcome by fear and guilt. By the end we are not sure exactly what happened in that room, but Harry does see the couple that he thought had been murdered going to the funeral of the real victim—his client, the woman's husband. When Harry gets back to his apartment, he gets a threatening phone call convincing him that the apartment is bugged; Harry futilely strips the rooms from wall to wall. The last scene shows Harry utterly alone and deracinated.

The film skillfully shows the specifics of surveillance, the everyday world of the eavesdropper who meticulously plies his trade. His obsession with his work to the exclusion of all else exposes the shallowness of his endeavor and its inability to encompass his deeper concerns and sensibilities. The one relationship that could have given him a grounding in reality is undermined. The only things that he can confess to his woman friend are innocuous misgivings. Harry has taken up eavesdropping as a way of life; he is simply a "medium" for sound and he spends his career trying to convince himself that is where his responsibility ends. He breaks down meaningful conversation into sound bits and then reconstructs them for his clients. However, he cannot maintain his neutrality; his conscience will not permit him to be an accomplice in another murder. But when he tries to undo what he has done, he becomes the target; the bugger becomes the bugged, in spite of all his precautions and defenses. In the end his world is reduced to a room stripped of significance, contextless, solipsistic. The relationship between means and ends that Borgmann sees as essential to focal experience is severed.

A second film reveals a positive sense of a focal concern: *Babette's Feast,* a 1987 Danish film, which could be subtitled, "the celebratory meal." It begins with a view of flatfish drying on a rack, part of the staple diet in a small Danish village in Jutland in the nineteenth century. Like its food, life here is simple and austere; led by the rather famous minister who presides over matters spiritual—which seems to include all aspects of life. His two daughters dutifully follow him, attending his services and ministering to the needs of the congregation. Although young and attractive, they seem content to spend their lives with their father and his flock.

Even when the father discourages the marriage of one of them to a young man of the village, there seems to be an acquiescence, even an affirmation of his wishes. And when another young admirer is disheartened by what he feels is the stifling atmosphere of the home, the daughter again passively appears to understand and accept his departure. The other daughter attracts a famous opera singer through her talent and demeanor and is offered a career in Paris; but she turns him away when she senses his carnal desire. And yet while the film could have emphasized the repression and renunciation of sensuality, it offers instead a sense of strength of conviction and commitment to a way of life, however difficult and at times troubling.

After the minister dies, Babette enters the film as a wayfarer, a refugee from the civil strife in France in the 1870s, in need of shelter and work. She offers to work for the sisters without pay—since they cannot afford and perhaps really don't want hired help—only for room and board. They hesitatingly accept. Immediately one can see how Babette is disappointed by the life of the village—especially the diet. The plainness of the food, the austerity of the life in general seems stifling and even depressing. Yet she endures, adapts, and eventually even thrives in the village. She stays for many years and becomes integrated into a life far from what she was accustomed to.

Then the hundredth anniversary of the deceased minister's birthday arrives and the sisters plan a "simple meal" and a commemorative service. However, by this time, the villagers have lost most of the community spirit and cooperativeness that were instilled by the minister and have fallen into petty bickering. A restoration is needed and it comes from the least likely source. Babette, who has just won a lottery, makes a strange request: she would like to prepare the meal for the occasion—a French meal. Of course, to the sisters and to the village at large, "French" connotes extravagance and sensuality, especially French spirits. In trepidation, the sisters agree; after all, this is the only thing that Babette has asked for over all these years.

The film excels in the presentation of the meal in terms of its physicality—each aspect of the preparation is shown in detail: carloads of the best wine, cheese, quail, truffles are hand shipped, along with a giant turtle. It is truly a sacrificial supper. Details of cleaning, cutting, cooking, basting, and seasoning are included. And the transformation is not limited to the foodstuffs; the meal literally transforms the participants as the physical and spiritual are joined in a celebration of life. Old enemies converse and deep secrets are revealed; the conviviality is wonderfully graphic. In a speech made by the man who once courted one of the sisters and remembered her the rest of his life, the triumphant theme of the meal is sounded—"all choices are redeemed"—even the ones that led to lost loves, lives, and careers. The scene brings to mind Stanley Cavell's suggestion that:

> [i]f it is part of the grain of film to magnify the feeling and meaning of a moment, it is equally part of it to counter this tendency, and instead to acknowledge the fateful fact of a human life that the significance of its moments is ordinarily not given with the moments as they are lived, so that to determine the significant crossroads of a life may be the work of a lifetime. (Cavell 1984, 11)

After the meal the sisters expect Babette to leave now that she has the money, but she tells them she has spent the entire sum on the meal and would like to remain with them. It is as if an entire life (or should we say lives) was lived through this focal occasion. Here means and ends embrace in a joyful celebration.

Finally, let us turn to *Local Hero* (1983, directed by Bill Forsyth), the story of a remote Scottish town and its encounter with a giant oil company. This film conveys a splendid sense of the significance of place, so important to Borgmann's reform of technology. The plot is simple and could have been portrayed simplistically, but instead we are able to see how a particular place can enchant a people and maintain its integrity against the greatest odds. Of course there must be a few ironic twists or, should we say, fortuitous happenings that somehow come together to provide a saving grace.

A large oil company wants to buy out the village near a sparkling bay and turn it into an oil-refining complex. The job seems to be an easy one for the organization because in the words of one manager: "It's not a weird third-world country where the people are idiots who can't speak English and have strange uncivilized customs. They're good Nordic Europeans who will be rational and accommodating." And at first this seems to be the case in that the spokesman for the town immediately begins to negotiate a good price. However, he needs time to talk to his constituents and up the ante. However, this "dead time" for the company representatives who have come to negotiate the deal slowly comes alive as they fall under the spell of the town, settle into its rhythms, and admire its terrain.

The fortuitous happenstance that moves the plot along is that the owner of the oil company is an eccentric amateur astronomer who is more interested in meteors than oil. When he hears of the spectacular meteor showers and northern light shows that have already fascinated his representative, he comes to the village himself and, after a long discussion with an old fisherman-astronomer who lives on the beach, decides to change the company's plans. The company will still buy the property, but now will build a scientific institute to study and preserve the area.

The emphasis of the film is not, as perhaps one would expect, on the townspeople's resistance, but on village life (obviously not fully appreciated by the people themselves) and the beauty of the landscape. Only the old fisherman feels a definite commitment to the land. And it takes another eccentric, the company owner, to fully respond to what the place has to offer. So in its very lightness, its whimsicality, the film shows how strong the hold of technology is, how far it reaches, and the difficulty of taking a stand against it.

Conclusion

In *TCCL,* Borgmann gives special recognition to what he calls "deictic discourse," which he contrasts with "apodeictic discourses," based on laws and specific conditions, i.e., scientific explanation (Borgmann 1984, 71–72). Deictic discourse "points out" what is significant, revealing the subject in its uniqueness and concreteness. He cites poetry as the purest form of the deictic and gives precedence to the literary arts in this disclosive capacity. Nowhere does he mention film. Yet film—especially in its realistic mode—has a unique ability to reveal things in their immediacy and specificity. One could argue that film is the supreme deictic art.

The films that I have discussed graphically support three of Borgmann's theses in an eloquent way. *The Conversation* demonstrates the debilitating effects of technology, the way in which a technological orientation undermines the means/ends relationship and destroys focal experience. The other films exemplify two focal concerns: *Babette's Feast* captures the festive meal as a significant core of a convivial lifestyle, while *Local Hero* reveals the compelling power of a specific location and how it can transform lives. These films complement Borgmann's project and enhance its persuasive power. They testify to the importance of the cinematic art in the promotion of the reform of technology.

References

Borgmann, Albert. 1984. *Technology and the Character of Contemporary Life: A Philosophical Inquiry.* Chicago: University of Chicago Press.

Cavell, Stanley. 1984. *Themes Out of School: Effects and Causes.* San Francisco: North Point Press.

Cornford, F. M. 1977. *The Republic of Plato*. New York: Oxford University Press.

Giannetti, Louis. 1990. *Understanding Movies*. Englewood Cliffs, N.J.: Prentice Hall.

Heidegger, Martin. 1977. *Basic Writings*. New York: Harper and Row.

Kracauer, Siegfried. 1971. *Theory of Film*. New York: Oxford University Press.

McConnell, Frank. 1975. *The Spoken Seen: Film and the Romantic Imagination*. Baltimore: John Hopkins University Press.

NINE

Farming as Focal Practice

Paul B. Thompson

Albert Borgmann developed his ideas of focal things and focal practices during the final stage of a dialectical argument that pits the obvious promise of technology against the increasing failure of that promise to be realized. In *Technology and the Character of Contemporary Life* (or *TCCL*) focal things and practices open and can fulfill the promise of technology. Focal things and focal practices unify and harmonize our often-fragmented experience of being in the postmodern world. In doing so, they overcome technology's tendency to overwhelm us with consumable commodities, commodities whose sheer commodiousness undercuts the ends to which consumption is presumed to be a means. Borgmann notes that fragmentation is not solely a phenomenon of late capitalism. Citizens of capitalist and communist societies alike are trapped in an order of life dictated by the terms of technological production, distribution, and consumption. Careful and philosophically informed appreciation of focal things can, Borgmann thinks, end this entrapment. Only then may technology truly be in the service of ends worthy of human lives.

This chapter will (mostly) stand on Borgmann's shoulders, simply accepting the analysis that leads to focal practices without comment. I will apply Borgmann's ideas to the practice of farming and will argue that the philosophical meaning of farming is unveiled when it is understood as a focal practice. However, Borgmann believes that many things might serve as focal things, and many practices might, for different groups and individuals, have the edifying, unifying, and salvific importance that he describes. Wilderness sports and activities figure prominently in his examples. David Strong has developed this theme from Borgmann's work in *Crazy Mountains* (Strong 1995). He argues that the encounter with wilderness is a focal practice of special moral and existential significance. In describing farming as focal practice I will defend a claim that is in some tension, at least, with Borgmann's pluralism and Strong's deep ecology. Farming is something like the quintessential focal practice, and farms, not wilderness, are the indispensable paradigm for focal things. This is not, of course, to say that everyone should farm. The claim is that the critique of technology and the notion of a focal practice fulfill their implicit logic in the contemplation of farming. Farming is thus central to the philosophy of technology as Borgmann and Strong develop it. Their failure to take up farming as a focal practice can be forgiven in light of intellectual fashions, but the philosophical drama plotted in *TCCL* and developed in *Crazy Mountains* demands a denouement on the farm. There are three parts to this argument. First, it is necessary to establish that farming could be a focal practice for at least some individuals or groups. Second, farming has a special status among focal practices, one that demands philosophical attention and reflection. Third, so what? What have we lost by neglecting farming?

FARMING AS A FOCAL PRACTICE

Borgmann makes the first part of the argument easy.[1] Virtually anything might serve as a focal thing, and any practices might be constructed as focal. Focal things differ from commodities, in Borgmann's sense, because one "decommodifies" them in taking a reflective and inquiring stance toward them. Practices become focal when they ground or center a range of activities in an ontologically deep way. Though Borgmann does not use the language of authenticity, the discussions of focal things and focal practices recall existentialist themes from Heidegger and Sartre, where the human condition is characterized in terms of the need to create one's own being through one's projects. For Sartre especially, modernity undercuts our ability to complete the fundamental ontological project of becoming by presenting us with an array of ready-made projects. Adopting the ready-made project conceals authentic being (still implicit within the adopting) and submits becoming to an order of being that is inherently at odds with its essence.[2] Borgmann leaves the link to authenticity implicit, pointing his readers toward the potential for authentic experience in many quarters of life: running, cooking, wilderness. The list is indefinitely long, and one interesting question is whether *anything* (Super Nintendo, perhaps) is irreconcilably off it. Borgmann shifts the burden of proof so that we ask why something could *not* be a focal thing, rather than why it could.

1. Indeed, Borgmann references Wendell Berry's *Farming: A Handbook* (1970) among a list of books that point their readers to focal concern with rowing, skiing, riding, and tennis. Strong (1995, 225) refers to Berry's poem "Horses" (Berry 1985), a paean to farmers who resist the commodious tractor and continue to use draft animals.
2. We know that this kind of philosophizing can continue at length, and to salutary effect. Works such as *Being and Time* or *Being and Nothingness* become focal things for their readers (or they become bookends).

For my purposes, this means that farming could be a focal project, but Borgmann does not point us toward essential characteristics of focal projects. He refers to George Sheehan's book *Running as Being* to articulate how running could be focal, and there is no shortage of authors who stand to farming as George Sheehan stands to running. Among the best are Gene Logsden, whose 1994 book *At Nature's Pace* is a successor to a series of books by Wendell Berry, beginning with *The Unsettling of America,* and including *The Gift of Good Land, Home Economics,* and *What Are People For?* Any argument that would articulate farming as focal practice would be an example of what Borgmann calls deictic discourse. It would require its own sense of timing and pace and could not be controlled in advance. Logsden and Berry have provided a deictic discourse on farming that is far more poignant than any I might offer here, and it would be presumptuous (and misleading) to summarize it in a few sentences.[3] In the spirit of beginning a different kind of argument, then, we can say that Logsden's account reveals good farming to be a set of practices and technologies that have evolved in response to a specific place. Good seed, for example, takes on its genetic characteristics as a result of having been planted in and harvested from the same soil, year in and year out. The traditional barn "has evolved through centuries of experience" (Logsden 1994, 90–91). Farming then is "the resolute and regular dedication"[4] to land, and not just any land or land in general, but to the land on which one farms.

One might, of course, be dedicated to a place without farming it. Many people are dedicated to the preservation of natural or historic places, or to the beautification of their homes. While these kinds of dedication might also create focal practices, farming unifies "achievement and enjoyment of mind, body, and the world" in a way that preservation and beautification cannot. Farming is and has been throughout history the preeminent practice (from a relatively small set of practices that include fishing and hunting and gathering) by which human beings bring forth the sustenance of their lives. Farming is productive. Farming must, of course, be supplemented by skills

3. It is furthermore a topic that I have taken up before: see Thompson 1986, 1988, 1990, 1998; Thompson and Madden 1987. I take up the agrarian foundations of the virtue of citizenship in Thompson 1992, 1995. Thompson 1993 contrasts the meaning of farm animals in traditional and industrial farming; and "Saving the Family Farm," is the final chapter of Thompson, Matthews, and van Ravenswaay 1994. My most recent go at this topic is Thompson 1998. None of these efforts makes explicit reference to Borgmann, focal practices, or the device paradigm; indeed I must confess that to my own loss, I read this important book only after it had been in print for nearly a decade.

4. "A focal practice, generally, is the resolute and singular dedication to a focal thing. It sponsors discipline and skill which are exercised in a unity of achievement and enjoyment of mind, body, and the world, of myself and others, and in a social union" (Borgmann 1984, 219).

and crafts that are not part of farming per se, but building, tool making, and the martial arts do not center, order, and unify "myself and others" in the way that farming does.

Farming demands the engagement of mind and body with the world. The evolution that Logsden praises is a mindful and social process whereby individuals and social groups evolve a way of being, of continuing to be, that is highly attuned to the "expectations" of the place in which they are situated. On the traditional farm, everyone depends upon everyone else; fields must be tilled and animals must be fed. Milk must be made into butter and cheese, crops must be thrashed, and the cows and pigs must be slaughtered and rendered. These practices establish roles for each member of the farm family, and, in some communities, for butchers or millers who undertake a specialized craft. There is no formula for these roles; every farm is different. The differences reflect different soils, different microclimates, different neighbors, and different social institutions. In every case, however, practices evolve under the weight of feedback mechanisms and impressive object lessons that communicate the interdependence of people with each other and with the land.

In traditional farming, farmers who are not attuned to the unique characteristics of their own situation or who are not singularly dedicated to the pursuit of livelihood that is farming's central norm fail. Communities that are not organized around the needs and capacities of their farmers fail. Those that are successful "guard in its undiminished depth and identity the thing that is central to" farming, namely the dialectic between land and livelihood.[5] This dialectic is perilously vulnerable to the diremption of which Borgmann speaks, whereby mere sustenance becomes the end, and land a mere means. Noting these singular features of farming is not to claim that farming is unique in its status as a focal practice, however. Clearly fishing unifies the practices of many coastal villages, as it did for North American tribes that followed the salmon, and some groups of Native Americans adopted the buffalo as focal thing, engendering a way of life highly attuned to nature and unsurpassed in freedom. The centrality of farming is contingent upon the actual history of European civilization, as well as on its role in providing (and organizing) life and livelihood. Farming represents a particularly significant focal practice because the history of European civilization is, in crucial respects, a history of the spread of

5. "It is certainly the purpose of a focal practice to guard in its undiminished depth and identity the thing that is central to the practice, to shield it against the technological diremption into means and end. Like values, rules and practices are recollections, anticipations, and, we can now say, guardians of the concrete things and events that finally matter" (Borgmann 1984, 209).

European plants, animals, and agricultural techniques across the entire expanse of temperate zones of the earth.[6]

FARMING'S SPECIAL STATUS

Focal things and focal practices have a salvific quality in *TCCL,* and *Crazy Mountains* emphasizes the saving character of wilderness with an extended meditation on the story of Job. Borgmann's own analysis insinuates a jeremiad where the integrity and wholeness of life suffer from diremption through the device paradigm. Both Borgmann and Strong acknowledge debts to Marx and Heidegger in constructing the analysis but both also offer an ameliorative, rather than totalizing, restoration of life's meaning through focal practices. Focal things fall into two groups: those, such as food or wilderness, whose lost meaning is recovered, and those, such as a run or a motorcycle, whose very availability is contingent upon technology.[7] Precisely because we think of ourselves as having come from a pretechnological past and heading into an uncertain technological future, it is the first kind of focal thing that lends itself most readily to themes of salvation and recovery, and clearly farmland falls into that group.

Strong cites Wendell Berry's discussion of plow horses to illustrate focal things in *Crazy Mountains* (Strong 1995, 119–20), but to see how farming can be saving, it is first necessary to see how it is fallen. Throughout Berry's writings the fall of farming is documented in very Borgmann-like language, emphasizing how modern farm technology is seductive in offering relief from toil. Industrial technology offers commodious solutions to the problem of farm production. If soils are poor or if insect or rodent pests are an annoyance, a package of chemicals can be pulled off the shelf of the local co-op. Farm machinery relieves the drudgery of farm work, and can be well utilized when farmers standardize field size and shape and when they plant crop varieties whose size and time of maturation have also been standardized through crop breeding. Berry also laments the farmer's loss of contact with the land that results from farming from behind the windshield

6. The links between the ecological history of farming and the history of civilization are the subject of an extensive literature that has been largely neglected by philosophers. In this context, see especially Crosby 1986 and Diamond 1997.

7. I respect the possibility for disagreement. Clearly, people have run for millennia. A few individuals (military messengers, for example) must have trained at running and must have experienced something akin to what George Sheehan documents in his book. Yet it does not seem to be an accident of history that Sheehan's book appears in a highly industrialized late twentieth-century culture. The point is interesting to the extent that it bears upon the historicity of focal practices.

Also, perhaps I am mischievous to mention motorcycles here, for Borgmann (1984, 160–62) is critical of Robert Pirsig's *Zen and the Art of Motorcycle Maintenance.*

of the computerized and comfort-controlled modern farm tractor (Berry 1977, 1981). The result is that land as place is replaced by a version of the device paradigm in which land presents itself as but one of many purchased inputs in the production process. Land is reduced to its chemical and physical characteristics, which are bought or leased along with machinery, seeds, and chemicals. Land as place recedes, and with it, perhaps, the social and moral connections to nature and to community that give land its focal character before the device paradigm takes over (Berry 1987, 1991).

Berry's vision of farming can be contested. His reference point is farming as it was done in North America from European settlement until approximately World War II, but there are few generalizations that hold true of farming over such a broad swath of time and geography. Berry is thinking of a form of community and family-based farming that still exists in many parts of the United States and Canada, but one that is unarguably on the wane. We must allow Berry a bit of latitude with respect to his agricultural history, and I will not pick nits here. It *is* important to remember that there is little constancy in the way that land reveals itself as focal throughout European history. Land is a focal thing in feudal and plantation agriculture, for example, but in a manner quite different from Berry's vision. If one examines the transition from feudal to either plantation- or community-based agriculture as a transformation of focal practice, agricultural history becomes a form of philosophical history. Berry's critique selects one dimension of that history, to be sure, and a different reading of history might stress how communitarian ideals become implicated in the exploitation of race, class, and gender. Berry's version of that history illustrates why industrial farming offers far less than even feudal or plantation agriculture does in its capacity to "guard in its undiminished depth and identity the thing that it is central to" (Borgmann 1984, 209).

Which history is better? Which history is true? In another context, we might consider whether modern agriculture is more or less repressive than admittedly repressive feudal and slave-based systems of food production or family farms with rigid gender roles. Here, we must notice how Berry and Strong call attention to the disappearance of place and the dissolution of community. Perhaps the moral and political accomplishments of industrial society compensate for this ontological loss, but this is not the appropriate comparison in the present context. Farming has a history in which the focal character of land has plasticity subject to biological, technological, and economic constraints. The economic and social circumstances of nineteenth- and early twentieth-century North America were conducive to farming as a focal practice, and given the liberal political culture of the United States and Canada, healthy farm communities were often the result. Because many people farm under this regime, virtually everyone is at only a few degrees of separation from the land. Farming can, in that sense, be a focal thing for society as a whole. It can unify the social practice even of those who do not farm. Contemporary industrial agriculture is not conducive to farming as a focal practice. Borgmann's device paradigm coupled with Berry's account of agricultural technology gives us the words we need to explain why.

In the industrial age, the actual practice of farming becomes wholly preoccupied with devices that stand between person and task, person and land. In a preindustrial dairy, the farmer knows each cow as an individual, and addresses the animal's health and productivity on a case-by-case basis. In fact, the size of the dairy is often constrained by the farmer's ability to manage complex information. The more cows the more things to remember or record. The farmer has a variety of tools—feed rations, forage management, nutrient cycling, genetics, veterinary medicine—that may be drawn upon in successful dairying. Some of these tools are relatively devicelike, animal drugs being a prime example, but most require skill and judgment in their application. In an industrialized dairy, cows wear bar code ear tags that are read by a computer as the cow walks through the automated milking stall. Drugs and feed rations are metered and milk production data are recorded automatically. When a cow ceases to be productive, the computer marks it for culling. Only the number of gigabytes limits the computer's "husbandry." The farm as a focal thing or dairying as a focal practice recedes behind the devices that manipulate animals, feed, and manure in an economically efficient production of milk.

Indeed, the philosophical history of farming itself comes to an end with industrial technology. From the beginning of agriculture until the present age, changes in culture mirrored changes in the land. Human communities that depended on agriculture (we must always remind ourselves that some did not) formulated basic patterns of thought and practice within the technological parameters of survival through farming. For many (including many among relatively elite classes) much of their waking lives was spent in attentiveness to land and in directly applying hand tools to land, crops, animals, and the things of the household. A change from itinerant cropping or pastoralism to permanent agriculture is accompanied by a huge change in the daily texture of existence for everyone, as is a shift from feudal to capitalist patterns of control over land and technology. If human life consists in the actual passing of a human lifetime, innovations such as the mold board plow restructure the way that time is passed and, in this respect, who and what people are to an extent that is virtually unimaginable today. Even in an age when innovations such as the computer introduce huge changes in the way that many people pass their hours, the depth and breadth of technology's impact on our being and that of the things around us pales in comparison to what occurs in an agrarian society. But there is a constant in the agrarian world. Throughout every episode of technological transformation, land dominates human experience, and until recently only simple tools mediate the experience of land. This is especially true for manual labor, but even the overseer or the plantation manager (indeed even the priest or army general who does not farm) is close to the land. Virtually everyone feels the give and take between land and food production, and for this reason, changes in that give and take are ontological changes. Yet how isolated are we from that give and take today? How little does a new method of harvesting, of planting, or of breeding affect the texture of our lives? The philosophical history of farming is over, or at least so it would seem.

I would submit, though the present format does not provide the opportunity to establish, that the philosophical history of land and farming should play a central role in any philosophy of technology, and especially in any built upon Marxian or Heideggerian foundation. It is impossible to make ontological sense out the peasant experience without an account of farming. Marx admired the communitarian foundations of feudal societies and, in his early thought, at least, problematized capitalism in terms of the individual's estrangement from those foundations through wage labor. Yet the communitarian foundations of feudal society consist not in the construction of cathedrals (cf. Borgmann 1984, 159–60) but in the peasant's bond to land, understood implicitly as a particular place. If we accept Wendell Berry's vision, the historical development of farming reached its apogee under capitalism and at a time when the industrial revolution was well under way. These circumstances were undercut by a gradual change in the technological constitution of agriculture. Farming's capacity to unify the world and to guard its essence has been drastically reduced. Today, farming might be focal in the way that running might be focal—as a hobby or life activity for a few. It no longer possesses a focal power to unify society and culture. Furthermore, the decline of farming is a technological phenomenon that cuts across economic systems.

I would argue that the ontology of peasant experience and the history of farming are even more fundamental to Heidegger's thought. Marx remained rather optimistic about the eventual effects of technology, after all, and seemed to think that what had been lost could be regained in a socialist society. The themes of ontological loss that pervade Wendell Berry's critique of agricultural technology are far more consistent with Heidegger's thought. Heidegger devoted his life to reconciling thinking with the Being that presents itself as grounded in a given place. The links between this central theme and farming will be explored a bit in the next section. Nevertheless these remarks on Marx and Heidegger must be taken as suggestions for a program of research in the philosophy of technology rather than as developed arguments.

THE NEGLECT OF FARMING

There are good reasons why Borgmann and Strong should not have devoted more of their argument to farming than they do, though the reasons have more to do with tactics than substance. Every author chooses how to get a point across, and getting read in the first place is a prerequisite to success. Despite the appeal of Wendell Berry farming has a negative psychosocial meaning that complicates this always-formidable task. As Logsden writes, "Most of us grew up in a society where farmer was often merely a synonym for moron" (Logsden 1994, 48). On the one hand, this phenomenon reflects a natural tendency for verbal and social skills to be more extensively developed among people who live in close quarters and in constant commerce with one another (and amid greater cultural diversity). A certain amount of mocking and derision reinforce the acquisition of these skills. Farming is like farting; people with refinement and breeding avoid mentioning it as assiduously as they avoid being caught doing it. Building farming into *TCCL* in a fundamental way might have doomed the book to obscurity.

On the other hand, farming's negative valence extends more deeply into intellectual than popular culture, suggesting that something more than lack of sophistication is being repressed. Rusticity is an acceptable literary topic (though more for its metaphorical power than in its own right), but the philosophy of agriculture simply does not come up. To bring up farming in a philosophical context is (or has been until recently) a faux pas—an intellectual fart—of the first order, likely to be followed by embarrassed silence. This phenomenon is all the more reason for Borgmann and Strong to have downplayed farming, but is also itself revealing and relevant to farming as a focal project. Again, the exhibition of this relevance must be limited to suggestive allusions, rather than full arguments, but it does provide a way to link farming more closely to Heidegger's thought, if only implicitly.

Western philosophy has increasingly become committed to the analytic, the universal, and the a priori. Indeed, to describe a topic as philosophical is, for many Western intellectuals, to say that understanding it involves analysis of thought and language as a nexus of concepts. In extreme logicist philosophies, commitment to the a priori takes an ahistorical form, but even those who recognize the role of time in history have neglected the role of place. By extension, they neglect the praxis of being in place. Universalizing philosophy has, in fact, commodified place, concealing the particularity of the exchange between place and consciousness. The commodified product is space or (ominously) *environment,* commodiously available for determination, first by things, then by devices.

Things, of course, might be quite particular. In Borgmann's treatment things admit of open-ended engagement and invite the creation of meanings. Heidegger's interpretation of Kant, however, identifies things with those dimensions of being that may be known in advance of their exhibition through presence. They form the basis for *techne,* which demands an anticipation of being (Heidegger 1968). However, it is precisely the stability of things that avails the commodity form, in Borgmann's sense, and, contingently, historically, the device paradigm. Are places simply spaces determined by things? That is, is a place simply an empty space that has been filled up with Kantian things, with objects that disclose themselves essentially in their ability to be known in advance? Borgmann's philosophy does not answer this question. With apologies to Kant scholars who will correctly note that Kant never said anything like this, I would like to call the view that the environment is simply the thing-determined space around a person, a community, or a set of practices "Kantian environmentalism."

One reading of Borgmann is that we need to recognize that things proper and devices are two different kinds of Kantian things. Both disclose themselves through a stable form of being that is knowable in advance of a particular encounter with them. But things have a richness that is absent in devices. Things open into a deep and extended encounter with being, while devices perform their predetermined task so quickly and faithfully that our moment of encounter with them is without ontological dimensionality altogether. Another way to say this is that in using, appreciating, or addressing things we are drawn into a mode of being where self and world are but moments of a unified praxis. Using a device, in contrast, permits one to remain absorbed in the particular construction of one's "self" that is dominant at any given moment and to forget or neglect that this self is even in a world (much less a contingent, fragile, and unstable construction). Is Borgmann a Kantian environmentalist? If the answer is yes, then the predicament he identifies in *TCCL* simply reflects the degree to which our environment is filled with (hence determined by) devices, rather than things. An environment completely given over to devices is psychologically perilous, for it is devoid of ends. The organism (or perhaps I should say the subject) itself becomes the only source of meaning.

The Kantian environment is not completely given over to devices, of course; there *are* other subjects, other people. But for Kant, things and devices appear to be equivalent with respect to their capacity for generating ends. If so, then the environment is sterile, save for the encounter with other subjects. The Kantian subject generates its own meaning (its end) in its attempt to recognize subjectivity in others—other subjects revealed as ends, not things. For Kant, time and history unfold through agency, as the subject struggles to act in ways that are free and meaningful, not merely the physio-chemical responses of an organism to its environmental influences. Farming is only one act among others in the Kantian quest for autonomy and authenticity, and likely to be relatively unfruitful by virtue of the relative paucity of subjects in the farmer's orbit. Borgmann may be read as enriching the Kantian environment by driving a wedge between things and devices. Perhaps new sources of meaning can be found in the space opened by this wedge, and perhaps even farming (as well as running or skiing) can be meaningful.

Here, Strong's analysis complements and extends Borgmann's. The recovery of a proper relationship occurs through an encounter with wilderness. Strong discounts the efficacy of ordinary nature recreation, however. In their common forms, fishing and camping are dominated by technologies that mute nature's voice. Graphite rods and high-tech gear allow us easy access to wilderness environments, but the ease comes at a price. It is also easy to neglect—even to negate—the way that things call out to us, the way in which things are (potentially) efficacious in forming the wilderness experience. Strong writes, "Nature, in my account of it, is not simply observed nature. It takes hold of us, *animating* and involving us. . . . Animating nature quickens us, nourishes us, gives us the air that makes our breath worth the breathing" (Strong 1995, 175). Like Borgmann, Strong attributes this power to focal things, but unlike Borgmann, many of Strong's focal things are in fact places—Cottonwood Canyon or Walden Pond.

This extension of Borgmann is potentially a move beyond Kantian environmentalism. If places are not determined either by things or devices, then being in a place will be contingent upon more than things arrayed at a given time. The place itself becomes fecund. It contributes to the being that subsists and is revealed in being there. On this view, the contingencies of a place cannot be known in advance, though some are revealed in their particularity when one is there. In this regard, places are quite unlike Kantian environments, populated by Kantian things whose peculiar feature is precisely their potential for being known in advance. One can come to know one's place, but not even the most exhaustive description of Kantian things and their relationships would yield such knowledge. One comes to know a place by being there over time, by habitation, by making it one's own. Clearly, the efficacy that Strong attributes to wilderness derives from its placial character (to use a word coined by Edward Casey [1992]). Strong's wilderness has fecundity, a generative power lacking in Kantian environments. If Borgmann is a Kantian environmentalist, Strong certainly is not.

Yet why would Strong think that wilderness has primacy among places for the recovery of focal practice? In particular, why should wilderness have primacy over farms? Although there are as many ways to inhabit a place as there are places to inhabit, farms are places where land is inhabited by farming. Farming is the way of habitation, of being there in the place, and of ownership (that is, making the place one's place). Though Strong talks about coming home and being at home in the wilderness, surely only those few (loggers or trappers) who make their productive lives in wilderness can truly be said to dwell there. Surely our sense of becoming beings who have a place must be derived from the way that a place contributes to that process of becoming, a process that must include the material as well as the spiritual work that sustains human life. This, I think, returns us to ponder again the ontological loss that occurs when the philosophical history of farming comes to an end. Both farms and farmers exist through farming. One is a painter by painting, but the sense in which one's material existence depends on painting is less rich and certainly less literal. Farmers produce not only themselves, but also all who eat through farming.

While it would be easy to trivialize the grammar of the formulation, one way of saying that farms and farmers exist through farming rejects Kantian environmentalism in favor of a philosophy that privileges production rather than action. The Kantian subject "acts" by choosing one action rather than another, but how does the Kantian subject subsist? This has seemed to be an important question to a generation of philosophers who have wondered whether the subject supervenes on brain activity, or has a kind of substance all its own. Yet Marx, Husserl, and Heidegger anticipated a generation of postmodernists in believing that the Kantian subject is an artifact of a certain kind of reflection and that long before that reflection there were praxial moments in which person and place are coproduced. Production, on this view, is an ontological process that yields the person as it shapes the place. To take this notion of production seriously in material terms demands a hard look at eating, which should point almost everyone living after the second millennium toward farming.

Strong's philosophy goes part of the way. He ends *Crazy Mountains* with a call to "build again" (Strong 1995, 205–10), but his fascination with wilderness stops short of an existential embrace with production. Hikers hike, but might do so anywhere. Wilderness itself does not need loggers or trappers, much less hikers, to realize its placial character. Farmers farm, and necessarily on farms. Perhaps this *is* an oddity of grammar (that would be the Kantian view), but in the spirit of Borgmann and Strong, let us entertain the possibility that in farming on farms, farmers participate in a metaphysical whole that is manifest in the particular moment of being that we call "the farmer" in one mode, "the farm" in another. Beings are produced at a particular place by being there. Being there can be abstracted into universal moments of subjectivity and spatial extension only through a turn of mind that erases the particularity inherent in any production of being. But why bother to perform the abstractions that spawn Cartesian and Kantian philosophy? Philosophy itself has been somewhat dogmatic and question begging with respect to this inquiry, but a perfectly good answer is readily available: because this program of abstraction yields technology. Technology, in turn, can be brought to bear on the problem of production. Technology is not then evil, nor is production. Each has its place.

But a philosophy that systematically obscures the contingency of places and that denigrates the modes of being that are dedicated to the cultivation of places might undo the conditions that gave rise to its own creation. Such a philosophy would be threatened by moral problems or by focal practices that depend upon the particular, and especially threatened by practices that challenge its fundamental abstractions. Such a philosophy can maintain its grip on the minds and practices of people only if enough of them deny tout court the possibility that places have philosophical significance. One principled reason for this denial resides in the belief that philosophies of place are implicated in nationalist, racist, and totalitarian ideologies, and of course they are. Yet ironically, the philosophies of universal autonomy seem to be equally implicated in the device paradigm so ably critiqued in Borgmann's book. Philosophers of technology, in other words, have their work cut out for them, and an adequate philosophy of agriculture could be more than a case study for accomplishing the larger tasks.

Conclusion

If Borgmann is read as a Kantian environmentalist, then farming is a focal project among others. That is well and good, a respectable philosophical liberalism. Strong and I prefer a different reading, one in which focal practices are intended to remediate a world where subjects act in response to environmental forces. On this view, focal practices ought not to be purely consumptive. Focal practices (including running and skiing) ought to involve habitation of places. The focal things to which they are attached may be things indeed, but they must be things that allow one to dwell in a place. The encounter with focal things would then produce being within a place, and the sterility of the Kantian environment would dissolve into a sea of contingency, surprise, and serendipity. Strong and I differ in our assessment of the solvent that accomplishes this task. For Strong it is wilderness, and the dissolution has qualities of aesthetic contemplation. For me it is productive work that best responds to place, and farming is the quintessential example. Technology would be employed within the tasks of production, and there would be no need to ask, "Production for what?"

In *Crossing the Postmodern Divide,* Borgmann takes up the moral dimensions of practice. He warns that "[t]he segregation of doing from making, of morality from production, goes back to Aristotle and has been carried forward into the modern era by Immanuel Kant" (Borgmann 1992, 110). The book ends with a shift from "focal things" to "focal realities," which we might interpret as a shift toward the placial. These developments in Borgmann's philosophy imply that he would not endorse a Kantian environmentalist reading of *TCCL* and that he would be very sympathetic to the kind of philosophy developed not only in the present essay, but in Strong's *Crazy Mountains.* My purpose in confining the present analysis to the 1984 book is not to hold Borgmann to a Kantian view that he does not hold. My aim is to note the possibility of reading the book in that way and to argue that attention to the philosophical dimensions of farming is a particularly effective way to develop a more congenial reading of Borgmann's philosophical project.

Yet I hasten to agree with another of Borgmann's points. Focal practices do not yield a utopia, and preindustrial society was no Eden. Focal practices ameliorate and remediate; they do not redeem.[8] They remediate because they are productive rather than consumptive. Focal practices may, of course, be productive in an immaterial way. Heidegger's account of thinking and John Dewey's account of art as experience represent paradigm cases of immaterial production of being. But these accounts of production are incomplete, for like Aristotle, Heidegger and Dewey take the material production of thought or experience for granted. In societies where land is a focal thing and where farming is a focal practice unifying the society it may be permissible to grant material production when philosophizing. Philosophy can stray only so far under such circumstances. But Borgmann and Strong have taught us of the dangers that arise when focal practices give way to the device paradigm. The world of consumption, where subjects

8. Heidegger's error may have consisted in attributing a redemptive rather than an ameliorating power to cultivating, building, and dwelling. Such an error would link Heidegger's thought to some dangerous elements of German National Socialism in a manner that is philosophically significant. While I do not join those who think that Heidegger's association with the Nazis tarnishes his philosophy irrevocably, it *is* precisely this link that should make those of us who are impressed by Heidegger's philosophy to undertake some serious self-examination.

choose commodities, is a world where places have been so polluted by devices that the project of being in a place seems entirely passive, a form of tourism. The experience of production as both means and end fades from view. Focal practices remediate device pollution when they take you from the milieu of choice and replace you in a milieu of producing being by being there (there rather than anywhere or nowhere).

When focal practices and things are interpreted in light of their role in the production of existence, they imply a hierarchy, and it is this hierarchy that leads one to farming, in my view. Practices and things that are more comprehensive of and fundamental to production have a correspondingly greater claim to significance. This is not to say that any given person must take up the most fundamental practices, but it does indicate why the philosophy of agriculture, for example, might be a more pressing topic than the philosophy of recreational skiing. It is in this sense that farming is a quintessential focal practice and a crucial topic for the philosophy of technology. Borgmann's book is a way of beginning the philosophy of technology. It leads its readers to a crucial juncture, and Strong's environmental ethics point beyond that juncture. In avoiding the totalizing tendencies of Marx and Heidegger, Borgmann presents an ameliorative response to the existential dilemma of the device paradigm that is equally consistent with Kantian liberal and with radical interpretations. Strong sets out resolutely in the direction of the radical alternative.

Strong's philosophy lays the foundation for a therapeutic aesthetic, one that will go some distance toward the remediation of device pollution. In this he also provides a rationale for wilderness preservation, a move that opens a path for an environmental ethic within the philosophy of technology. Yet if he fails to embrace the productive moment, if he retreats to the contemplative, Strong may have narrowed the range of praxis, as Marx, Heidegger, and Borgmann understood it. In this he would be mirroring other environmental ethicists' tendency to overlook the material prerequisites of spiritual being, focusing only on the burden that those prerequisites place on wilderness preservation. *Crazy Mountains* does not entail such a narrowing, though it is consistent with it.

I would prefer to look toward farming as the quintessence of our being in the world. It is a moment of praxis in which most of us participate at great remove and with utter dependency on a technology that shields us too thoroughly from farming's material and spiritual dimensions. Farming understood as focal practice contains many lessons both from our history and for our future. While we may not become farmers in contemplating farming as focal practice, perhaps we will better understand the place that we have cultivated in this world, and how to guard it.

REFERENCES

Berry, Wendell. 1970. *Farming: A Handbook.* New York: Harcourt, Brace, and Jovanovich.

———. 1977. *The Unsettling of America: Culture and Agriculture.* San Francisco: Sierra Club Books.

———. 1981. *The Gift of Good Land.* San Francisco: North Point Books.

———. 1985. *Collected Poems.* San Francisco: North Point Press.

———. 1987. *Home Economics.* San Francisco: North Point Books.

———. 1991. *What Are People For?* San Francisco: North Point Books.

Borgmann, Albert. 1984. *Technology and the Character of Contemporary Life: A Philosophical Inquiry.* Chicago: University of Chicago Press.

———. 1992. *Crossing the Postmodern Divide.* Chicago: University of Chicago Press.

Casey, Edward. 1992. *Getting Back into Place.* Bloomington: Indiana University Press.

Crosby, A. W. 1986. *Ecological Imperialism.* New York: Cambridge University Press.

Diamond, Jared. 1997. *Guns, Germs and Steel: The Fates of Human Societies.* New York: W. W. Norton.

Heidegger, Martin. 1968. *What Is a Thing?* Trans. W. B. Barton Jr. and Vera Deutsch. Chicago: H. Regnery.

Logsden, Gene. 1994. *At Nature's Pace: Farming and the American Dream.* New York: Pantheon Books.

Sheehan, George. 1978. *Running and Being: The Total Experience.* New York: Simon and Schuster.

Strong, David. 1995. *Crazy Mountains: Learning from Wilderness to Weigh Technology.* Albany: SUNY Press.

Thompson, Paul B. 1986. "The Social Goals of Agriculture." *Agriculture and Human Values* 3, no. 4: 32–42.

———. 1988. "The Philosophical Rationale for U.S. Agricultural Policy." In *U.S. Agriculture in a Global Setting: An Agenda for the Future.* Ed. M. A. Tutwiler. Washington: Resources For the Future.

———. 1990. "Agricultural Ethics and Economics." *Journal of Agricultural Economics Research* 42, no. 1: 3–7.

———. 1992. *Sacred Cows and Hot Potatoes: Agrarian Myths and Policy Realities.* W. Browne, J. Skees, L. Swanson, P. B. Thompson, and L. Unnevehr. Boulder, Colo.: Westview Press.

———. 1993 "Animals in the Agrarian Ideal." *Journal of Agricultural and Environmental Ethics* 6 (special supplement 1): 36–49.

———. 1994. *Ethics, Public Policy, and Agriculture.* P. B. Thompson, Robert Matthews, and Eileen van Ravens
waay. New York: Macmillan.

———. 1995. *The Spirit of the Soil: Agriculture and Environmental Ethics.* London: Routledge.

———. 1998. *Agricultural Ethics: Research, Teaching and Public Policy.* Ames: Iowa State University Press.

Thompson, Paul B., and J. Patrick Madden. 1987. "Ethical Perspectives on Changing Agricultural Technology in the United States." *Notre Dame Journal of Law, Ethics, and Public Policy* 3, no. 1: 85–116.

TEN

Design and the Reform of Technology: Venturing Out into the Open

Jesse S. Tatum

The reform of technology seems an impossible dream. Virtually monolithic in its expression of the device paradigm, modern technology looms as an overwhelming colossus, its interrelated parts reinforcing one another, and its sheer and seamless face offering no visible hand- or foothold for the reformer. Even the nature of the fundamental flaw itself is concealed by the wondrous flow of gifts modern technology bestows.

Yet careful attention at the margins of public attention reveals advanced reform efforts expressive of newly emerging ways of being in the world. The home power movement in the United States, for example, now boasts in excess of one hundred thousand homes equipped with photovoltaic (i.e., solar cell), microhydroelectric, and small wind electric power systems. This movement has emerged from efforts to recover or invent orienting practices centered in a reinvigorated sense of community, an enriched work life, and a reconfiguration of relationships with the natural environment. It has involved the development of new small-scale user-owned and -operated energy supply technologies, and a full range of energy use systems from superefficient refrigerators and LED lighting fixtures to horizontal axis washing machines and highly efficient well pumps—all as a product of participatory collaborations among ordinary people committed to a redirected way of life. While radical efficiencies at the point of energy use reduce the economic costs of near total reliance on expensive renewable energy sources, home power systems are *not* cost-effective by any traditional standard. The home power movement rests quintessentially on what Borgmann describes as the willingness to give up without resentment the "affluence" of the device paradigm in the name of the "wealth" of engagement (Borgmann 1984, 223–31; Tatum 1994).

Developments such as the home power movement can be heartening as we grapple with the formidable task of a reform of technology. They can also be a source of concrete initial guidance, helping to confirm, for example, the notion that engineering and economic "efficiencies" will need to be set aside as governing desiderata as we recognize that efficiency in the production of "affluence" is not the same thing as efficiency in the promotion or preservation of the "wealth" of engagement. (As noted in greater detail below, efficiency, in the traditional technical and economic senses, cannot finally govern the reform of technology, for it is caught up in what is to be reformed.) In more general terms, developments like home power confirm the need to distance ourselves from the assumptions and rules of thumb that have dominated technological "advance" in the past. In our reform efforts we will indeed be "ventur[ing] out into the open" (Borgmann 1992, 4) to some degree, suspending past notions of our location and destinations in the world.

Exploring both real and hypothetical examples, this chapter cultivates lessons of this sort and offers a new, if preliminary, set of guidelines for the move out into the open that any genuine reform of technology will imply. My discussion of reform proceeds through a new approach to the *design* process. Suggestions will be offered for enhancing engagement both where technology is intended to *center* a focal practice, and where technology is primarily intended to *enable* us elsewhere to cultivate such a practice. The chapter will also identify some of the specific obstacles that may regularly impede our efforts and offer initial strategies for overcoming them.

WHERE DO WE STAND AND WHERE DO WE WISH TO GO?

I take our social situation to be pervasively darkened by our embrace of the device paradigm, but it is at least latently brightened by the hope inherent in a stubborn residual awareness that devices are not enough to sustain us. On the one hand, we abandon not only "things" for devices, but we abandon the most fundamental *processes of choice* in favor of the device of expert analysis. We accept, for example, energy technologies selected for us through the most simplistic (expert) calculations involving market prices and interest on money, allowing our way of being in the world to be profoundly shaped for us in terms of relationships both with the environment and with other people (Tatum 1995). Disburdened of the social interactions and of the material subtleties and complexities involved in actively choosing how we are to live, we allow ourselves to become increasingly cut off from the engagement with others and with the world that would accompany a more genuine politics.

Yet, on the other hand and in sharp contrast, we continue to burden ourselves quite deliberately in areas such as music making and sports, with little or no thought of producing anything in the usual sense. The simple machines that could be built to propel a tennis ball back and forth across a net with essentially total predictability and control would be thought quite preposterous. At some level, we remain aware that "life cannot be delegated," and we draw back from the "deep hostility to life" (Mumford 1964) unmasked as we begin to push the device paradigm to its logical limits.[1] As Borgmann suggests, "there is a wide and steady, if frequently concealed, current of focal practices that runs through the history of this country. It is the other American mainstream" (Borgmann 1984, 219). And this is the hope that brightens a situation otherwise darkened by the pervasive device paradigm.

Interestingly, our present situation generates a certain restiveness among those of us raised on the promise of technology and long accustomed to waiting, not too patiently, for the next "technological fix" (Weinberg 1990). Loneliness, the sense of incapacitation, and the simple emptiness and boredom of a world of devices give rise to "sullenness" and "hyperactivity" (Borgmann 1992). These reactions offer some confirmation of our sense of where we stand: they are precisely the responses we would expect. But they offer further signs of hope as well: present displays of restiveness could mature into constructive efforts at reform.

In any case, we need to move beyond mere sullenness and hyperactivity to a fundamental reform, tapping every sign of restiveness and acting in new ways to push back the darkness of the ruling device paradigm. Our task will be to set aside the false (and in the end unattractive) images of ever more complete and effortless prediction and control, and learn more vigorously to "take up life" again (Borgmann 1984, 246). For all their value, the methods of science and the advances of technology are not sufficient to sustain us. We must venture out into the open again.

How is this to be accomplished? Where do we wish to go from here? What Borgmann has proposed does indeed amount to a move out into the open. In place of neat social and political devices like cost-benefit analysis, he urges deictic discourse, a form of discourse that "illuminates what concerns me and, if successful, provides you with an understanding that [may] move you to act as I have been moved" (1984, 180–81). In place of the uniform automatism of material devices, he calls for a move toward the engagement of things. He also offers a helpful reassurance that eases the way somewhat: while some of the "affluence" of the consumption made possible by devices may need to be traded away in favor of the "wealth" of engagement, this should be possible without resentment as the hollowness of a commodity existence is moderated by the orientation of

---

1. C. S. Lewis (1944) has also provided an apt description of the highly undesirable state of affairs that would likely flow from carrying something like the device paradigm to its logical limits.

things. Specifically, because he seeks more than a set of technologies better suited to the delivery of what we value as a collection of commodities, it is not going to be enough to reform technology *within* the device paradigm. Neither is he looking for a reform *of* the device paradigm in the sense of trying to replace it or eliminate entirely the disburdening benefits of technology. The object of reform is to "prune back the excesses of technology and restrict it to a supporting role" (Borgmann 1984, 247). Thus, he seeks the recognition and restraint of the device paradigm as the basis for a more profound commerce with reality that gives focal things and focal practices a central place in our lives.

From this start, though, how precisely are we to proceed in this new and unmapped terrain? How are we to avoid feeling simply lost out in the open? These are the questions I wish to address below by trying to build from Borgmann's work to sketch a few additional guidelines for a reform of technology.

TECHNOLOGY BY DESIGN

A genuine reform of technology must, I believe, carry Borgmann's "intelligently selective attitude toward technology" (Borgmann 1984, 211) beyond a selection among technologies already made available under the dominance of the device paradigm. It must reach into the design process itself. Our design objective, after all, is no longer what it has been under the sway of the device paradigm. It is no longer enough simply to disburden the user. Nor do we seek simply to make a particular commodity easily and ubiquitously available for consumption. Neither is there any inherent interest in making the mechanism of production recede from view or disappear.[2] To be thorough and effective in the deracination of old and flawed objectives we must, in some sense, start from first principles.

As the reform of technology reaches into design, it is useful to draw distinctions in our thinking between two types of technology at opposite ends of a continuum: one in which the technology itself is to be the focal thing, and the second in which something more like, though better guided than, a device is sought. Along with "technology as thing," then, we can think of a second group of "enabling technologies," intended primarily

2. It can be argued that there has long been a *commercial* interest in making the mechanism of production recede from view or disappear. To the extent that the "consumer" leaves the generation and distribution of electricity to the utility, for example, and is satisfied simply to flip the switch at home, the utility is justified in charging a fee for its services. To the extent that the "consumer" remains uninformed of environmental and other implications of his/her electricity use, the benefits provided by the utility may appear larger than in fact they are, rates of consumption will tend to be higher, and the utility's financial benefit is likely to be greater.

to make the engagement of other focal things and practices possible. Opportunities for reform exist in both categories of technology.

TECHNOLOGY AS THING

Where the design process is intended to evoke technology as thing, preserving or enhancing engagement is a first-priority principle and the shift in design objectives that lies at the heart of the matter may best be conveyed through a concrete example. Let us look at the design of technology for personal transport, first in terms of the problems of the traditional device approach, then with a more positive alternative in the sense of technology as thing. Some of our most striking departures from classical objectives and design processes will be encountered in this illustration of a move toward technology as thing.

If we wish to design a machine to move us about not as a device but as a thing, the popular image of an electric car could hardly be more *wrong* as a design idea. The standards of success for this car call for us to match or improve upon the performance of our present transportation devices: effortless acceleration and control; high-speed travel and extended range; a comfortable interior isolated from the heat, cold, noise, and pollution of the outside world; and an invisibly reliable mechanism virtually instantaneously refueled from an equally invisible and, by all appearances, undepletable source of energy. These are indeed classical objectives, not only the easing of ordinary burdens, but the "annihilation" of time and space (Borgmann 1984, 191).[3] But where are we when we are without time and space, without weather, and without consequence (e.g., observable resource depletion) in consumption?

If we were to design a machine for human transportation as a "thing" rather than a device, it would probably bear little resemblance either to existing gasoline-fueled automobiles or to their electric reincarnations. In fact, it might look more like an electric moped. It might be designed for little more than a twenty-mile range and a maximum speed of twenty-five miles per hour, declining as the batteries are depleted. Linking a lightweight motor to a battery, it could be configured as a retrofit, bolting directly onto a bicycle and leaving the pedals in place and operational. Let's equip it also with a small solar panel so that it can be recharged while parked all day in the sun (or not, if it's cloudy). Now we will know time and space and

3. Lewis Mumford (1970, 172–73) has also suggested that an implicit goal of technological society has been the removal of all natural barriers. There is the appearance of "only one efficient speed, *faster;* only one attractive destination, *farther away;* only one desirable size, *bigger;* only one rational quantitative goal, *more.*"

distinguish uphill from down, hot weather from cold. And we will discipline and pace ourselves from an awareness of the sunshine. The roadway will no longer be one homogeneous blur but show us blades of grass and perhaps even wildflowers in the cycle of the seasons. Now we know better where we stand.

Traditional technologists (under the sway of the device paradigm) would scoff at our electric moped. "Zero to twenty in thirty-three seconds!" they would laugh. But where technology is intended as a thing to provide a focus and orientation in our lives, and to reveal the world in its essential dimensions, it makes no sense to measure it by the standards of the device. Where an electric moped or an ordinary bicycle is at the center of a focal practice, replacing it with a new automobile, electric or otherwise, would defeat our central purpose and elevate technology from its role as means to an end in itself.

As this illustration suggests, technology *as* thing is likely to carry our reform efforts into the most striking departures from standards that are implicit in the device paradigm and now second nature to most of us. It will require an approach to design in which the prevailing criteria of technical and economic efficiency, for instance, neither uniformly apply nor necessarily prevail. If technology A delivers a particular commodity like "warmth" or "electricity" more cheaply or more efficiently than technology B, this will no longer be sufficient reason to choose A over B. Other desiderata will regularly be granted equal or greater weight and would be freely offered and freely heard in the course of deictic discourse in which individuals and groups bear witness to any focal concerns that they may see are affected.

In its shifts toward genuine participation and an openly political exchange, the process of deciding among technologies—indeed the design process itself—would be as much affected as technical standards. 'Citizens' would replace 'consumers' as the uneasiness typical of our usual implication in and complicity with the device paradigm is aired and becomes converted into a basis for altering our course of action. Politics itself would, in this sense, move well beyond its present status as a mere "metadevice of the technological order" under which one might "as well call for participation in pocket calculators" (Borgmann 1984, 113). Citizen designers would be assisted, rather than eclipsed, by technical experts as the primary task of the latter came to be to make clear the limits, range, and mechanisms of technical possibility. Thus, the design of technology as thing will unquestionably move us out into the open both *technically,* in the sense of distancing ourselves from classical design standards of technical and economic efficiency, and *sociopolitically,* in the sense of a shift from the closely limited circle of expert design professionals to the open participation of deictic discourse.[4]

New technical guidelines of a general nature will remain difficult to formulate in the first instance, but are also less necessary where focal things and practices are to displace commodity production as our objective. On the other hand, citizen design processes along the lines described above are already beginning to emerge. One of the best illustrations, again, may rest in the home power movement (Tatum 1996). Other illustrations can be drawn from the "science shops," parliamentary offices of technology assessment, and other developments occurring in Europe (e.g., Vig 1992). Such processes are also well modeled in once marginalized practices from the past such as those of the Highlander Research and Education Center in New Market, Tennessee (Horton 1990; Adams 1975).

ENABLING TECHNOLOGY

Some kind of an electric automobile may yet be valued in some situations as an "enabling" technology. One who finds engagement even as an avid runner, for example, may still find it appropriate to get to and from work in a car. If we reorient our design process, however, classical standards will still be fundamentally affected.

The primary object with enabling technologies will be the disburdenment that makes other focal practice possible; but there will be no call in general to make the goods procured by such technology appear to be free or to hide from view the mechanism of their procurement. Under a reform of technology, our electric vehicle still would *not* be designed strictly within the device paradigm like the ubiquitous resort to mobility—the automobile—that now itself shapes our modern lives; rather it would be designed as a tool for facilitating and enhancing engagement *directly* as well as with other focal things. Even where we are to rely on batteries and power plants, or on gasoline in tanks, for example, these need not be made invisible. If users should find elements of the functioning of their revised automobile discomforting—e.g., observing the weight and bulk of batteries, or the volume of fuel that goes into a twenty-gallon gasoline tank—they could choose to make design adjustments ranging

4. I have pursued the critically important issue of participation, especially in the design process, elsewhere (Tatum 1996, 1999) in detail. Benjamin Barber's work (1984) is also very helpful in this connection, including his notion of "political talk," which closely complements Borgmann's "deictic discourse." In the "political talk" of Barber's "strong democracy," "no voice is privileged, no position advantaged. . . . Every expression is both legitimate and provisional, a proximate and temporary position of a consciousness in evolution" (Barber 1984, 183).

from a more fuel-efficient auto to a reduced reliance on auto travel. But these adjustments would not be made simply to mask actual connections in the world.

The reform of technology envisioned here would, in other words, redirect the design of this second group of enabling technologies as much as it does the design of technology as thing.[5] Unlike devices, enabling technology would seek deliberately at least to *disclose* rather than obscure significant realities, at least insofar as this does not actively interfere with their central purpose as the practical means to other focal concerns. Even where it might supply food, warmth, or other needs as smoothly and easily as the device, enabling technology would make the mechanism of procurement as visible, understandable, and accessible as possible. Whether or not we chose to follow them, the trails leading from enabling technology into the larger world would remain well marked and invite entrance rather than end abruptly at chained gates.

In many instances, it is possible to press the design of enabling technology even farther along the continuum toward technology as thing without significantly undermining other focal things or practices its design may support. To pursue a new example, someone whose practice is centered in the culture of the table might seek to make time available for the preparation and sharing of meals by purchasing a furnace. As an enabling technology, that furnace can itself be variously designed. Under the device paradigm, its mechanism and the origins of its fuel are hidden from view and its operation is automatic. Under a reform of technology, we might first make its mechanism and the flow of its fuel supply more visible. We might provide for individualized control of temperatures by room, and for clearer indications of energy use based on outdoor conditions and thermostat settings. But we might also choose to press further in the direction of technology as thing. We might choose to build a superinsulated[6] home requiring so little heat, in fact, that we might continue to enjoy some of the engagement of a wood stove; through careful design, the gathering, cutting, and splitting of fuel wood might then be accomplished without sacrificing so much time as to undermine devotion to the culture of the table. Alternatives of this sort would not amount to the usual technological fix. Not designed simply to disburden consumers as devices do, they would, in the very process of easing

5. I depart, here, from a strict construction of Borgmann's work in which existing devices are not to be rethought insofar as they are used in support of other focal practices.

6. This is a technical term indicating a combination of heavy insulation and extraordinary measures to control infiltration. Fresh air is introduced actively using air-to-air heat exchangers that warm incoming fresh air with outgoing interior air. Carefully designed superinsulated homes require essentially no heating even through the winter in Canada.

burdens to make room for other focal engagements, deliberately disclose and preserve awareness of significant realities.[7]

DEALING WITH INTERRELATED TECHNOLOGIES

Past practice and patterns of thought will, of course, pose major difficulties in the reform of technology. It will remain possible, for example, to get carried away with the design of enabling technologies and drift back into something like the device paradigm.

> Engagement, however skilled and disciplined, becomes disoriented when it exhausts itself in the building, rebuilding, refinement, and maintenance of stages on which nothing is ever enacted. People finish their basements, fertilize their lawns, fix their cars. What for? The peripheral engagement suffocates the center, and festivity, joy, and humor disappear. (Borgmann 1984, 222)

As a safeguard, it may be appropriate to cut back on enabling technology categorically.

Here we encounter another obstacle, however, that is likely to be felt not only here but throughout the reform effort: present technologies in all of their interrelationships take on the appearance of necessity. It will be difficult to remove or redesign any single piece, any particular technology in isolation from the rest. In the case of the present example, we find that the modern suburb

> is technological through and through. It is a pretty display of commodities resting on a concealed machinery. There is warmth, food, cleanliness, entertainment, lawns, shrubs, and flowers, all of it procured by underground utilities, cables, station wagons, chemical fertilizers and weed killers, riding lawn mowers, seed tapes, and underground sprinklers. (Borgmann 1984, 189)

And it is difficult to see how or where one might intervene.

Warmth, food, cleanliness, and other practical needs can, however, be supplied by enabling technologies designed at least not to conceal significant realities. And this can be done without adding such a load of practical burdens as to displace other focal engagement. In this particular instance, the practices of the home power movement, which rest in part on the

7. As Borgmann suggests, care will be required in the design of (enabling) technologies to minimize their interference with focal concerns they do not directly support. Where possible the avid runner's automobile should not, for example, be permitted to crowd the bicycle or electric moped off the roadways as commonly occurs under the device paradigm (cf. Winner 1986, 48). Conflicts should be eased, however, by the "kinship [that exists] among eloquent and focal things" (Borgmann 1984, 195).

systematic addition of aggressive energy efficiency measures to renewable supply systems, offer concrete indications of such an alternative course. An element of skill and discipline is required in the implementation and use of these systems. And they inevitably preserve some awareness of the links they establish between their users and the larger world. As indicated earlier, these specific alternatives actually seem to have emerged essentially as enabling technologies in defense of particular focal concerns, including a sense of community, the desire for meaningful work, and the expression of different relationships with the natural environment. They stand as a shining example of how *multiple* technological substitutions can be made more or less simultaneously in reforming a fabric of devices even where single substitutions might be impossible or unduly burdensome.

The design task is undoubtedly more formidable in cases such as this where one must ultimately redesign the whole before each of the parts can again make sense. The job can be done, however, as it has been in the case of the home power movement. Such difficult cases will take time, of course, especially where microelectronic and other advances now appear to place technology beyond direct intervention or modification by anyone. *Yet differently directed design efforts alter the evolutionary environment,* as it were. And design processes that are participatory rather than dominated by experts (specialists)—processes with engagement rather than narrowly specified commodity procurement as their aim—make a gradual reshaping even of interlinked enabling technology possible in ways that greatly enhance prospects for engagement.

Breaking from the Assumption of Affluence

A second difficult task in the design process under a general reform of technology will be to escape one of the most pervasive legacies of the device paradigm: design for affluence. The technologies now in place in the typical suburb, for example, are not only designed to *provide* affluence in the sense of ubiquitous and easily available commodities, they are designed *for* affluence—i.e., they are designed under the assumption of an affluent pattern of life. The typical home, the typical automobile—indeed, the whole patterning of our lives around technology—assumes the availability and exploitation through technology of a veritable glut of "resources" in the sense of capital, materials, and other "factors of production." Under this assumption it is difficult even to obtain a quart of milk, for example, without hauling a couple of tons of steel (our automobile) to town and back again with us;[8] caught up in affluence, that automobile absorbs time and

---

8. I owe this observation to an architecture professor I had at MIT in the mid-1970s, Shawn Wellesley-Miller.

resources and may well do more to displace than to facilitate or "enable" focal things and practices.

Simpler and much more affordable enabling technologies do not tend to be available under the pervasive assumption of affluence. But this need not be so. The automobile itself is a necessity only in a society built around the automobile. As our own history and current patterns of mobility in Europe and other parts of the world suggest, societies can be configured in other ways. Such things as community gardening, less resource intensive transportation systems, even actual reconfigurations of work, residence, and consumption sites, can be developed or accomplished over time.

The assumption of affluence, as it is embodied in virtually all of the technological components from which we might otherwise assemble new patterns of life, will constitute a major obstacle to reform. As we move away from affluence in the name of the wealth of engagement, it will be important here again to recognize that individual redesign efforts may well seem ineffectual until they begin to join systematically, overcoming this obstacle on a broad front.[9]

VENTURING OUT INTO THE OPEN

Perhaps I am too optimistic in all of this. Yet much of what has been described is already happening to one degree or another. With modest effort, it is now possible to find and purchase an electric moped, build a superinsulated home, and embrace a full range of slightly more burdensome but highly engaging home power systems,[10] all of which emerged, if informally, from precisely the kinds of concerns Borgmann has raised. Home power technologies in particular show signs of moving systematically beyond the assumption of affluence. Running and other prospective focal practices, while surely fads to a degree, show signs of an enduring core not shared by the hula hoop or the latest turn in popular music or dance. Changes in the political process also appear to be in the works with respect to technology, both in theory (Barber 1984) and in practice (Tatum 1996).[11] And finally, a preference for the simple life, along with other signs of an

9. In some instances, necessity may prove the mother of invention. For example, in India basic transportation is increasingly being provided by inexpensive homemade vehicles simply by placing small gasoline engines on what amount to wagons (Mitra 1995). Necessity, in other words, may at times leave formal designs behind in their unresponsiveness to working realities.

10. Electric mopeds much like the one described earlier in the text are available, for example, along with a range of home power technologies, from a variety of retail sources. See any current issue of *Home Power Magazine* (PO Box 520, Ashland, OR 97520).

11. A number of ongoing practical developments in this country are being spearheaded by the work of Dr. Richard Sclove of the Loka Institute (Amherst, Massachusetts).

as yet largely inchoate struggle toward reform, seems to refuse to disappear from individual and popular consciousness.[12]

Not surprisingly since they are impossible to comprehend in terms of traditional concerns and categories, developments such as home power do remain somewhat obscure and subject to misrepresentation.[13] Nor, since the terms and procedures of a more flexible and inclusive deictic discourse are not yet widely available to us, should we be surprised under the present regime of compromised communications when reform efforts that appear to be motivated by focal concerns are instead defended, even by those pursuing them, in classical terms of cost and efficiency. There is evidence, nevertheless, in areas touched upon here of solid new ventures out into the open where the old rules do not apply and where we learn, share, and communicate in different ways.

To bring focal concerns into practice, it is useful to concentrate on a reform of technology directly. It is helpful to distinguish the degree to which a particular artifact or arrangement is intended to *enable* as opposed to constitute or *center* a focal practice, recognizing that the satisfactions of engagement with things can be enhanced over the whole range of this continuum. It is necessary as well to identify and address such recurring obstacles as the interrelationships among technologies and the assumption of affluence. Given, for example, our obstinate intuitive commitment to sport and to music in our lives, and the vast unexplored realms of technical possibility toward which the device paradigm has for so many years turned a blind eye, our prospects are brighter than they first appear.

REFERENCES

Adams, F. (with Myles Horton). 1975. *Unearthing Seeds of Fire: The Idea of Highlander.* Winston-Salem, N.C.: John F. Blair.

Barber, Benjamin. 1984. *Strong Democracy: Participatory Politics for a New Age.* Berkeley and Los Angeles: University of California Press.

Borgmann, Albert. 1984. *Technology and the Character of Contemporary Life: A Philosophical Inquiry.* Chicago: University of Chicago Press.

———. 1992. *Crossing the Postmodern Divide.* Chicago: University of Chicago Press.

Elgin, Duane. 1981. *Voluntary Simplicity: Toward a Way of Life That Is Outwardly Simple, Inwardly Rich.* New York: William Morrow.

"Here Comes the Sun." 1993. *Time,* October 18, 84

12. Indications to this effect range from scholarly attention to the so-called voluntary simplicity movement (Elgin 1981) to notice of it in the popular press (e.g., Zachary 1995; Pooley 1997).

13. Recent press attention, for instance, tends to present home power either as a newly attractive alternative under traditional economic standards ("Here Comes the Sun" 1993) or in terms of its attractiveness among reactionary survivalists (Weiss 1995). More perceptive treatments do, on the other hand, occasionally appear (e.g., Rierden 1991).

Horton, Myles (with J. Kohl and H. Kohl). 1990. *The Long Haul.* New York: Doubleday.

Lewis, C. S. 1944. *The Abolition of Man.* New York: Macmillan.

Mitra, Barun S. 1995. "India's Informal Car." *Wall Street Journal,* January 31, A18.

Mumford, Lewis. 1964. "Authoritarian and Democratic Technics." *Technology and Culture* 5:1–8.

———. 1970. *The Myth of the Machine.* Vol. 2, *The Pentagon of Power.* New York: Harcourt, Brace, Jovanovich.

Pooley, Eric. 1997. "The Great Escape: Americans Are Fleeing Suburbia for Small Towns." *Time,* December 8, 52–64.

Rierden, Andi. 1991. "Homeowners Revive Interest in Solar Power." *New York Times,* August 25, sec. 12, p. 1.

Tatum, Jesse S. 1994. "Technology and Values: Getting Beyond the 'Device Paradigm' Impasse." *Science, Technology, and Human Values* 19 (winter): 70–87.

———. 1995. *Energy Possibilities: Rethinking Alternatives and the Choice-Making Process.* Albany: SUNY Press.

———. 1996. "Home Power: A Model for Participatory Democracy in Technology Decision Making." Report of work under National Science Foundation Grant Number SBR-9511857.

———. 1999. *Muted Voices: The Recovery of Democracy in the Shaping of Technology.* Bethlehem, Penn.: Lehigh University Press.

Vig, Norman J. 1992. "Parliamentary Technology Assessment in Europe: Comparative Evolution." Paper delivered at the Annual Meeting of the American Political Science Association, September 3–6, Washington, D.C.

Weinberg, Alvin. 1990 [1966]. "Can Technology Replace Social Engineering?" In *Technology and the Future,* 5th ed. Ed. Albert Teich. New York: St. Martin's Press.

Weiss, Philip. 1995. "Off the Grid." *New York Times Magazine,* January 8, 24–33, 38, 44, 48–52.

Winner, Langdon. 1986. *The Whale and the Reactor.* Chicago: University of Chicago Press.

Zachary, G. Pascal. 1995. "When Shopping Sprees Pall, Some Seek the Simple Life." *Wall Street Journal,* May 24, B1, B10.

ELEVEN

Nature by Design

Eric Higgs

. . . Therefore am I still
A lover of the meadows and the woods,
And mountains; and of all that we behold
From this green earth; of all the mighty world
Of eye, and ear,—both what they half create,
And what perceive; . . .
　　　　　　—William Wordsworth, "Tintern Abbey"

I shall be telling this with a sigh
Somewhere ages and ages hence:
Two roads diverged in a wood, and I—
I took the one less traveled by,
And that has made all the difference.
　　　　　　—Robert Frost, "The Road Not Taken"

From my office at the Palisades Research Center, a few yards from the
location of the workshop that gave rise to this book, I look out on Roche
Bonhomme to the east. Walking less than half a mile toward the peak I
come to the Athabasca River, a river that drains into the McKenzie River
and from there to the Arctic Ocean. First, however, I must cross the railway
tracks and railway communication lines, then an abandoned access road
to the center and a railway service road, a transmountain gas pipeline,
a T1 fiber optic cable, the Yellowhead trans-Canadian highway, several
trails, and a roadside picnic area. These intrusions are at the heart of the
ecologically rich montane ecoregion of Jasper National Park, the largest of
the Rocky Mountain national parks and one of Canada's most celebrated
wilderness areas. It was the juxtaposition of so-called wilderness and human
congestion, a clash between perception and reality, that led me several years

ago to begin research on the ecological history of human influence in Jasper.[1] I want to know, at present and in decades to follow, how we can respect this place.

Our understanding of wild nature—wilderness—is changing. It is no longer constituted of people-less places, of mountain vistas and remote, inaccessible valleys. These caricatures are drawn from Euro-American cultural values that have produced a view of nature-as-wilderness, an Edenic place in the receding distance (Nabhan 1997; Cronon 1995; Blackburn and Anderson 1993). Accordingly, with a growing awareness that such values are indeed rooted in part in cultural projections (Soulé and Lease 1995), the subject of ecological management keeps changing form. What are we after in a place such as Jasper? Should we allow natural and cultural processes to proceed without regulation? Should we use management practices to mimic or amplify natural processes, for example in the increasing use of prescribed fire? Should we be setting long-term goals based on negotiations about desired landscapes, and then design our practices to achieve these ends? At least one matter is clear: cultural beliefs, threaded through a labyrinth of institutions and shielded increasingly from direct experience, impinge on ecological management.

Design in a technological culture is of interest here because it embodies simultaneously the greatest hope and peril in working with nature. It impels us with the best of intentions to think creatively and constructively, and, at the same time, it courts terrific problems for the quality of our designs; design seems tainted by the very technological patterns and impulses we are trying to avoid. Nevertheless, since marks will continue to be etched in nature, we should think about how to remove or make such marks less damaging. (We must, of course, continue chanting and acting upon the critical and essential message of ecological responsibility: avoid damage in the first place.) *Ecological restoration* provides a fruitful vantage point for examining the technological constitution of our work with nature. A rapidly growing movement worldwide, restorationists are working to repair human disturbance to ecological patterns and processes. Most operate under the belief that what they are doing is an unequivocal good, yet we must inquire whether any deliberate intervention can escape the very problems that produced the need for restoration in the first place. If so, what conditions encourage such thoughtful intervention, not just in theory, but more so in practice?

Ecological restoration can be a salutary practice, I will argue, on the basis

1. For information on the Culture, Ecology and Restoration research project, please consult our homepage: http://www.arts.ualberta.ca/~cerj/cer.htm.

of Albert Borgmann's device paradigm and reform proposals via focal things and practices. That this kind of practice is possible does not mean it will be enacted. It is more likely that the best intentions will be washed away by a much more powerful movement of technological intervention: restoration as a technoscientific apologia. The device paradigm explains how a thing, such as wilderness, and a practice, such as ecological restoration, can be rendered commodities and stripped of sinuous connection with social and natural processes. Reform must come from an outside awareness of the consequences of the device paradigm, a perspective that grows more difficult in proportion to the rise of the paradigm.

In effect, two paths describe the future trajectory of ecological restoration, and my point is to show how the gentler one—the "one less traveled by"—can be found and conserved. For many, restoration is about the perfection of technique. For others, it is about building communities in relation to natural processes and patterns. While these two approaches are not mutually exclusive, they have begun to form a decisive fork. Having traveled a dusty, bumpy road, we are faced with a choice: to the right is a wide road paved with efficiency and along the way are manicured rest stops and regular services; to the left is a meandering path lined with unforeseeable focal experiences and shaped by ongoing processes. Fewer walk this one; it is less efficient and predictable, but finally more engaging. The means are similar, but the ends quite different. The former, what I term *technological restoration,* is connected to the patterns of technological culture; the latter, *focal restoration,* is constituted by engaged relationships between people and ecosystems.

Borgmann inspires this image of the two paths. Suffused in his writing is the need to make a choice between living a life filled with *devices* and one that achieves orientation through the power of *things*. In these terms the issue is: Will restoration become a practice that turns out ecosystems as predictable commodities, in perfect order, according to principles of technical expertise? Alternatively, will it remain a heterogeneous ambition, one imbued with community intelligence and scientific modesty?

THE PROBLEMS OF ECOLOGICAL RESTORATION

Ecological restoration is the term used to refer to practices that aim at returning integrity to ecosystems disturbed by human activity. A restoration project typically involves assessing change or damage, and comparing this with predisturbance historical site conditions. A variety of therapies are deployed, ranging from removal of particular insults (e.g., surface water pollution) to removal of invasive species to rebuilding, literally, the conditions for flourishing of an integral ecosystem. Restoration operates at all scales, from microsites (e.g., a few square meters) to broad, landscape-wide, process-oriented initiatives. Most fall in the middle range where it is possible to monitor closely the changes to specific habitats. The Society for Ecological Restoration promotes an inclusive definition: "Ecological Restoration is the process of assisting the recovery and management of ecological integrity. Ecological integrity includes a critical range of variability in biodiversity, ecological processes and structures, regional and historical context, and sustainable cultural practices."[2] This definition allows some room to move without either riveting restoration to historical accuracy, which would result in an austere and exclusive practice, or giving up historical commitments entirely. Ecological restoration has become a covering term for a wide variety of practices all aimed at returning a measure of predisturbance ecological (and sometimes cultural) integrity.

Because ecosystems are dynamic and perfect knowledge of predisturbance conditions is virtually impossible to obtain, restoration requires design according to specific goals, rather than exact replication (as one would attempt in the restoration of a work of art). Some projects involve greater amounts of site treatment. Others require long schedules to return vegetation communities. The technical challenges vary considerably from ecoregion to ecoregion. Budget constraints are a major factor in limiting the extent to which proponents can meet restoration goals. These factors, and others, produce diverse ambitions; it is not always possible to return to predisturbance conditions, or even to come terribly close. This matter is further complicated when other kinds of value are created or highlighted in the process of restoration, for example the elevation of ecological awareness and activity in a community, or a refuge for a threatened species (Higgs 1997).

In Jasper, for example, dense, closed-canopy lodgepole pine forests now cover the valley bottoms. Based on an analysis of historical survey photographs from 1915, we know that the valley was relatively open with forest patches interspersed with grasslands and savannas. The proximate cause for such dramatic change in vegetation cover in eighty years is the prevention and suppression of wildfire. As in many other jurisdictions in the mountainous west of North America, the effective elimination of large-scale fire events, at least to the greatest extent possible, has led to significant changes in forested ecosystems for which fire is a crucial process (Graber 1995; Pyne 1995).

Restoration is appealing to park managers, and others, because it offers the means, in the form of flexible techniques, and an end, the return to pre-

---

2. The definition was approved by the Board of the Society for Ecological Restoration in 1996 and presented on the society's web site, http://www.ser.org.

disturbance conditions. However, as this initial optimism fades managers are left with three difficult questions: What were predisturbance conditions? Which conditions, assuming a historical range of variability, are most appropriate? Are historical conditions the best guide in an uncertain cultural and environmental climate? Suppose Jasper managers choose the option of greater patchiness in the Athabasca Valley. Does it make more sense to return the overall conditions to those that obtained just before the formation of the park in 1907? Why stop there? There were over a hundred native peoples living agriculturally in the valley at the turn of the century. Before that, fur trade activities had a remarkable effect on wildlife populations. This takes us back to the turn of the nineteenth century, when paleoecological evidence suggests slightly different climatic conditions. This is the historical problem in ecological restoration: the difficulty of fixing a historical epoch for a system in motion. The outcome is likely to be as arbitrary as leaving the system alone. And, such deliberate intervention in ecological processes and patterns, this time under the banner of ecological integrity, introduces another layer of human intervention.

Some philosophical critics of restoration address this last point. They argue that restoration itself is a product of a commodified relationship to nature, one that provides technological fixes to damaged ecosystems at the expense of the important matter of environmental preservation (Katz 1992a, 1992b). Despite good intentions, people are remaking nature, often in their own image. Ecological restoration is a preeminent device for managing nature; it encourages by its very constitution the deliberate manipulation of nature. Unlike other environmental practices that typically are intended either to remove human insults or protect places from abusive activities, restorationists oblige people to pick up shovels, plant, seed, weed, burn, and selectively use biocides. For critics, restoration is a wave of ultimately lamentable meddling; it conforms to the same destructive patterns that produced the problems to which restorationists are now responding. As Robert Eliot points out, restoration becomes an end in itself, distracting us from more significant aims (Eliot 1997). Philosophers such as Katz and Eliot offer provocative challenges.[3] They highlight the instrumental qualities of restoration, and point toward the commodification of practice.

3. I am deliberately avoiding a detailed discussion of the arguments made by Katz and Eliot against ecological restoration. While they demand an accounting of why it is that restoration is, or could be, a good thing, they have engaged little, if at all, with restoration practitioners, a tack that weakens the practical utility of their arguments. Restorationists, so far in any case, have largely ignored the philosophical literature on restoration. A productive and sustained conversation between restorationists and philosophers would be valuable to both parties.

However, I worry that they, and other critics, have underestimated the diversity of contemporary restoration practice, and in so doing ignored the possibility of a genuinely salutary type of restoration.

Contrary to what we may think about wilderness management and environmental protection, cultural involvement through focal practices is a crucial factor in the success of ecological restoration. Contingent meanings for wilderness, and nature more generally, must be negotiated, not merely assumed. It is an apparent irony, as I shall argue, that for restorations to be durable and salutary as regards ecological integrity, they must also manifest cultural integrity. Ecological restoration is easily absorbed by the device paradigm unless there is a conscious, sustained local participation in decision making and practice.

This points to a theoretical distinction between *technological restoration* and *focal restoration*. Technological restoration is that which results from commodified practices in a hyperreal setting. The reference points for such restoration are conditioned by the device paradigm. Focal restoration resists the device paradigm by centering on reality and the precarious resource-fulness of local participation and focal practices. The former describes the wide, paved road to the future, the latter the meandering and less traveled one.

However, the reasons for oversight might be plain: focal restoration is frustrated in all areas of contemporary life by the rise and rule of the device paradigm. Glamorous distractions are produced by a dizzying machinery of illusion that inveigles us to consume packaged, digestible cultural products, and these morsels appeal to deeply held beliefs about community, family, adventure, achievement, and nature. Similarly, we are witnessing the rise of technological representations of nature and wilderness that are proving alluring, and they condition our awareness of the real thing. Our cultural beliefs affect our practices, and spectacles such as the Disney Corporation's Wilderness Lodge are renovating our beliefs about wilderness. The message is that nature is more pliable and congenial than previously thought, and this undermines the potential for locally engaged ecological restoration. Can we bring focal restoration to prominence and articulate its character in a society given over to technological spectacle? Are our imaginative capacities becoming attenuated such that we are less able to conceive what might be possible as regards salutary engagement with reality?

The Commodification of Nature

The more we study, the more we understand the landscape in Jasper is the re-sult of decades of cultural belief and practice at work: shifting management philosophies, types and modes of visitation, national-level parks policy, and larger cultural dispositions to nature and wilderness. Nature is continuously processed through the projectors of cultural institutions, and interpreted multifariously through the lenses of individuals and communities. When the ground upon which our beliefs about nature shifts, as it is doing rapidly through the advent of what Borgmann terms *hyperreality*, the power of nature to hold moral and spiritual beliefs weakens. Nature becomes a pliable device.

This power to "half create," following Wordsworth, suffuses the modern era and inspires a fundamental ambiguity by which our knowledge of nature and wilderness is formed. We understand two seemingly inconsistent verities about things: that there is nature out there that lies beyond our ability to cocreate, and that our forms of perception make it resemble what we choose. Over the past decade, it has become apparent that "wilderness" specifically and "nature" more generally are culturally conditioned terms (Cronon 1995; Hayles 1995). There is a line between an essentialist "what you see is what you get" epistemology anchoring one end and a belief that all of nature is constructed on our experience suspended from the other. Soulé (1995) argues that in moving too far along that taut line, away from an essentialist idea of nature towards a constructed one courts the very hazard of themed nature. Reality, that gritty, tangible experience of a place such as Jasper, evaporates in a mist of virtuality.

The creative ambiguity that arises here is a blessing because it reveals to us the extent and thickness of the cultural layers we impose on top of wild nature. It offers a new way of seeing wilderness, one that admits of human practice in its myriad forms. At the same time, such critical constructivism is tempered by the presence of wild places that still exist beyond the beaten path in remote, inaccessible, or forgotten places, and possess what Borgmann calls "commanding presence and telling continuity" (1995, 38). Natural processes have a way of poking through in any case, lending hubris to those who believe that nature can be fully ensnared.

Ambiguity is a curse too. One can read the peopled quality of the Athabasca Valley as evidence of the disappearance of wilderness and as a license to reconfigure it according to contemporary desires. After all, runs the argument, if people have been present all along, using and transforming the place, why not perpetuate the tradition? This view makes a crucial mistake in necessarily justifying current practice on the basis of former activities. It is almost certainly the case that the context, and often the intensity and scale, of human activities in the past was different. Equally troubling is the threat to the notion of ecological integrity. Integrity is a core value for ecological restorationists and most conservation biologists and environmental managers. Robust accounts of integrity depend on some hard realities: historical reference conditions, presence of keystone species, species diversity and abundance, absence of weedy or exotic species, and so on. Sophisticated definitions of integrity also allow for long-standing, typically traditional, cultural practices. In hovering too close to a constructivist idea of wilderness we court the loss of ecological integrity as well as a misreading of historical human activities. Restoration is unleashed from conventional constraints and a freer commerce is permitted with popular notions of nature and wilderness.

The ambiguity of our epistemic commitments to wilderness is compounded directly by a more general shift in our allegiance to reality. Borgmann argues that reality is giving way to hyperreality, a kind of reality that is detached from direct experience and context. In losing an authentic engagement with things, we lose sight also of moral commitments to those things. Images become the currency of morality, but images lack stability and resonance. The rise of electronic communications and scientific approaches to image management and marketing, fusing advertising and propaganda, has increased the rate of change of cultural images and produced a uniformity of perception. The potency and pervasiveness of such imagery makes local, vernacular conditions less attractive, and compels their replacement with sophisticated commodities. The globalization of imagery collides with local views and creates confusion over what to believe and when it is appropriate to believe one thing over another. As Jennifer Cypher and I suggest, the pervasiveness and intent of image generation constitutes a "colonization of the imagination," or a reconfiguring of people's imaginative capacities (1997).

Nowhere is this system of colonization more advanced than in the products of the Disney Corporation. For forty years, Disney has pumped out film and television images that have shaped the imaginations of millions of viewers around the world. Wild animals are anthropomorphized and domesticated. The boundaries between wild and tame are redrawn, and primary experience of wild things is displaced by voyeuristic and mediated experiences. The works of culture industries such as Disney accelerates the reception of a constructed nature.

People are flocking to Disney's Wilderness Lodge, for example, as an escape to a land far away in time and space. The fact that this large, luxurious resort it is a simulacrum seems not to disturb most people. And most people are apparently unperturbed by the presence in Florida of redwood trees, northwest coast Native American artifacts, bison, and western log construction. A development such as the lodge builds on ingrained public ideas about wilderness, which is to be expected. In the hands of an organization as powerful as Disney, it has the potential to reshape meaning by imparting its ideological message to the visitor (or viewer) as though it were part of the natural order of things. It is the Disney version of nature that becomes a primary referent for experiences in real nature, not the other way around. How long will it be before we are searching for those elusive mouse ears carved on the wall of the Grand Canyon?

In colonizing the imagination what the lodge and similar projects are accomplishing is a friendly takeover of the reality that underlies themed experience. By turning wilderness into a conceptual product, one that is adaptable, delimitable, endlessly pliable and available, Disney is also creating a new reality in which to experience it. Then, in recursive fashion, consumption conditions our understanding of reality; nature outside of the empire becomes subject to the interpretations of the empire. Of course, Disney's products are converging with and abetting other simulations. The worry is that this takeover of reality to produce a world filled with hyperrealities will displace reality as a moral center. But isn't hyperreality supposed to produce a world richer in opportunity and experience? Shouldn't this be preferable? An abidingly intractable question is whether or not authentic nature (i.e., reality that has commanding presence and telling continuity) possesses attributes vital to the health of people? Are the rough edges on reality important? In the absence of limits and boundaries as imposed by reality, nature is opened to endless manipulation, not only in the style of domination that has become so familiar to us in the modern era, but now in a thematic sense that creates a theme out of concepts such as ecological integrity.

For ecological restorationists, questions of historical fidelity and ecological integrity will be reset within a context of artificiality. The goals that are chosen may resemble manufactured images instead of carefully negotiated ones rooted in participation and faithful articulation of locale. Inclined this way, and possessed of sophisticated skills, the restorationist is able to specify, say, a tradeoff between forest cover and openings more congenial to the touring public. Historical authenticity could drive restoration goals, but it may be that in a changed landscape of roads and utility corridors, Disney and niche tourism, corridors for threatened wildlife species could be placed conveniently along safe, unobtrusive "watchable wildlife" areas. From here, the theming of a national park begins in earnest, satisfying the latest in cultural views about nature and wilderness.

As Tad Friend points out in a recent article on Disney's newest and most ambitious (one billion dollars) attraction, Animal Kingdom, "As I experienced Disney's treats, it became clear that in an important sense this park isn't about animals at all. It's about us, about our wishes and needs.

For how we behave toward animals taken from their natural surroundings reveals us to ourselves" (1998, 103). A logic of justification forms against this backdrop: if we don't think it really matters to the ecosystems, but it does matter to us, then implement a design based on our desires and values. Yes, ecological integrity still counts, but even this concept becomes a commodity to be rendered more efficient (how many grizzly bears *are* necessary?). Restoration becomes part of a thematic endeavor and is pushed along the multilane production highway of the future and away from the gentler choice at the fork in the road. This is technological restoration.[4]

Through an elaborate system of simulation and image management, corporations such as Disney produce commodities that change the meanings of nature and wilderness. This complicates the task of restoration. *Modern* restoration, the style still favored by many practitioners, meant returning an ecosystem as closely as possible to its predisturbance condition. *Postmodern* restoration, which appears ascendant, means adapting to a variety of contingent meanings and ambiguity. By and large this creates a healthy reflexivity for the practitioner, but it also entails wrestling simultaneously with scattered purpose and technological ambition. This is a more specific version of the ambivalence I described above. Now that restoration has become a diverse activity, ranging from natural urban gardens to whole river basin megaprojects, and now that historical fidelity is relative, what restoration is, exactly, is difficult to discern. When this uncertainty is compounded by a culture of hyperreality, the hazard of it falling to the pattern of the device paradigm is a real one. Commitments to authentic engagement with reality, to things, are unhinged. Ecosystems become devices as the rush begins to (re)produce commodities in the form of restorations that meet the interests of those who pay the bills. The commodification of nature and wilderness, therefore, diverts the project of restoration along a technological path. The more pervasive technological restoration becomes, the less easy it is to articulate and justify thingly restoration—the path less traveled. What will restorationists of the future restore: things or devices, reality or hyperreality?

The Commodification of Practice

There is a further way in which ecological restoration is becoming more devicelike: via the commodification of practice. To understand fully the

---

4. Perhaps people will begin to flock to artificial experience—my favorite is the thought experiment, SimJasper, in which a paying client can wrestle with a grizzly bear—and leave wild place alone. It strikes me that this kind of thinking misjudges the significance of engaged human involvement in ecosystems and merely forestalls the consumption of real places such as Jasper.

implications of the device paradigm for ecological restoration requires examination of the commodification of *nature,* which I accomplished in the previous section, as well as the commodification of *practice.* To commodify a practice means to change the locus of attention from things to devices and to transform it to an exclusive professional enclave geared to efficiency (Higgs 1991). This is a well-known phenomenon identified under a number of labels: professionalization, specialization, a decisionistic society (pace Habermas), a culture of expertise. That we should detect it as a trend in ecological restoration is hardly surprising.

Part of the reason for its assimilation surely has to do with its seeming win-win qualities. Reacting against zero-sum thinking in which all compete for the slices of the same unchanging pie, dispute resolution, policy, and business specialists turned to a new paradigm in the 1980s and 1990s. If the size or quality of the pie could be changed, then decisions, presumably more creative and usually more profitable ones, were possible. In the landmark book that kicked off much of this discussion, Fisher and Ury's *Getting to Yes* (1981) proposed that seemingly intractable problems could be resolved by demonstrating, sometimes with considerable difficulty, that a mutually advantageous solution was possible. Ecological restoration opened a potent new opportunity for corporate and governmental environmental management: ecosystems could be rebuilt or reconfigured, thereby augmenting and extending increasingly costly commitments to environmental protection. During an era of environmental loss of innocence, restoration represents hope for converting past destructive practices; it has tremendous symbolic authority. Government agencies, sometimes in concert with corporate partners, are providing more support for restoration projects, in some cases sponsoring enormous endeavors such as the Kissimmee River initiative (Cummins and Dahm 1995).

Corporations have taken up the cause, supporting restoration through modified development projects, grants, and awards. Jonathan Perry (1994) reported a surge of interest beginning in the 1970s in mollifying the effects of corporate development, typically office complexes, and improving environmental profile. Such projects serve to "naturalize the presence of the corporation" and lend the appearance of solidity to a (likely) transitory local commitment, to create a history for exurban sprawls seeking identity, and to provide a calming experience to the corporate world. The uses of restoration in corporate environments serve to justify the political-economic interests of the firm as much as or more than the ecological interests of the site. Another controversial practice concerns ecosystem mitigation (Higgs 1993). In areas of intense development pressure, notably along the eastern seaboard of the United States, property developers gaze longingly at parcels that are protected by local, state, or federal environmental statutes. A popular approach is to compensate, or mitigate, the effects of development on, say, at coastal wetland, with purchase, dedication, and restoration of another property of equal ecological value.

Mitigation is a clear example of the commodification of restoration; restored ecosystems are converted to tradable units for consumption. Mitigation also illustrates the commodification of practice. Even though most restorationists in my experience view mitigation projects as a crass commercial endeavor that ought to be avoided, mitigation is on the rise and producing a cadre of professionals skilled at such arrangements. If ecosystems can be bought, sold, and traded, does this mean they will also be subject to economically analogous processes of disposal and recycling? Will professional practice be codified and restricted? Will certain techniques become proprietary? Will ecosystem designs be franchised? Such questions seem far-fetched today. They point, however, directly at larger trends in the commodification of experience and the production of a hyperreal environment. When we cease to find such questions peculiar, that is when the commodification of restoration will have reached a zenith.

FOCAL RESTORATION

If what ails ecological restoration is the device paradigm as manifest in the commodification of nature and restoration practice, joined with a colonization of the imagination, then the prescription is focality. The challenge is to develop effective, resilient, durable focal practices, ones that when combined with shared practices and economic reforms produce authentic communities centered on matters of concern greater than mere consumption. Repairing damage by designing interventions that reconstitute ecological integrity, in the inclusive sense mentioned above, requires treating ecosystems as things rather than devices. For the ecological restorationist, this entails focal restoration: practices that configure a stronger relationship between people and natural process, a bond that is reinforced by the varieties of communal experience. A focal restoration is one that centers the world of the restorationist, expresses the commanding presence of nature and demonstrates continuity between that particular act of restoration and other activities on the landscape. Focal restoration is mindful restoration.

Focal things and practices are central to reforming the device paradigm. On this account, the debilitating effects of a life centered in consumption can be countered by embracing focal things: things in our lives that have significance through the definition and delineation of context. Focal things and practices help distinguish technological restoration such as corporate projects and hyperreal commodities (e.g., Disney's Wilderness Lodge) from focal restoration activities of grassroots restorationists. In this way I have found the theory of the device paradigm helpful in clarifying the extent and means of assimilation of restoration into technological culture. Fortunately restoration continues to produce a clear stream of practices and commitments that resist the desiccating effects of technology, and through this we glimpse the most significant hope for restoration as a way of integrating ecological concern and cultural practices. What we need is not less intervention; perhaps we need even more intervention, but such deliberate involvement must center on things, not devices. Focal restoration offers the promise of being a main force against incursions of the device paradigm into nature and wilderness.

Ecological restoration is blessed with many fine writers who have eloquently mapped a salutary view of restoration that connects the restoration of ecosystems to the restoration, regeneration, or reinhabitation of human values and spirit. In dozens of essays and case studies, writers such as Stephanie Mills, Bill Jordan, and Gary Paul Nabhan, not to mention Aldo Leopold and his lyrical midcentury classics, demonstrate the correlational coexistence of practitioners and dynamic ecosystems and the strength of focal practices to maintain the attachments of people to places. *Correlational coexistence* was coined by David Strong (1995) to characterize the distinctive mutual relationships that occur between people and things. A thing is enlarged by care, and a person is rewarded with a more profound understanding of existence and responsibility. This is what happens in ecological restoration when lives both human and natural are inspired.

What does ecological restoration look like as a focal practice? I recall how my understanding of short- and tallgrass prairies changed when I helped create the Robert Starbird Dorney memorial garden at the University of Waterloo. My studies of ecology and countless hours in the field were no match for the knowledge required to arrange an integral assembly of organisms that would ultimately work together in a fashion closely resembling what must have once occupied this site. It was humbling to know this hard work was but a slight contribution to the autonomous ecological processes that took over the moment planting had ceased. I discovered myself making countless tiny decisions—the outer boundaries of a *Monarda fistulosa* planting—rooted in the integrity of organisms and their relationships, and my art. Hence, reciprocity formed between me and the garden that opened up an appreciation of things that I had long taken for granted. Ten years and twenty-five hundred miles distant, the Dorney garden has left deep contours in my understanding of restoration. In the words of Gary Nabhan, I was "building *habitat* as well as memories" (1997, 87).

Ecological restoration is seldom a solitary pursuit; it works beautifully as a demonstration of communal focal practices. For example, the "Bagpipes and Bonfire" festival in Lake Forest, Illinois developed out "of the yearly act of burning all the exotics and weedy non-natives" (Christy 1994, 123) removed from the Lake Forest Preserve. On a Sunday afternoon in the fall more than a thousand people participated in a festival that includes "family entertainment, period actors, hot-air balloons, food and drink . . . [and] . . . at dusk, a 100 piece Scottish piping band [that] emerges from the prairie, solemnly circles the brush pile, and plays traditional airs" (Holland 1994, 123). Holland suggests such a ceremony as this "invite[s] participation by society." Moreover, "with its bonfire ritual, [the festival] renews the spirit of a community sharing in the regeneration of a native ecosystem" (1994, 122). Historical connections are invoked and respected. The strict normative division between culture and nature is constructively blurred, and people are brought into closer connection with natural processes and cultural patterns. Communities are strengthened through the gathering of energy and commitment.

In a recent article, Andrew Light and I argue that ecological restoration has inherent democratic capacity (1996). The qualities of restoration practice promote community engagement, experimentation, local autonomy, regional variation, and a level of creativity in working along with natural patterns and processes. It is the combination of offering value to nature and value to community that give it the capacity to enhance a participatory politics. To argue that ecological restoration has participatory capacity does not mean that it will be participatory. Commitment and fortitude are required to maintain community-based focal restoration practices, and to ensure that the political terrain remains hospitable for this more embracing view of restoration. Thinking of restorations as focal practices is the surest way of maintaining such openness.

The Limits of Reform

Are focal practices sufficient for the reform of the device paradigm? In particular, if we accept the implications of the device paradigm on ecological restoration through the commodification of nature and practice, then engagement with focal things (ecosystems) is the best place to begin a concerted program of reform. Change must begin by individuals' (re)appropriating things that matter in their lives and developing practices that uphold and protect the significance of these things. Borgmann wants us to undertake two important challenges: first, to clarify our understanding of those things that have final significance, a list comprising widely agreeable civic, physical, and character virtues; and second, to acknowledge the way in which things serve clearly and unambiguously as the center of our lives. The next stage of reform involves the translation of personal focal practices to a civic level. Communal focal practices—athletic gatherings, public events, participation in local decision making—are not merely an aggregation of individual practices, but an awareness of the importance of maintaining a vital communal life.

There is a third and final stage that involves the establishment of a two-tier economic system. At one level is an artisanal economy comprising locally autonomous practices. At the other would be economic productions that would issue desirable goods deemed too complicated and intensive to be produced through decentralized processes. By this, Borgmann means consumable goods such as refrigerators, automobiles, computers, and so on. Governmental actions to encourage an artisanal sector would simultaneously lessen the grip of large industrial activity. People committed increasingly to focality and against the device paradigm would resist as much as possible the use of culturally and ecologically destructive goods. Pragmatic choices would be made to lessen our dependence on mass-produced goods, seeking to use and support local ones.

I have been puzzled by Borgmann's economic reforms since first reading them in 1984. They seem inadequate against the size and effectiveness of institutions involved in what Edward Herman and Noam Chomsky have termed "manufacturing consent" (1988). The continued spread of television, increasingly stealthy and inventive approaches to advertising and marketing, corporate concentration of the media and entertainment industries, and the confluence of information technologies beginning with the world wide web have changed the character of the economy. Borgmann's theory of the device paradigm accounts for the implications: as the pattern of decomposition of machinery and commodities becomes more pervasive, the more difficult it will be to imagine a world other than one captivated by this pattern of manufactured consent. The manufacture of consent becomes easier and easier, making it more difficult to resist the incursion of technology. The result is a hyperreal economy, and one that is more resistant to conventional strategies of resistance. In my mind, arresting the device paradigm is a race between the formation and stability of focal practices and the manufacture of consent.

These dour observations are moderated by the endless resourcefulness and motivation of individuals and communities who remain steadfast in their support of focal practices; there is nothing, fortunately, that can remove the possibility of focal concerns and practices, in the same way that spirit and private conviction survive through extreme deprivation. This is good news for ecological restorationists who have thrived mostly on local ambition, experimentation, and humility. The fact that restoration *is* a grassroots movement, by and large, bodes well for the near future. What concerns me is the longer-term trajectory. The warning signs of commodified practice within a hyperreal nature are close enough to witness, and the energy required to steer toward the gentler path is much less now than it will be in a decade.

Borgmann's quiet politics of technology, especially his associated economic reforms, may prove inadequate against hyperreality. I think that more active resistance to the device paradigm is required. The opening I see is the interest in local and bioregional economies coupled with the development of left ecological politics,[5] especially variations on a theme of libertarian socialism or communal forms of anarchism. Is there a theory of political resistance, more radical than what he proposes in *TCCL*, that would fit comfortably alongside Borgmann's political beliefs? Is there a coherent political economic theory that would protect and elevate personal and communal focal practices and resist more effectively the corrosion of choice through manufactured consent? Can we shield and support focal restoration?

A PATH OF GREATER RESISTANCE

Back in Jasper, the issue of focal practices and the device paradigm are far from the minds of most managers and visitors. The struggle between "wise use" and preservation results in partially erected barricades to further development, barriers that are neither sufficient to meet the needs of the preservationists nor flexible enough to permit creative interventions through focal restoration. The park is conceived as a refuge under threat, a view that makes good sense of reality. However, this view lacks imagination; it assumes that whatever is present now will continue to be eroded by time like the mountains themselves. What is badly needed is a conscious design for the park and the region, one that takes into account the value and fragility of the region with the changing interests of visitors and residents. Such a design should stretch at least one hundred years into the future, a time slightly longer than the park has been in existence. Thinking this far into the future is marvelous tonic. It builds simultaneously an understanding of the scale and dynamism of changes that have taken place

5. This is what I proposed in my doctoral dissertation, "Planning, Technology and Community Autonomy" (University of Waterloo, 1988). Having studied the decline of community political and social autonomy in Bruce County, Ontario, I proposed a radical political reform along the lines of Bookchin's "radical municipalism" or some variation of libertarian socialism, i.e., an oppositional movement against top-down management and governance.

over the last hundred years and the kind of wisdom required to make good, humble design decisions about the future. The restoration of a place must look both forward and backward.

I have shown two parallel processes of commodification at work which when combined produce conditions necessary for technological restoration: one that renders nature into products for consumption, and the other that sets up the practice of restoration as a technical one bound by matters of efficiency. The distinction between technological and focal restoration eases the transition to a postmodern nature and sharpens an understanding of the corrosive character of technology as regards reality and nature. Focal restoration involves restorationists working in communities, blending knowledge about local nature with social needs and cultural awareness. This level of engagement is a necessary condition for the realization of a new kind of relationship with nature, one that enforces humility and respect. Traditional views of nature as something entirely other than us, brought to its greatest height in the idea of wilderness, separate us from a direct and profound understanding of natural structure and process. Of course, in advocating this I am not proposing that license be given to indiscriminate meddling; local bounds are placed on this through focal awareness. Nor am I suggesting that wild nature is an irrelevant concept. We need places such as Jasper where human activity is limited by access to show us the measure of technology (Strong 1995). Even in Jasper, however, restoration is needed to guide large and small decisions in the future.

Fortunately, restoration practice is rooted in focal things, but these roots are being cut away by the forces of hyperreality and commodification. Albert Borgmann's theory of the device paradigm reveals a fork in the road. Philosophical articulation of the issues underlying ecological restoration helps us understand what the choices are and how we might promote the conditions for thoughtful intervention in natural processes. A major theoretical challenge lies in developing a politics of resistance to the commodification of reality and the accompanying colonization of the imagination that makes the meandering path, the one less traveled, a viable one. After all, resistance is fertile.[6]

REFERENCES

Blackburn, T. C., and K. Anderson, eds. 1993. *Before the Wilderness: Environmental Management by Native Californians.* Menlo Park, Calif.: Ballena Press.

Borgmann, Albert. 1984. *Technology and the Character of Contemporary Life: A Philosophical Inquiry.* Chicago: University of Chicago Press.

6. Inspiration for this chapter came from collaboration with Jennifer Cypher.

————. 1995. "The Nature of Reality and the Reality of Nature." In *Reinventing Nature? Responses to Postmodern Deconstruction.* Ed. Michael Soulé and Gary Lease. Washington: Island Press.

Christy, Stephen. 1994. "A Local Festival." *Restoration and Management Notes* 12:123.

Cronon, William. 1995. "The Trouble with Wilderness: Or Getting Back to the Wrong Nature." In William Cronon (ed.) *Uncommon Ground.* New York: W. W. Norton.

Cypher, Jennifer, and Eric Higgs. 1997. "Colonizing the Imagination: Disney's Wilderness Lodge." *Capitalism, Nature, Socialism* 8, no. 4: 107–30.

Cummins, Kenneth W., and Clifford N. Dahm. 1995. "Restoring the Kissimmee." *Restoration Ecology* 3, no. 3: 147–48.

Elliot, Robert. 1997. *Faking Nature: The Ethics of Environmental Restoration.* London: Routledge.

Fisher, R., and W. Ury. 1981. *Getting to Yes: Negotiating Agreement without Giving In.* Boston: Houghton-Mifflin.

Friend, Tad. 1998. "Please Don't Oil the Animatronic Warthog." *Outside,* May, 100–108.

Graber, David M. 1995. "Resolute Biocentrism: The Dilemma of Wilderness in National Parks." In *Reinventing Nature? Responses to Postmodern Deconstruction.* Ed. Michael Soulé and Gary Lease. Washington: Island Press.

Hayles, Katherine. 1995. "Searching for Common Ground." In *Reinventing Nature? Responses to Postmodern Deconstruction.* Ed. Michael Soulé and Gary Lease. Washington: Island Press.

Herman, Edward S., and Noam Chomsky. 1988. *Manufacturing Consent: The Political Economy of the Mass Media.* New York: Pantheon Books.

Holland, Karen. 1994. "Restoration Rituals: Transforming Workday Tasks into Inspirational Rites." *Restoration and Management Notes* 12:121–25.

Higgs, Eric. 1991. "A Quantity of Engaging Work To Be Done: Restoration and Morality in a Technological Culture." *Restoration and Management Notes* 9:97–104.

————. 1993. "The Ethics of Mitigation." *Restoration and Management Notes* 11:138–43.

————. 1997. "What Is *Good* Ecological Restoration?" *Conservation Biology* 11, no. 2: 338–48.

Katz, Eric. 1992a. "The Big Lie: Human Restoration of Nature." *Research in Philosophy and Technology* 12:231–43.

————. 1992b. "The Call of the Wild." *Environmental Ethics* 14:265–73.

Light, Andrew, and Eric Higgs. 1996. "The Politics of Ecological Restoration." *Environmental Ethics* 18:227–47.

Nabhan, Gary Paul. 1997. *Cultures of Habitat: On Nature, Culture, and Story.* Washington: Continuum.

Perry, Jonathan. 1994. "Greening Corporate Environments: Authorship and Politics in Restoration." *Restoration and Management Notes* 12:145–47.

Pyne, Stephen. 1995. *World Fire: The Culture of Fire on Earth.* New York: Henry Holt.

Soulé, Michael. 1995. "The Social Siege of Nature." In *Reinventing Nature? Responses to Postmodern Deconstruction.* Ed. Michael Soulé and Gary Lease. Washington: Island Press.

Soulé, Michael, and Gary Lease, eds. 1995. *Reinventing Nature? Responses to Postmodern Deconstruction.* Washington: Island Press.

Strong, David. 1995. *Crazy Mountains: Learning from Wilderness to Weigh Technology.* Albany: SUNY Press.

PART FOUR

Extensions and Controversies

The chapters in this part press Borgmann's theories on fundamental issues and develop alternative positions. Borgmann cautioned us, as Phillip Fandozzi pointed out in chapter 8, that "the peril of technology lies not in this or that manifestation, but in the pervasiveness and consistency of its pattern." The pervasiveness of devices threatens to exclude all counterbalancing forces. Does technology tend toward this exclusivity? While none of the contributors to this volume pursues this question directly, how one answers it will color one's attitude toward devices and hyperreality in general. If one answers this question affirmatively, as Borgmann does, then the task of meeting the problem of the device paradigm becomes finding, fostering, and maintaining counterbalances to this tendency of technology toward exclusion. However, if one answers this question negatively or if the question is of less concern, then new possibilities may open up. One may be more concerned about finding good ways of having both devices and focal things without having to worry about devices displacing focal things in the totalizing way Borgmann is concerned about. It may seem from this perspective that Borgmann's highlighting of the negative consequences of devices is exaggerated, for he claims that it is a mistake to believe that devices can positively enrich our lives. Is Borgmann right about this last claim? Surely a good case can be made, but are there important exceptions to this rule?

In accord with his concern about the peril of technology, Borgmann argues that fundamental material decisions, whether or not to substitute devices for things, are the most consequential for individuals and communities. On this view, devices are most genuinely promising when held at bay to a supporting role for focal things and practices. In her recasting of Borgmann's theory, Diane Michelfelder in chapter 12, "Technological Ethics in a Different Voice," appreciates the way focal things may counterbalance devices, but she finds that Borgmann's evaluation of the device paradigm does not always bear out for individual devices. Simply because a technological object can be classified as a device does not necessarily mean that it will have the negative effects on engagement and human relationships that Borgmann's theory predicts; some devices actually foster these values, she argues, illustrating her points with a study done on women's use of telephones. "The machinery that clouds the story of a device does not appear to prevent that device from playing a role in relationship building." If so, devices under some conditions may be more promising than Borgmann thinks; Michelfelder finds that devices can themselves support focal practices if they are used in a context of narrative and tradition. Enhancing the features of ordinary life guides her reform of technology so that it poses more promise than threat to democratic life.

Douglas Kellner takes on the pessimism of Borgmann's view of "hyper-modernism" as depicted in Borgmann's later work, *Crossing the Postmodern Divide.* Kellner is less sure that where we are headed will lead to this extreme. In chapter 13, "Crossing the Postmodern Divide with Borgmann, or Adventures in Cyberspace," Kellner argues for a more moderate position on the changes we are now experiencing in material culture. On the one hand, he objects to Borgmann's claim that we really are at a point of crossing from modernity to postmodernity, arguing instead that our present cultural position is much more complex and confusing than Borgmann makes it out to be. Also, while he agrees with Borgmann's claim that technology is a major force in postmodernity, Kellner argues that technology is not the only shaping force in postmodernism. We also must pay attention to the forces of capitalism. On the other hand, Kellner, too, seeks the counterbalance of something like Borgmann's postmodern realism, where focal things weigh against devices, but his deconstruction of Borgmann's hyperreal/real distinction finds that Borgmann sells technology short for the positive part it may play in this counterbalancing relationship. Kellner argues that as technology supplements face-to-face contact and encounter with things, it can actually help to form a richer life for those who avail themselves of it. He illustrates his argument with examples from cyberspace.

In her recasting of Borgmann's thought, Mora Campbell in chapter 14 makes use of the idea of temporal ambiguity, the condition of discordant synchronous events in one's life. Imagining a world that could be otherwise would involve resolving the deeper problems that Borgmann shows with technology. Here Campbell finds attractive the notion that undesirable kinds of ambiguity can be eliminated through focal things and practices. But reform needs to go further to meet her additional concerns with social changes, addressing gender, cross-cultural differences, and a greater appreciation for the continuity between humans, other living beings, and the natural world. For instance, while Borgmann speaks well of traditional focal things and practices, the function of gender never complicates this picture. Moreover, from the standpoint of temporal ambiguity, focal things and practices are too limited since, on her reading of Borgmann, they are caught up within the private home, leisure time only, and the Gregorian calendar. "Unless focal practices serve to shift this overall pattern, they cannot, in temporal terms, significantly reorient the context of our lives."

Campbell argues that Donna Haraway's more positive stance toward ambiguity helps us to see gender, cultural differences, and continuity as problems. But Haraway's resolutions to these problems fall short by Haraway's own standards when seen in light of Borgmann's device paradigm and temporal ambiguity. We end up ironically with a kind of "cyborg narcissism." Borgmann and Haraway together, supplemented by the notion of temporal ambiguity, get us further along in our efforts to create a world within which we can listen to our bodies and to others instead of enacting patterns of domination upon them. However, Campbell cautions that really reimagining such a world will take larger-scale cultural changes.

What is the relationship between technology and capitalism? Many readers of Borgmann wonder how the forces of capitalism influence technology, questioning whether the device paradigm misses the root of the problem. Isn't consumerism after all the result of capitalism and not technology? For Borgmann, consumerism is the result of the structure of the device itself, the pervasiveness of devices, and the promise of technology. According to his theory, the very structure of the device calls forth consumption of the commodity it provides and calls for nothing more. When our human-built environment becomes pervaded with these devices, we can expect only consumption on our part. What initiates this process and maintains it, however, is the unquestioned and nearly unquestionable belief in the promise of technology: that technology can procure for us a free and prosperous life. This underlying belief, Borgmann argues, cuts across socioeconomic classes. Importantly, for instance, for him it is the preexistence of this belief that advertisements address rather than create (at least at a fundamental level). So for him, deeper than the market forces is this underlying belief in what technology can provide. Hence, much of his critique turns on challenging it.

In chapter 15, "Trapped in Consumption: Modern Social Structure and the Entrenchment of the Device," Thomas Michael Power argues instead that consumerism is the result of market forces and is maintained by those forces. Market forces constrain our choices of what is realistically possible for most of us to choose. Blindness to this fact will only ensure that consumerism remains the dominant way of life in modern societies, regardless of how many people may wish to live in alternative ways. However, Power's reform tactic is not to do away with the market system, but rather to bring into relief how the market constrains (rather than enhances) choice to consumption and how the market depends on a human-crafted social context without which it would be brutal and inefficient. He also emphasizes how we are already intervening to constrain the market from interfering with aspects of our well-being. Bringing these factors into the foreground will enable us to reform the market that now practically forces us to live the kind of life Borgmann critiques.

How might a contemporary philosophy of technology, sharing similar concerns with Power about capitalism and the findings of the social sciences, engage the work of Borgmann and address the reform of technology? For the past decade Andrew Feenberg has been developing a philosophy of technology that competes with Borgmann's. In chapter 16, "From Essentialism to Constructivism: Philosophy of Technology at the Crossroads," he directly criticizes Borgmann's theory in light of his own.

As Diane Michelfelder notes in chapter 12, Borgmann contends that devices or hyperreality need to be "counterbalanced," not eliminated. On this view, to restrain devices appropriately, we need to counterbalance them with focal things (which can be in the normal sense technological, such as a flute) whose very meaning would be ruined if procured by a device (the flute replaced by a stereo). The device does not need to be redesigned so much as restrained in light of something non-devicelike, that is, in the strict technical sense of Borgmann's theory, something "nontechnological."

Feenberg understands this appropriation of technology as "a spiritual movement of some sort." He argues that the unifying powers of Borgmann's and Heidegger's "essentialist theories" need to be mitigated with an awareness of the significant differences between various technological designs and developments. The sophisticated developments of modern technology, on his account, allow for a "subversion" of their design for purposes that are more fully engaging and contextual. Roughly, Feenberg's two-level theory shows how this subversion can take place between the essentialist theories of philosophers and the attention to "differences" of the social sciences. From this standpoint, Feenberg argues that both Heidegger and Borgmann really characterize the essence of technology ethnocentrically from within a capitalist context; hence, many of the negative features of modern technology that they point to really belong to the influences of capitalism on technological development and design.

Finally, we know what Borgmann's vision of technology and postmodernism are, but what is his vision of philosophy itself? His works are saturated with philosophy and yet he has not produced a book developing his general philosophy. On the basis of Borgmann's two works on technology, David Strong, in chapter 17, "Philosophy in the Service of Things," tries to step beyond them and characterize Borgmann's philosophy. He understands this general vision of philosophy to mirror Borgmann's thinking about devices. For Borgmann, Strong argues, philosophy's strengths and limitations can best be understood in the light of things. He challenges Borgmann to work out his general philosophy, especially his idea that there is a kind of symmetry between humans and things. Strong finds that Borgmann's most significant philosophical advance has to do with his careful analysis of the physical characteristics of devices and things and of the physical transformation of Earth and our built environment. Here Strong questions whether some of the received religious elements of Borgmann's books are consistent with the radical nature of his general philosophy in its concern with physical things.

TWELVE

Technological Ethics in a Different Voice

Diane P. Michelfelder

The rapid growth of modern forms of technology has brought both a threat and a promise for liberal democratic society. As we grapple to understand the implications of new techniques for extending a woman's reproductive life or the spreading underground landscape of fiber-optic communication networks or any of the other developments of contemporary technology, we see how these changes conceivably threaten the existence of a number of primary goods traditionally associated with democratic society, including social freedom, individual autonomy, and personal privacy. At the same time, we recognize that similar hopes and promises have traditionally been associated with both technology and democracy. Like democratic society itself, technology holds forth the promise of creating expanded opportunities and a greater realm of individual freedom and fulfillment. This situation poses a key question for the contemporary philosophy of technology. How can technology be reformed to pose more promise than threat for democratic life? How can technological society be compatible with democratic values?

One approach to this question is to suggest that the public needs to be more involved with technology not merely as thoughtful consumers but as active participants in its design. We can find an example of this approach in the work of Andrew Feenberg. As he argues, most notably in his recent book *Alternative Modernity: The Technical Turn in Philosophy and Social Theory*, the advantage of technical politics, of greater public participation in the design of technological objects and technologically mediated services such as health care, is to open up this process to the consideration of a wider sphere of values than if the design process were to be left up to bureaucrats and professionals, whose main concern is with preserving efficiency. Democratic values such as personal autonomy and individual agency are part of this wider sphere. For Feenberg, the route to technological reform and the preservation of democracy thus runs directly through the intervention of nonprofessionals in the early stages of the development of technology (Feenberg 1995).

By contrast, the route taken by Albert Borgmann starts at a much later point. His insightful explorations into the nature of the technological device—that "conjunction of machinery and commodity" (Borgmann 1992b, 296)—do not take us into a discussion of how public participation in the design process might result in a device more reflective of democratic virtues. Borgmann's interest in technology starts at the point where it has already been designed, developed, and ready for our consumption. Any reform of technology, from his viewpoint, must first pass through a serious examination of the moral status of material culture. But why must it start here, rather than earlier, as Feenberg suggests? In particular, why must it start here for the sake of preserving democratic values?

In taking up these questions in the first part of this paper, I will form a basis for turning in the following section to look at Borgmann's work within the larger context of contemporary moral theory. With this context in mind, in the third part of this paper I will take a critical look from the perspective of feminist ethics at Borgmann's distinction between the thing and the device, a distinction on which his understanding of the moral status of material culture rests. Even if from this perspective this distinction turns out to be questionable, it does not undermine, as I will suggest in the final part of this paper, the wisdom of Borgmann's starting point in his evaluation of technological culture.

PUBLIC PARTICIPATION AND TECHNOLOGICAL REFORM

One of the developments that Andrew Feenberg singles out in *Alternative Modernity* to back up his claim that public involvement in technological change can further democratic culture is the rise of the French videotext system known as Teletel (Feenberg 1995, 144–66). As originally proposed, the Teletel project had all the characteristics of a technocracy-enhancing device. It was developed within the bureaucratic structure of the French government-controlled telephone company to advance that government's desire to increase France's reputation as a leader in emerging technology. It imposed on the public something in which it was not interested: convenient access from home terminals (Minitels) to government-controlled information services. However, as Feenberg points out, the government plan for Teletel was foiled when the public (thanks to the initial assistance of computer hackers) discovered the potential of the Minitels as a means of communication. As a result of these interventions, Feenberg reports, general public use of the Minitels for sending messages eventually escalated to the point where it brought government use of the system to a halt by causing it to crash. For Feenberg, this story offers evidence that the truth of social constructivism is best seen in the history of the computer.

Let us imagine it does offer this evidence. What support, though, does this story offer regarding the claim that public participation in technical design can further democratic culture? In Feenberg's mind, there is no doubt that the Teletel story reflects the growth of liberal democratic values. The effect generated by the possibility of sending anonymous messages to others over computers is, according to Feenberg, a positive one, one that "enhances the sense of personal freedom and individualism by reducing the 'existential' engagement of the self in its communications" (Feenberg 1995, 159). He also finds that in the ease of contact and connection building fostered by computer-mediated communication, any individual or group of individuals who is a part of building these connections becomes more empowered (Feenberg 1995, 160).

But as society is strengthened in this way, in other words, as more and more opportunities open up for electronic interaction among individuals, do these opportunities lead to a more meaningful social engagement and exercise of individual freedom? As Borgmann writes in *Technology and the Character of Contemporary Society* (or *TCCL*): "The capacity for significance is where human freedom should be located and grounded" (Borgmann 1984, 102). Human interaction without significance leads to disengagement; human freedom without significance leads to banality of agency. If computer-mediated communications take one where Feenberg believes they do (and there is little about the more recent development of Internet-based communication to raise doubts about this), toward a point where personal life increasingly becomes a matter of "staging . . . personal performances" (Feenberg 1995, 160), then one wonders what effect this has on other values important for democratic culture: values such as self-respect, dignity, community, and personal responsibility.

The Teletel system, of course, is just one example of technological development, but it provides an illustration through which Borgmann's concern with the limits of public participation in the design process as a means of furthering the democratic development of technological society can be understood. Despite the philosophical foundations of liberal democracy in the idea that the state should promote equality by refraining from supporting any particular idea of the human good, in practice, he writes, "liberal democracy is enacted as technology. It does not leave the question of the good life open but answers it along technological lines" (Borgmann 1984, 92). The example we have been talking about illustrates this claim. Value neutral on its surface with respect to the good life, Feenberg depicts the Teletel system as encouraging a play of self-representation and identity that develops at an ever-intensifying pace while simultaneously blurring the distinction between private and public life. The value of this displacement, though, in making life more meaningful, is questionable.

To put it in another way, for technology to be designed so that it offers greater opportunities for more and more people, what it offers has to be put in the form of a commodity. But the more these opportunities are put in the form of commodities, the more banal they threaten to become. This is why, in Borgmann's view, technical politics cannot lead to technical reform.

For there truly to be a reform of technological society, Borgmann maintains, it is not enough only to think about preserving democratic values. One also needs to consider how to make these values meaningful contributors to the good life without overly determining what the good life is. "The good life," he writes, "is one of engagement, and engagement is variously realized by various people" (Borgmann 1984, 214). While a technical politics can influence the design of objects so that they reflect democratic values, it cannot guarantee that these values will be more meaningfully experienced. While a technical politics can lead to more individual freedom, it does not necessarily lead to an enriched sense of freedom. For an object to lead to an enriched sense of freedom, it needs, according to Borgmann, to promote unity over dispersement, and tradition over instantaneity. Values such as these naturally belong to objects, or can be acquired by them, but cannot be designed into them.

To take some of Borgmann's favorite examples, a musical instrument such as a violin can reflect the history of its use in the texture of its wood (Borgmann 1992b, 294); with its seasonal variations, a wilderness area speaks of the natural belonging together of time and space (Borgmann 1984, 191). We need to bring more things like these into our lives, and use technology to enhance our direct experience of them (as in wearing the right kinds of boots for a hike in the woods), for technology to deliver on its promise of bringing about a better life. As Borgmann writes toward the end of *TCCL,* "So counterbalanced, technology can fulfill the promise of a new kind of freedom and richness" (Borgmann 1984, 248).

Thus for Borgmann the most critical moral choices that one faces regarding material culture are "material decisions" (Borgmann 1992a, 112): decisions regarding whether to purchase or adopt a technical device or to become more engaged with things. These decisions, like the decisions to participate in the process of design of an artifact, tend to be inconspicuous. The second type of decision, as Wiebe E. Bijker, Thomas P. Hughes, and Trevor Pinch have shown (1987), fades from public memory over time. The end result of design turns into a "black box" and takes on the appearance of having been created solely by technical experts. The moral decisions Borgmann describes are just as inconspicuous because of the nature of the context in which they are discussed and made. This context is called domestic life. "Technology," he observes, "has step by step stripped the household of substance and dignity" (Borgmann 1984, 125). Just as Borgmann recalls our attention to the things of everyday life, he also makes us remember the importance of the household as a locus for everyday moral decision making. Thus Borgmann's reflections on how technology might be reformed can also be seen as an attempt to restore the philosophical significance of ordinary life.

BORGMANN AND THE RENEWAL OF PHILOSOPHICAL INTEREST IN ORDINARY LIFE

In this attempt, Borgmann does not stand alone. Over the course of the past two decades or so in North America, everyday life has been making a philosophical comeback. Five years after the publication of Borgmann's *TCCL* appeared Charles Taylor's *Sources of the Self*, a fascinating and ambitious account of the history of the making of modern identity. Heard throughout this book is the phrase "the affirmation of everyday life," a life characterized in Taylor's understanding by our nonpolitical relations with others in the context of the material world. As he sees it, affirming this life is one of the key features in the formation of our perception of who we are (Taylor 1989, 13). Against the horizons of our lives of work and play, friendship and family, we raise moral concerns that go beyond the questions of duties and obligations familiar to philosophers. What sorts of lives have the character of good lives, lives that are meaningful and worth living? What does one need to do to live a life that would be good in this sense? What can give my life a sense of purpose? In raising these questions, we affirm ordinary life. This affirmation is so deeply woven into the fabric of our culture that its very pervasiveness, Taylor maintains, serves to shield it from philosophical sight (Taylor 1989, 498).

Other signs point as well to a resurgence of philosophical interest in the moral dimensions of ordinary life. Take, for example, two fairly recent approaches to moral philosophy. In one of these approaches, philosophers such as Lawrence Blum, Christina Hoff Summers, John Hartwig, and John Deigh have been giving consideration to the particular ethical problems triggered by interpersonal relationships, those relationships among persons who know each other as friends or as family members or who are otherwise intimately connected. As George Graham and Hugh LaFollette note in their book *Person to Person,* these relationships are ones that almost all of us spend a tremendous amount of time and energy trying to create and sustain (Graham and LaFollette 1989, 1). Such activity engenders a significant amount of ethical confusion. Creating new relationships often means making difficult decisions about breaking off relationships in which one is already engaged. Maintaining interpersonal relationships often means making difficult decisions about what the demands of love and friendship entail. In accepting the challenge to sort through some of this confusion in a philosophically meaningful way, those involved with the ethics of interpersonal relationships willingly pays attention to ordinary life. In the process, they worry about the appropriateness of importing the standard moral point of view and standard moral psychology used for our dealings with others in larger social contexts—the Kantian viewpoint of impartiality and the distrust of emotions as factors in moral decision making—into the smaller and more intimate settings of families and friendships.

Another, related conversation about ethics includes thinkers such as Virginia Held, Nel Noddings, Joan Tronto, Rita Manning, Marilyn Friedman, and others whose work has been influenced by Carol Gilligan's research into the development of moral reasoning among women. I will call the enterprise in which these theorists are engaged feminist ethics, since I believe that description would be agreeable to those whom I have just mentioned, all of whom take the analysis of women's moral experiences and perspectives to be the starting point from which to rethink ethical theory.[1] Like interpersonal ethics, feminist ethics (particularly the ethics of care) places particular value on our relationships with those with whom we come into face-to-face contact in the context of familial and friendly relations. Its key insight lies in the idea that the experience of looking out for those immediately around one, an experience traditionally associated with women, is morally significant, and needs to be taken into account by anyone interested in developing a moral theory that would be a satisfactory and useful guide to the moral dilemmas facing us in all areas of life. Thus this approach to ethics also willingly accepts the challenge of paying philosophical attention to ordinary life. This challenge is summed up nicely by Virginia Held: "Instead of importing into the household principles derived from the marketplace, perhaps we should export to the wider society the relations suitable for mothering persons and children" (Held 1987, 122).

On the surface, these three paths of ethical inquiry—Borgmann's ethics of modern technology, the ethics of interpersonal relationships, and feminist

---

1. I am not using "feminist ethics" in a technical sense, but as a way of referring to the philosophical approach to ethics that starts from a serious examination of the moral experience of women. For philosophers such as Alison Jaggar, the term feminist ethics primarily means an ethics that recognizes the patriarchal domination of women and the need for women to overcome this system of male domination. Thus she and others might disagree that the ethics of care, as I take it here, is an enterprise of feminist ethics.

ethics—are occupied with different ethical questions. But they are united, it seems to me, in at least two ways. First, they are joined by their mutual contesting of the values upon which Kantian moral theory in particular and the Enlightenment in general are based. Wherever the modernist project of submitting public institutions and affairs to one's personal scrutiny went forward, certain privileges were enforced: that of reason over emotion, the "naked self" over the self in relation to others, impartiality over partiality, the public realm over the private sphere, culture over nature, procedural over substantive reasoning, and mind over body. In addition to the critique of Kantian ethics already mentioned by philosophers writing within a framework of an ethics of interpersonal relationships, feminist ethics has argued that these privileges led to the construction of moral theories insensitive to the ways in which women represent their own moral experience. Joining his voice to these critiques, Borgmann has written (while simultaneously praising the work of Carol Gilligan), "Universalism neglects . . . ways of empathy and care and is harsh toward the human subtleties and frailties that do not convert into the universal currency. . . . The major liability of moral universalism is its dominance; the consequence of dominance is an oppressive impoverishment of moral life" (Borgmann 1992a, 54–55).

A second feature uniting these relatively new forms of moral inquiry is a more positive one. Each attempts to limit further increases in the "impoverishment of moral life" by calling attention to the *moral* aspects of typical features of ordinary life that have traditionally been overlooked or even denied. The act of mothering (for Virginia Held), the maintenance of friendships (for Lawrence Blum) and the loving preparation of a home-cooked meal (for Borgmann) have all been defended, against the dominant belief to the contrary, as morally significant events.[2]

Despite the similarities and common concerns of these three approaches to moral philosophy, however, little engagement exists among them. Between feminist ethics and the ethics of interpersonal relationships, some engagement can be found: for instance, the "other-centered" model of friendship discussed in the latter is of interest to care ethicists as part of an alternative to Kantian ethics. However, both of these modes of ethical inquiry have shown little interest in the ethical dimensions of material culture. Nel Noddings, for example, believes that while caring can be a

2. For example, Virginia Held has written: "[Feminist moral inquiry] pays attention to the neglected experience of women and to such a woefully neglected though enormous area of human moral experience as that of mothering. . . . That this whole vast region of human experience can have been dismissed as 'natural' and thus as irrelevant to morality is extraordinary" (Held 1995, 160).

moral phenomenon when it is directed toward one's own self and that of others, it loses its moral dimension when it is directed toward things. In her book *Caring,* she defends the absence of discussion of our relations to things in her work: "as we pass into the realm of things and ideas, we move entirely beyond the ethical. . . . My main reason for setting things aside is that we behave ethically only through them and not toward them" (Noddings 1984, 161–62).

And yet in ordinary life ethical issues of technology, gender, and interpersonal relationships overlap in numerous ways. One wonders as a responsible parent whether it is an act of caring to buy one's son a Mighty Morphin Power Ranger. If I wish to watch a television program that my spouse cannot tolerate, should I go into another room to watch it or should I see what else is on television so that we could watch a program together? Is a married person committing adultery if he or she has an affair with a stranger in cyberspace? Seeing these interconnections, one wonders what might be the result were the probing, insightful questioning initiated by Borgmann into the moral significance of our material culture widened to include the other voices mentioned here. What would we learn, for instance, if Borgmann's technological ethics were explored from the perspective of feminist ethics?

In the context of this paper I can do no more than start to answer this question. With this in mind, I would like to look at one of the central claims of *TCCL:* the claim that the objects of material culture fall either into the category of things or devices.

FEMINISM AND THE DEVICE PARADIGM

As Borgmann describes them, things are machines that, in a manner of speaking, announce their own narratives and as a result are generous in the effects they can produce. For example, we can see the heat of the wood burning in the fireplace being produced in front of our eyes—the heat announces its own story, its own history, in which its relation to the world is revealed. In turn, fireplaces give us a place to focus our attention, to regroup and reconnect with one another as we watch the logs burn. In this regard, Borgmann speaks compellingly not only of the fireplace but also of wine: "Technological wine no longer bespeaks the particular weather of the year in which it grew since technology is at pains to provide assured, i.e. uniform, quality. It no longer speaks of a particular place since it is a blend of raw materials from different places" (Borgmann 1984, 49).

Devices, on the other hand, hide their narratives by means of their machinery and as a result produce only the commodity they were intended to produce. When I key the characters of the words I want to write into my portable computer they appear virtually simultaneously on the screen in front of me. I cannot see the connection between the one event and the other, and the computer does not demand that I know how it works in order for it to function. The commodity we call "processed words" is the result. While things lead to "multi-sided experiences," devices produce "one-sided experiences."[3] Fireplaces provide warmth, the possibility of conviviality, and a closer tie to the natural world; a central heating system simply provides warmth.

What thoughts might a philosopher working within the framework of feminist ethics have about this distinction? To begin with, I think she would be somewhat uneasy with the process of thinking used to make decisions about whether a particular object would be classified as a thing or a device. In this process, Borgmann abstracts from the particular context of the object's actual use and focuses his attention directly on the object itself. The view that some wine is "technological," as the example described above shows, is based on the derivation of the wine, the implication being that putting such degraded wine on the table would lead to a "one-sided experience" and further thwart, albeit in a small way, technology's capability to contribute meaningfully to the good life. In a feminist analysis of the moral significance of material culture, a different methodology would prevail. The analysis of material objects would develop under the assumption that understanding people's actual experiences of these objects, and in particular understanding the actual experiences of women who use them, would be an important source of information in deciding what direction a technological reform of society should take.

The attempt to make sense of women's experience of one specific technological innovation is the subject of communication professor Lana Rakow's book *Gender on the Line: Women, the Telephone, and Community Life* (1992). As its title suggests, this is a study of the telephone practices of the women residents of a particular community, a small midwestern town she called, to protect its identity, Prospect.

Two features of Rakow's study are of interest with regard to our topic. One relates to the discrepancy between popular perceptions of women's use of the telephone, and the use revealed in her investigation. She was well aware at the beginning of her study of the popular perception, not just in Prospect but widespread throughout American culture, of women's use of the telephone. In the popular perception, characterized by expressions such as "Women just like to talk on the phone" and "Women are on the phone all the time"; telephone conversations among women appear as "productivity

---

3. The term "multi-sided experiences" is used by Mihaly Csikszentmihalyi and Eugene Rochbert-Halton in their work *The Meaning of Things,* discussed in Borgmann 1992b.

sinks," as ways of wasting time. Understandably from this perception the telephone could appear as a device used for the sake of idle chatter that creates distraction from the demands of work and everyday life. This is how Borgmann sees it:

> The telephone network, of course, is an early version of hyperintelligent communication, and we know in what ways the telephone has led to disconnectedness. It has extinguished the seemingly austere communication via letters. Yet this austerity was wealth in disguise. To write a letter one needed to sit down, collect one's thoughts and world, and commit them laboriously to paper. Such labor was a guide to concentration and responsibility. (Borgmann 1992a, 105)

Rakow's study, however, did not support the popular perception. She found that the "womentalk" engaged in by her subjects was neither chatter nor gossip. Rather, it was a means to the end of producing, affirming, and reinforcing the familial and community connections that played a very large role in defining these women's lives. Such "phone work," very often consisting of exchanges of stories, was the stuff of which relations were made: "Women's talk holds together the fabric of the community, building and maintaining relationships and accomplishing important community relations" (Rakow 1992, 34).

Let me suggest some further support for this view from my own experience. While I was growing up, I frequently witnessed this type of phone work on Sunday afternoons as my mother would make and receive calls from other women to discuss "what had gone on at church." Although these women had just seen each other at church several hours before, their phone calls played exactly the role that Rakow discovered they played in Prospect. At the time, they were not allowed to hold any positions of authority within the organizational structure of this particular church. The meaning of these phone calls would be missed by calling them idle talk; at least in part, these phone visits served to strengthen and reinforce their identity within the gendered community to which these women belonged.

Another interesting feature of Rakow's study was its discovery of how women used the telephone to convey care:

> Telephoning functions as a form of care-giving. Frequency and duration of calls . . . demonstrate a need for caring or to express care (or a lack of it). Caring here has the dual implication of caring *about* and caring *for*—that is, involving both affection and service. . . . While this [care-giving role] has been little recognized or valued, the caring work of women over the telephone has been even less noted. (Rakow 1992, 57)

As one of the places where the moral status of the care-giving role of women has been most clearly recognized and valued, feminist ethics is, of course, an exception to this last point. Rakow's recognition of the telephone as a means to demonstrate one's caring for speaks directly to Nel Noddings's understanding of why giving care can be considered a moral activity (Noddings 1984). In caring one not only puts another's needs ahead of one's own, but, in reflecting on how to take care of those needs, one sees oneself as being related to, rather than detached from, the self of the other. In commenting that not only checking on the welfare of another woman or phoning her on her birthday but "listening to others who need to talk is also a form of care" (Rakow 1992, 57), Rakow singles out a kind of caring that well reflects Nodding's description. More often one needs to listen to others who call one than one needs to call others; and taking care of the needs of those who call often involves simply staying on the phone while the other talks. As Rakow correctly points out, this makes this particular practice of telephone caring a form of work. Those who criticize the ethics of care for taking up too much of one's time with meeting the needs of individual others might also be critical of Rakow's subjects who reported that

> they spend time listening on the phone when they do not have the time or interest for it. . . . One elderly woman . . . put a bird feeder outside the window by her telephone so she can watch the birds when she has to spend time with these phone calls. "I don't visit; I just listen to others," she said. (Rakow 1992, 57)

As these features of telephone conversations came to light in the interviews she conducted with the women of Prospect, Rakow began to see the telephone as "a gendered, not a neutral, technology" (Rakow 1992, 33). As a piece of gendered technology, the telephone arguably appears more like a thing than like a device, allowing for, in Borgmann's phrase, the "focal practice" of caring to take place. Looking at the telephone from this perspective raises doubts about Borgmann's assessment of the telephone. Has the telephone in fact become a substitute for the thing of the letter, contributing to our widespread feelings of disconnectedness and to our distraction? Rakow's fieldwork provides support for the idea that phone work, much like letter writing, can be "a guide to concentration and responsibility." By giving care over the phone, the development of both these virtues is supported. Thus, on Borgmann's own terms—"The focal significance of a mental activity should be judged, I believe, by the force and extent with which it gathers and illuminates the tangible world and our appropriation of it" (Borgmann 1984, 217)—it is difficult to see how using the telephone as a means of conveying care could not count as a focal concern.

Along with the question of whether a particular item of our material culture is or is not a device, looking at the device paradigm from a feminist point of view gives rise to at least two other issues. One is connected to an assumption on which this paradigm rests: that the moral significance of an object is directly related to whether or not that object is a substitute for the real thing. This issue is also connected to the idea that because technological objects are always substitutes for the real thing, the introduction of new technology tends to be a step forward in the impoverishment of ordinary life.

Certainly technological objects are always substitutes for *something or another*. A washing machine is a substitute for a washing board, dryers are substitutes for the line out back, krab [*sic*] is often found these days on salad bars, and so forth. In some cases, the older object gradually fades from view, as happened with the typewriter, which (but only as of fairly recently) is no longer being produced. In other cases, however, the thing substituted for is not entirely replaced, but continues to coexist alongside the substitute. In these cases, it is harder to see how the technological object is a substitute *for the real thing*, and thus harder to see how the introduction of the new object threatens our sense of engagement with the world. While it is true that telephones substitute for letter writing, as Borgmann observes, the practice of letter writing goes on, even to the point of becoming intertwined with the use of the telephone. Again, from Rakow:

> The calls these women make and the letters they send literally call families into existence and maintain them as a connected group. A woman who talks daily to her two nearby sisters demonstrated the role women play in keeping track of the well-being of family members and changes in their lives. She said, "If we get a letter from any of them (the rest of the family) we always call and read each other the letters." (Rakow 1992, 64)

Perhaps, though, the largest question prompted by Rakow's study has to do with whether Borgmann's distinction itself between things and devices can hold up under close consideration of the experiences and practices of different individuals. There are many devices that can be and are used as the women in this study used the telephone. Stereos, for example, can be a means for someone to share with someone else particular cuts on a record or songs from a CD to which she or he attaches a great deal of personal significance. In this way, stereos can serve as equipment that aid the development of mutual understanding and relatedness, rather than only being mechanisms for disengagement. The same goes, as Douglas Kellner points out in chapter 12, for the use of the computer as a communicative device. Empirical investigations into the gendered use of computer-mediated communications suggest that while women do not necessarily use this environment like the telephone, as a means of promoting care, they do not "flame" (send electronic messages critical of another individual) nearly as much as do men, and they are critical of men who do engage in such activity.[4]

In particular, from a feminist perspective one might well wonder whether, in Borgmann's language, the use of those "conjunctions of machinery and commodity" inevitably hamper one's efforts at relating more to others and to the world. Borgmann argues that because devices hide their origins and their connections to the world, they cannot foster our own bodily and social engagement with the world. But as I have tried to show here, this is arguably not the case. Whether or not a material object hides or reveals "its own story" does not seem to have a direct bearing on that object's capacity to bind others together in a narrative web. For instance, older women participating in Rakow's study generally agreed that telephones improved in their ability to serve as a means of social support and caregiving once their machinery became more hidden: when private lines took the place of party lines and the use of an operator was not necessary to place a local call. To generalize, the machinery that clouds the story of a device does not appear to prevent that device from playing a role in relationship building.

DEVICES AND THE PROMISE OF TECHNOLOGY

While a child growing up in New Jersey, I looked forward on Friday evenings in the summer to eating supper with my aunt and uncle. I would run across the yard separating my parents' house from theirs to take my place at a chair placed at the corner of the kitchen table. The best part of the meal, I knew, would always be the same, and that was why I looked forward to these evenings. While drinking lemonade from the multicolored aluminum glasses so popular during the 1950s, we would eat Mrs. Paul's fish sticks topped with tartar sauce. With their dubious nutritional as well as aesthetic value, fish sticks are to fresh fish as, in a contrast described eloquently by Borgmann, Kool Whip is to fresh cream (Borgmann 1987, 239–42). One doesn't know the seas in which the fish that make up fish sticks swim. Nearly anyone can prepare them in a matter of minutes. Still, despite these

---

4. See, for example, Susan Herring, "Gender Differences in Computer-Mediated Communication: Bringing Familiar Baggage to the New Frontier" (unpublished paper).

considerations, these meals were marked by family sociability and kindness, and were not hurried affairs.

I recall these meals now with the following point in mind. One might be tempted by the course of the discussion here to say that the objects of material culture should not be divided along the lines proposed in *TCCL* but divided in another manner. From the perspective of feminist ethics, one might suggest that one needs to divide up contemporary material culture between relational things, things that open up the possibility of caring relations to others, and nonrelational things: things that open up the possibility of experience but not the possibility of relation. Telephones, on this way of looking at things, would count as relational things. Virtual reality machines, such as the running simulator Borgmann imagines in *CPD,* or golf simulators that allow one to move from the green of the seventeenth hole at Saint Andrews to the tee of the eighteenth hole at Pebble Beach, would be nonrelational things. One can enjoy the experiences a virtual golf course makes possible, but one cannot in turn, for example, act in a caring manner toward the natural environment it so vividly represents. But the drawback of this distinction seem similar to the drawback of the distinction between things and devices: the possibility of using a thing in a relational and thus potentially caring manner seems to depend more on the individual using that thing and less on the thing itself. Depending on who is playing it, a match of virtual golf has the potential of strengthening, rather than undoing, narrative connections between oneself, others and the world.

But if our discussion does not lead in this direction, where does it lead? Let me suggest that although it does not lead one to reject the device paradigm outright, it does lead one to recognize that while any device does use machinery to produce a commodity, the meaning of one's experience associated with this device does not necessarily have to be diminished. And if one can use technology (such as the telephone) to carry out focal practices (such as caregiving), then we might have cause to believe that there are other ways to recoup the promise of technology than Borgmann sees. As mentioned earlier, his hope is that we will give technology more of a supporting role in our lives than it has at present (Borgmann 1984, 247), a role he interprets as meaning that it should support the focal practices centered around focal things. But if devices can themselves support focal practices, then the ways in which technology can assume a supporting role in our lives are enhanced.

But if the idea that devices can support focal practices is in one way a challenge to the device paradigm, in another way it gives additional weight to the notion that there are limits to reforming technology through the process of democratic design. When they are used in a context involving narrative and tradition, devices can help build engagement and further reinforce the cohesiveness of civil society. Robert Putnam has pointed out the importance of trust and other forms of "social capital" necessary for citizens to interact with each other in a cooperative manner. As social capital erodes, democracy itself, he argues, is threatened (Putnam 1995, 67). While this paper has suggested that devices can under some conditions further the development of social capital, it is difficult to see how they can be deliberately designed to do so. In thinking about how to reform technology from a democratic perspective, we need to remember the role of features of ordinary life such as narrative and tradition in making our experience of democratic values more meaningful. Borgmann's reminder to us of this role is, it seems to me, one of the reasons why *TCCL* will continue to have a significant impact in shaping the field of the philosophy of technology.

REFERENCES

Bijker, Wiebe E., Thomas P. Hughes, and Trevor Pinch, eds. 1987. *The Social Construction of Technological Systems.* Cambridge: MIT Press.

Borgmann, Albert. 1984. *Technology and the Character of Contemporary Life: A Philosophical Inquiry.* Chicago: University of Chicago Press.

———. 1987. "The Invisibility of Contemporary Culture." *Revue internationale de philosophie* 41:234–49.

———. 1992a. *Crossing the Postmodern Divide.* Chicago: University of Chicago Press.

———. 1992b. "The Moral Significance of the Material Culture." *Inquiry* 35:291–300.

Feenberg, Andrew. 1995. *Alternative Modernity: The Technical Turn in Philosophy and Social Theory.* Berkeley and Los Angeles: University of California Press.

Graham, George, and Hugh LaFollette, eds. 1989. *Person to Person.* Philadelphia: Temple University Press.

Held, Virginia. 1987. "Non-contractual Society: A Feminist View." In *Science, Morality and Feminist Theory.* Ed. Marsha Hanen and Kai Nielsen. Calgary: University of Calgary Press.

———. 1995. "Feminist Moral Inquiry and the Feminist Future." In *Justice and Care: Essential Readings in Feminist Ethics.* Ed. Virginia Held. Boulder, Colo.: Westview Press.

Noddings, Nel. 1984. *Caring.* Berkeley and Los Angeles: University of California Press.

Putnam, Robert D. 1995. "Bowling Alone: America's Declining Social Capital." *Journal of Democracy* 6:65–78.

Rakow, Lana F. 1992. *Gender on the Line: Women, the Telephone, and Community Life.* Urbana: University of Illinois Press.

Taylor, Charles. 1989. *Sources of the Self.* Cambridge: Harvard University Press.

THIRTEEN

Crossing the Postmodern Divide with Borgmann, or Adventures in Cyberspace

Douglas Kellner

In his major works, Albert Borgmann has explored in depth and detail the role of technology in contemporary life and provided compelling critical, philosophical perspectives. In this study, I primarily discuss *Crossing the Postmodern Divide* (or *CPD;* 1992) in relation to the themes of his earlier *Technology and the Character of Contemporary Life* (1984). While appreciating Borgmann's attempt to produce distinctions between modernity and postmodernity as historical epochs, I challenge his particular interpretation of a postmodern divide and sketch out an alternative conception of technology that critically engages some of Borgmann's positions. My argument will be that while technology threatens democracy, community, individual sovereignty, and other values many of us hold in common, it also furnishes the potential for a positive reconstruction of social life and an enhancement of human life. My provocation will be to deconstruct what I take to be a too sharp distinction in Borgmann's text between a "hyperreal" technosphere contrasted with a "real" world of concrete human interaction and focal activities. I attempt to show that some of Borgmann's own positive values can be realized in the cyberspaces of the new technologies and offer some examples. These reflections will compel us to rethink the concepts of the public sphere, democracy, community, and technology.

TECHNOLOGY AND MODERNITY: BORGMANN'S PERSPECTIVES

In his introduction to *CPD,* Borgmann refers to "the expatriate quality of public life," writing: "We live in self-imposed exile from communal conversation and action. The public square is naked. American politics has lost its soul. The republic has become procedural, and we have become unencumbered selves. Individualism has become cancerous. We live in an age of narcissism and pursue loneliness" (1992, 3). He then submits many eloquent examples of contemporary sullenness and hyperactivity and unfolds a polemic assault against our present condition. Referring back to the themes of his earlier book, I would agree that technology is at least partly responsible for the problems that Borgmann eloquently evokes, but that it also provides possible solutions—that the poison provides part of the cure.

There is no doubt that technology is a major constituent of our contemporary world and requires sophisticated philosophical perspectives to theorize its nature and effects. Technology has been acknowledged by Borgmann and many others as both a fundamental component of the modern world and a possible lever to a new postmodernity. Moreover, both its proponents and critics agree that technology has been of momentous importance in constituting modernity and most theorists of postmodernity argue that it is new technologies that are largely responsible for taking us over the postmodern divide. But before taking this trip I want to critically engage Borgmann's thoughts on modernity and to focus on the role of technology in the modern adventure.

Borgmann's argument is that modernity is largely characterized by aggressive realism, a methodical universalism, and an ambiguous individualism. The modern project for Borgmann involves the use of science and technology to dominate nature; constructing a method that is largely technical to ground knowledge that will enable us to control nature and construct a technological society; and the deployment of technology and its fruits to satisfy individual needs and ends in which technology is embedded in a means-ends nexus serving primarily the interests of atomized individuals.

Borgmann argues that philosophy is a seismographic register of epochal changes and reads off the features of modernity from the development of modern thought, of which he is sharply critical. There have been numerous discourses on modernity in recent years and I believe that Borgmann goes too far in privileging philosophy as definitive of modernity. If one looks at modernity from the perspective of critical social theory one gets a somewhat different picture of the modern/postmodern divide (I also believe that modernity looks different from the vantage of the arts, or economics and politics, but this is a story for another day). From the standpoint of classical social theory—the tradition of Marx, Weber, and Durkheim, et al.—modernity is interpreted as an epochal rupture involving momentous changes in the economy, polity, social order, culture, and everyday life. Within social theory, there are still passionate arguments concerning whether the capitalist economy, a democratic polity, an Enlightenment cultural revolution, the Protestant ethic à la Weber, urbanization and social differentiation, or—as I would argue—a concatenation of all of these factors is primarily responsible for the origins, development, and trajectory of the modern world (see Antonio and Kellner 1992).

Classical social theorists, like Habermas in our day but for different reasons, saw a positive potential in modernity and unrealized promises that could, however, be fulfilled in the future. Modernity thus looks different from the perspectives of democratic theory, which see the rise of democracy, of a liberal public sphere, of bills of rights and constitutions as defining features. Likewise, if one looks at modernity from the perspective of classical social theory one might find the new forms of association, communities, cooperation, communication, and other forms of social interaction, as well as the benefits of modern cities. And modernity from the perspective of modernism in the arts sees a rich variety of innovative aesthetic movements and works transcending the limitations and conventions of traditional art (see Berman 1982).

All of these features of economic, social, political, and cultural modernity are admittedly ambiguous and we could have many books and conferences on the benefits and disasters of the capitalist economy; the nature, substance, benefits, and recurrent obstacles to realizing democracy; the gains and erosion of the positive forms of social association from earlier modernity; and the vicissitudes of modern culture. Indeed, I would conclude that modernity is a highly ambiguous phenomenon with tremendous achievements and potentials and copious problems and disasters. But I would argue that we need transdisciplinary perspectives to theorize modernity and I am skeptical as to whether it is in fact over and that we have crossed the postmodern divide—as I will argue throughout this paper.

Of course, one can read modernity from the perspectives of philosophy, as Borgmann does, and this optic might very well illuminate key features of the modern world, but such a largely philosophical optic misses, as I am suggesting, some of the more positive, but also ambiguous and enduring features of modernity. And one also might occlude from a purely philosophical perspective the embeddedness of technology in social relations and within a specific socioeconomic system, which I believe the perspectives of the sort of critical social theory developed by Marx, Weber, Lukács, and the Frankfurt School affords. Consequently, I will argue that this tradition provides a better optic to illuminate our present situation and the role of technology in the contemporary world than the philosophical traditions and this theme will be a primary focus of much of the rest of my paper.[1]

1. For my own perspectives on the critical theory of the Frankfurt School, see Kellner 1984, 1989a, 1989b; Bronner and Kellner 1989; Best and Kellner 1991, 1997; and Kellner 1995. Many of my articles on critical theory are collected at the Illuminations website: http://www.uta.edu/huma/illuminations.

A POSTMODERN DIVIDE?

The perspectives of critical social theory become relevant when one analyzes Borgmann's concept of a postmodern divide and we address the question of how we can best characterize our contemporary moment. Borgmann primarily sketches the postmodern divide in terms of a philosophical critique of modern ideology, or if you prefer, the modern mind set, the framework (Heidegger's *Gestell*), or what Borgmann sometimes calls the "modern project" or "modernism." He writes that "[a]n epoch approaches its end when its fundamental conviction begins to weaken and no longer inspires enthusiasm among its advocates. That is true of each of the three parts of the modern project: realism, universalism, and individualism" (1992, 48). It is here that Borgmann privileges "the seismographic significance of modern philosophy" that registers this shift. Borgmann elicits Richard Rorty in his *Philosophy and the Mirror of Nature* as a paradigmatic critic of the modern project of securing representational knowledge to dominate nature grounded in a foundation of truth and certainty that would provide the tools to dominate nature—and who advocates a shift to a more modest concept of philosophy as conversation and, I might add, interpretation.

Yet it is Horkheimer and Adorno in *Dialectic of Enlightenment* ([1947] 1972) who—decades before Rorty—provide a powerful critique of the Enlightenment project of the domination of nature and radically deconstruct the foundations, premises, and systems of modern thought that lead from the slingshot to the hydrogen bomb—to use a later phrase of Adorno. Moreover, while Adorno and Horkheimer, Rorty, Borgmann, myself, many philosophers of technology, and some postmodern theorists are prepared to break with the aggressive mind set that sees nature as the stuff of domination and that conceives of science and technology as instruments of domination, this is still a dominant optic among the hegemonic economic, political, and intellectual elites. Further, such "realism" is probably shared by all too many sectors of the public as well, thus it is not clear that there is a postmodern break with the dominant philosophical orientation of modernity, though there have certainly been a plethora of critiques and proposed alternatives—including Borgmann.

Likewise, while many critical theorists —as well as feminists like Carol Gilligan, whom Borgmann cites—reject the concept of a universalist method of truth or set of techniques to dominate and control nature, nonetheless neopositivistic concepts of science, faith in technocracy and technical solutions according to correct method, and desire for rigor and quantitative "hard" science, which guarantee truth and objectivity, continue to constitute the dominant mind set in leading intellectual and academic circles. Borgmann writes that "universalism has been dethroned in almost every field of contemporary culture, from mathematics, by way of physics and biology to anthropology, the law, and literature. It is now seen as an anxious and pretentious yet ultimately futile effort to enforce rigor and uniformity in an unruly and luxuriant world" (1992, 55). It is, however, largely avant-gardes in these fields who are contesting the modern paradigm, and even in philosophy belief in rigor, foundations, and correct techniques and modes of argumentation continue to prevail—at least this is the impression I get from philosophy conventions, department tenure meetings, and perusals of mainstream philosophy journals.

Finally, although the possessive individualism criticized by Borgmann has led to social anomie, destruction of community, and downright despair, unfortunately there is little evidence that significant numbers of people are turning away from a limited and destructive conception of the individual— a point that Borgmann himself makes:

> Despite its beneficence, the transformative power of postmodernism is in doubt because it has failed to resolve the ambiguity of individualism. The latter term designates the human condition that has lost its premodern communal bonds. But we lack a unified and positive understanding of the person who would answer to the term. The individual was thought to be the beginning and end of the modern project, its author and beneficiary, but this coherence was an illusion. (1992, 79)

And so while some philosophers see the limitations of the modern mind set and paradigms, I am afraid that the features that Borgmann ascribes to modernity still hold sway, although they are admittedly and in many ways deservedly under attack. Yet Borgmann concludes his interrogations of modernity with an uncharacteristically positive and optimistic note:

> The intellectual, artistic, and economic developments of the past generation have led us beyond the broad and once fertile plains of modernism to a point where, looking back, we can see that we have risen irreversibly above the unworried aggressiveness, boundlessness, and unencumberedness of modernism. The latter now seems brash and heedless to us, if not downright arrogant and oppressive. The transition from modernism to postmodernism is reflected in many kindred shifts of sympathy: from the belief in a manifest destiny to respect for Native American wisdom, from male chauvinism to many kinds of feminism, from liberal democratic theory to communitarian reflections, from litigation to mediation, from heroic medical technology to the hospice

movement, from industrialism to environmentalism, from hard to soft solutions. (1992, 78)

Borgmann then notes countervailing tendencies—"shifts from light to darkness as well: from Enlightenment to dogmatism, from tolerance to ethnic strife, from liberalism to self-righteousness, from freedom to censorship"—but concludes that "the shifts to the good have prevailed" (1992, 78). Here we could endlessly argue whether or not this paradigm shift has or has not taken place in these specific issues, and whether shifts to the good or the bad or somewhere in between have occurred. But the crux of the matter is that Borgmann does not really convincingly theorize a postmodern *divide*. He does produce a model, or ideal type, of the modern project, shows that it is under assault in certain sectors of contemporary philosophy and life, but does not provide a convincing set of criteria to distinguish the modern from the postmodern, or arguments and evidence that we have actually crossed the postmodern divide to reach the other side.

To be sure, Borgmann attempts himself to articulate an alternative project that he calls "postmodern realism," but I have my doubts whether this notion is adequate to articulate a divide between modern and postmodern thought. Indeed, I would argue that perspectivism and social constructivism are at the heart of the postmodern critique, which is radically at odds with the sort of realism that Borgmann professes.[2] Indeed, Borgmann consistently maintains a realist ontology and philosophy of nature and there is a covertly normative realism embedded in Borgmann's concepts of focal things and practices as well (Borgmann 1984, 1992). In effect, Borgmann is telling us to attend to the really real, the authentic, in organizing our life. That, for instance, we should engage in the "real" activities of eating, gardening, running, and the like, and avoid fragmentation and dispersion in superficial practices and unreal consumer or media fantasies—or, as I shall discuss and contest—the realm of cyberspace.

Two senses of realism are thus collapsed in Borgmann: an ontological conception of reality and a normative celebration of that which is deemed to be really real and authentic: e.g., focal things and practices. But for postmodern theory, "reality" is a construct and notions of the "authentic" or really real are regularly deconstructed. Thus, I would argue that the main lines of the postmodern critique are directed against such forms of realism and that the notion of a "postmodern realism" runs against the main lines of postmodern thought. Moreover, even if there has been some shift in philosophical paradigms, I'm not sure this alone would support the sort of epochal shift from one historical period to another that Borgmann

2. See the discussions of postmodern theory in Best and Kellner 1991, 1997.

evokes with his concept of a "postmodern divide." My own view is that while there are what might be called emergent tendencies that might lead to overcoming the modern mind set or project as described by Borgmann and others, these emergent and allegedly "postmodern" phenomena, at present, nevertheless cannot bear the burden of articulating a postmodern divide from the modern or serve as evidence of the transcendence of the modern—although there is arguably evidence of a postmodern turn in many circles.[3]

Indeed, I would argue that for the concept of the postmodern to have force and substance, we need to distinguish between modernity and postmodernity as historical epochs, modernism and postmodernism as forms of art, and modern and postmodern theory (see Best and Kellner 1991, 1997); I believe that Borgmann collapses these distinctions. For the most part, Borgmann folds his conception of a postmodern paradigm shift in philosophy into an epochal conception of postmodernity, although in the latter part of Borgmann's discussion of the postmodern (1992, chap. 3), he discusses architecture and what he calls a postmodern economy rooted in information processing. In these evocations of postmodern culture and society, there are more promising candidates to theorize a postmodern divide, yet I would argue that we need to shift from philosophy to social theory and cultural critique, or to transdisciplinary perspectives, to theorize these divisions. This is, I believe, the unarticulated direction in which Borgmann is going, but currently he tends to overly privilege philosophy— whereas I would argue that we need to focus more intently on the economy, culture, and society to see the paradigm shifts from the modern to the postmodern in action and producing a rather different mode of life, rooted, as we shall see, in the dramatic impact of new technologies—as well as the global restructuring of capitalism (see Best and Kellner 1997 and forthcoming).

In other words, I am arguing on a metalevel to go beyond philosophy to transdisciplinary social theory as was developed by the Frankfurt School and some versions of postmodern theory, and is happening today in some forms of feminism, cultural studies, and critical multiculturalism (see Kellner 1995). Thus, while I don't think that Borgmann has demonstrated a paradigm shift in philosophy and culture from the modern to the postmodern, I do believe that such distinctions can be made (see Best and Kellner 1997) and that there are undoubtedly manifestations of a postmodern shift in architecture and other cultural forms, as Borgmann suggests, although

3. For my own views of the paradigm shift from the modern to the postmodern, see Kellner 1989a, 1989b, 1989c, and 1994; and Best and Kellner 1991, 1997, and forthcoming.

he does not systematically engage the transition from modernism to post-modernism in the arts à la Jameson (1984, 1991), who correlates the sort of absence of affect, fragmentation, pastiche, and implosion in the arts with the experiences of postmodern subjects, arguing that postmodernism has become a "cultural dominant" and new mode of subjectivity and experience. While I do not agree with Jameson that postmodernism in culture is already a cultural dominant, it is certainly an important emergent force that may very well help register a shift under way more broadly from modern to postmodern paradigms.[4]

Borgmann does address recent trends in the economy and in particular information processing, and I would agree here that a postmodern paradigm shift is evident, described in terms of the emergence of a postindustrial society, information society, post-Fordism, postmodern globalization, and various other conceptions related to the restructuring of capitalism. Indeed, it is evident that immense changes are occurring in the economy and sooner or later these will affect, for better and worse, every aspect of life; when capital mutates, effects ripple from one end of the globe to the other, from one social domain to another. Thus, I agree with Borgmann that new modes of information processing and new computer and media technologies—as well as virtual reality, simulation, and other exotic high-tech instruments—*are* producing a set of dramatic changes in the economy, society, culture, and everyday life. In order to avoid technological determinism, however, I would propose theorizing the current developments in technology within the context of a global restructuring of capitalism, engaging the ways that new syntheses of capital and technology are dramatically transforming every aspect of our life (Best and Kellner, forthcoming).

This discussion brings us again to Borgmann's thematic of technology and contemporary life and here I will polemically engage Borgmann's views on technology and our contemporary moment, as well as Borgmann's philosophical perspectives.

POSTMODERN/FOCAL REALISM VS. HYPERREALITY

In *CPD,* Borgmann claims that two contrasting options confront us as we encounter the present situation, which he describes as the choice between an "instrumental hyperreality" and "hypermodernism" on the one hand and "postmodern realism" and "focal realism" on the other. Hypermodernism describes an intensification of the worst features of modernity and is constituted by a "hyperreality," "hyperactivity," and "hyperintelligence"

---

4. For my perspectives on Jameson, see Kellner 1989c; and for more on the shift from modernism to postmodernism in the arts, see Best and Kellner 1997, chap. 4.

(Borgmann 1992, chap. 4), which he contrasts with "focal realism," "patient vigor," and "communal celebration" (chap. 5). Borgmann thus deploys sets of triadic schemata to contrast the modern with two ways of living the postmodern condition, one of which intensifies modernism (i.e., hyper-modernism) and one that breaks more radically with it and thus constitutes a genuinely postmodern alternative. In Borgmann's words, "The alternative tendency is to outgrow technology as a way of life and to put it in the service of reality, of the things that command our respect and grace our life. This I call postmodern realism" (82).

Borgmann thus deploys a crucial normative distinction between a hy-permodernism contrasted to a postmodern focal realism, which provides, as Andrew Light has argued (1995), a remapping of Borgmann's earlier dis-tinction between devices and focal things and practices. These distinctions delineate what Borgmann's dislikes and likes in the contemporary moment, thus the two sets of terms function in his thought as positive and negative markers that critically counterpoise real experience to hyperreal experience. For instance, as the following examples indicate, Borgmann persistently attacks hyperreal experience and practice, excoriating "the cancerous growth of video culture" (Borgmann 1992, 10) and claiming that "[t]his middle region of physical reality is divided today by the line between the real and the hyperreal. On the one side are things of commanding presence, continuous with the world; on the other, disposable and discontinuous experiences" (118). Borgmann also writes that "[h]aving left modernism [e.g., as a mode of theory and intellectual paradigm] behind us, we now have to decide whether to proceed on the endless and joyless plain of hypermodernism or to cross over to another more real world" (126).

Borgmann thus recommends that we "outgrow technology as a way of life," put technology in the service of reality, and make focal things and practices the fundaments of our life (1992, 82 ff.). While I agree that we should not blindly worship technology, that we should put it in the service of enhancing our lives and serving our most cherished values, I disagree with Borgmann's distinction between hyperreality and reality and his assigning technological modes of experience or interaction to the hyperreal as opposed to real interaction with nature, objects, and human beings—which he in turn privileges over the hyperreal technosphere and its seductions. In the remainder of this paper, I argue that new technological modes of experience and interaction are just as real and life enhancing as conversation, gardening, taking a hike in the wilds, or caring for animals—examples positively valorized by Borgmann. I believe that Borgmann's distinction between the real and hyperreal and his denigration of hyperreality are problematic, that we need to deconstruct such oppositions, and that we should see how new technologies make possible the sort of focal, life-enhancing experiences and activities that Borgmann himself calls for.

Of course, the new technologies and technological mode of experience have some of the downsides that Borgmann points to and there are losses as well as gains in the uses of the new computer and information technologies that I will focus on. But the point that I would stress is that we need a dialectical optic on technology and, crucially, we need to focus our energies on the devising of uses for new technologies that will enhance our lives and serve the values that we hold in common. Consequently, I share the practical and political concerns of those philosophers of technology who want the philosophy of technology to concern itself with how technology can enhance our lives, can be put into progressive political uses—or that counters destructive and life-negating technologies and applications.

In the last chapter of *CPD,* Borgmann sketches out his own political vision and perspectives, and it is revealing that positive interactions with technology do not play a role in his deliberations. On one level, I appreciate the last sections of Borgmann's book, which sketch out a political and normative postmodern agenda with the discussions of focal realism, patient vigor, and communal celebration—this is for me far preferable to the apolitical and nihilistic versions of postmodernism that are circulating. Against such cynical and nihilistic versions of postmodern theory (Baudrillard and some of his followers come to mind), Borgmann provides a constructive and positive version. But he seems to leave out experiences with and uses of new technologies in those activities; the values and activities that he celebrates are counterpoised to "bad" "hyperreal" experiences and activities, betraying aspects of a technophobic refusal to see more positive uses of technology.

In fact, I found it curious that in his discussion of information processing, Borgmann describes the computerization of the economy with examples from business, but does not describe the ways that computer technologies are becoming embedded in the fabric of our own everyday lives and practices. Here—not only to supplement Borgmann's positive program but also to question and deconstruct his "real" vs. "hyperreal" dichotomy—I want to describe some concrete examples of how use of new technologies can be something like Borgmann's own focal practices and can help produce new modes of communication, writing, and art, which make possible the sort of positive postmodern experiences and activities that Borgmann himself desires. I argue that these technologies and their uses are just as real as our interactions in other dimensions of experience, though there are novelties and positive and negative features that need to be attended to, discussed, and engaged. I also argue that the new cyberspaces of media and computer technology produce new public spheres that might help overcome the individual isolation, apathy, and sullenness that have alienated large sectors of the public from our polity and from other people.

New Technologies and New Focal Things and Practices

Often very effectively, Borgmann does philosophy with examples and I will attempt to emulate his procedure here. Let us begin with e-mail and computer communication. For some years, I have used e-mail every day to communicate with people all over the world. This has facilitated a lot of business communication, saving a lot of the time and bother sometimes involved in planning trips, writing articles, and communicating instrumentally with colleagues. But computer communication also arguably nourishes personal and professional relationships that may lead to enduring friendships and productive relationships. Thus, technologically mediated communication can nourish significant social interaction and relationships as well as serve merely instrumental purposes.

Indeed, I would argue that e-mail and computer communication also have the features of establishing new forms of connectiveness and interaction. As noted, it enables me to connect with colleagues and friends all over the world, but also makes possible networking with people whom I have not met, thus expanding my network of friends and colleagues. Moreover, it is even possible to communicate in real time with people on a one-to-one basis via split screen or in new audio-video interactive spaces, as well as with groups of people in MOOs and MUDs, or computer conferences. There are admittedly losses and downsides to this mode of communication as well as benefits. While sometimes face-to-face communication is preferable, and telephone conversation is preferable to e-mail, it is not always possible to engage in such interaction. Thus, minimally, electronic computer-mediated communication supplements and expands one's connections and interactions.

Moreover, computer conferencing and MOOs and MUDs make possible interaction with a wider range and variety of individuals than is usually possible in face-to-face communication and without the burdens of travel. In past years, I have participated in virtual conferences in the PMC and IATH MOOs. The later conference was organized in Florida and during a Saturday afternoon session in 1995, we logged on to a MOO in Virginia. The interaction made possible a much more intense exchange of ideas than is usually possible at conferences, where the exchange of ideas is limited since individuals can only speak one at a time, often participants do not say much of interest, and, as we know too well, some speakers tend to monopolize the conversation. In MOO conferences, by contrast, individuals can textually multiply their ideas, one can interact in an intense way with more individuals and ideas, and observers can comment on the proceedings—with the entire event captured and archived for later scrutiny and further discussion and debate.

There are, of course, also losses in this sort of electronic conference. The traditional face-to-face conference can elicit exciting dialogue and interpersonal interaction (though after thirty years of conference going I would say that this is the exception rather than the rule). One can cultivate more personal and interactive relationships in face-to-face encounters, though computer-mediated communication can also create relationships that can be enhanced in real life. In computer conferences, one sacrifices the joys of travel, but also avoids the hardships. One does not have to disrupt the fabric of everyday life and can interact with people without leaving one's home, saving time (and money) often wasted on travel; one can spend these savings on more rewarding focal activities.

While in computer-mediated communication, there are undeniably losses due to absence of concrete presence, voice, personal interaction, and other semiotic features of interpersonal interaction, I would not create a hierarchy of interpersonal face-to-face versus electronic communication in any absolute register. Rather, I would argue for a logic of both/and rather than either/or, seeing these modes of communication and interaction as complementary and supplementary rather than mutually exclusive. Face-to-face communication can facilitate manipulation and domination as well as positive interaction; it can be boring and time-consuming and trap individuals in situations they do not really want to be in. Moreover, I certainly do not see what is intrinsically harmful about computer communication if it is a supplement to concrete social interactions, although I realize there is a danger of getting lost in virtual worlds, losing interpersonal communicative and interactive skills and the ability to socially interact with others.

In terms of research and writing, which are also an important focal domain of academic and intellectual life, I would claim that computer databases, the Internet, and web surfing provide tremendous resources and supplementation to standard academic practices. Accessing computer databases and web sites often saves research time, provides a wealth of information and access to alternative sources, and opens up debates to a wider range of views—though one might also benefit from traditional modes of library research.[5] In fact, new modes of communication, research,

5. The most one-sided and often inane attacks on e-mail and computer-based research and communication that I know is Stoll 1995. While Stoll points to limitations in the new modes of communication and interaction, and benefits in older institutions and practices, his polemic is relentlessly one-sided and occludes the obviously positive benefits of new

and writing are a supplement to traditional forms and should not be seen as their replacement.

Indeed, new technologies like CD-ROMs and other forms of multimedia provide new possibilities for creative work and education. For instance, I worked on a CD-ROM on Emile de Antonio's film *Painters Painting* with Voyager, editing over seven hundred pages of de Antonio's interviews with artists, providing biographies of the artists, and contextualizing the film and art history, thus generating over a thousand pages of text, too much for a book, but easily accommodated on CD-ROM. While Borgmann says that art has been subverted by technology (1992, 136), new technology in fact furnishes new possibilities for aesthetic creativity. I work with cyberartists Pat Lichty and Jon Epstein to develop illustrations for my books and web sites, and I have become excited concerning the aesthetic potential of new media. In addition, museum and other web sites make accessible to the entire world the heritage of world art that might not otherwise be accessible to many individuals. To be sure, such electronic reproduction of art, like slides and print reproduction, lacks the aura of the presence of the art work, but it furnishes supplementary experience and textual information that can enhance eventual museum experience of the works themselves.

When we turn to more social and political uses of new computer technologies, the concept of an information superhighway highlights the need to have an Internet system that is free, is open to all, and yields public spaces for diverse purposes and interaction. Given the extent to which capital and its logic of commodification have colonized ever more areas of everyday life in recent years, it is somewhat astonishing that cyberspace is by and large decommodified for large numbers of people—at least in the overdeveloped countries like the United States. In the United States, government and educational institutions, and some businesses, provide free Internet access and in some cases free computers, or at least workplace access. With flat-rate monthly phone bills (which I know do not exist in much of the world), one can thus have access to a cornucopia of information and entertainment on the Internet for free, one of the few decommodified spaces in the ultracommodified world of technocapitalism.

The metaphor of frontier signifies the adventure of computer explorations and quests that, in conjunction with the concept of an information superhighway, evokes images of a journey, a trip, postmodern adventures in information processing, communication, aesthetic creativity, and exploration—replicating some of the adventures of early modernity

---

technologies. For a wide range of positions on technology, society, and education, see my web site: http://www.gseis.ucla.edu/courses/ed253a/253WEB1.htm.

in the new cyberspaces. We are entering new frontiers, new modes of communication and interaction, new sources of knowledge and creativity, and new forms of social interaction. There are, of course, dangers that one can get lost in this world, and there is no doubt that some of our students and fellow citizens are getting enmeshed in the sometimes problematic worlds of cyberspace that include mindless games, stupid chitchat, and problematic activities such as pornography and gambling.

And while metaphors of the net and web point to connectedness, rhizomatic and multilayered levels of experience and texture, the same metaphors also signify, more negatively, that one can become trapped inside an artificial world, lost in the funhouses of cyberspace, unable to escape to worlds and relations outside. Yet many political organizations are using the Internet to advance their struggles and to connect people with real-world issues. Many labor organizations are also beginning to make use of the new technologies. Mike Cooley (1987) has written of how computer systems can reskill rather than deskill workers, while Shoshana Zuboff (1988) has discussed the ways in which high tech can be used to "informate" workplaces rather than automate them, expanding workers' knowledge and control over operations rather than reducing and eliminating it. The Clean Clothes Campaign, a movement started by Dutch women in 1990 in support of Filipino garment workers has supported strikes throughout the world, exposing exploitative working conditions (see the web site at http://www.cleanclothes.org/1/index.html). In 1997, activists involved in Korean workers' strikes and the Merseyside dock strike in England used web sites to gain international solidarity (for the latter see http://www.gn.apc.org/labournet/docks).[6]

Labor organizations, such as the North South Dignity of Labor group, note that computer networks are useful for coordinating and distributing information, but cannot replace print media that is accessible to more of its members, face-to-face meetings, and traditional forms of political struggle. Thus, the trick is to articulate one's communications politics with actual political movements and struggles so that cyberstruggle is an arm of political battle rather than its replacement or substitute. The most efficacious Internet organizing has indeed intersected with real movements

6. For an overview of the use of electronic communication technology by labor, see Moody 1988; Waterman 1990, 1992; and Brecher and Costello 1994. Labor projects using the new technologies include the U.S.-based Labornet, the European Geonet, the Canadian L-Net, the South African WorkNet, the Asia Labour Monitor Resource Centre, Mujer a Mujer, representing Latina women's groups, and the Third World Network, while PeaceNet in the United States is devoted to a variety of progressive peace and justice issues. On technopolitics and progressive political uses of new technologies, see Kellner, forthcoming.

ranging from campaigns to free political prisoners, to boycotts of corporate projects, to actual political conflict, as noted above.

Hence, to capital's globalization from above, cyberactivists have responded by attempting to carry out globalization from below, developing networks of solidarity and circulating struggle throughout the globe. To counteract the capitalist international of transnational corporate globalization, a Fifth International of computer-mediated activism is emerging, to use Waterman's phrase (1992), that is qualitatively different from the party-based socialist and communist internationals. Such networking links labor, feminist, ecological, peace, and other progressive groups providing the basis for a new politics of alliance and solidarity to overcome the limitations of postmodern identity politics (on the latter, see Best and Kellner 1997 and forthcoming).

Moreover, a series of conflicts around gender and race are also mediated by new communications technologies. After the 1991 hearings in the United States on Clarence Thomas's fitness to be Supreme Court justice, Thomas's assault on claims of sexual harassment by Anita Hill and others, and the failure of the almost all-male Senate to reject the obviously unqualified Thomas prompted women to use computer and other technologies to attack male privilege in the political system in the United States and to rally women to support women candidates. The result in the 1992 election was the election of more women candidates than in any previous election and a general rejection of conservative rule.

Many feminists have now established web sites, mailing lists, and other forms of cybercommunication to organize support for their cause. Likewise, African American insurgent intellectuals have made use of broadcast and computer technologies to advance their struggles. John Fiske (1994) has described some African American radio projects in the technostruggles of the present age and the central role of the media in recent conflict around race and gender. African American "knowledge warriors" are using radio, computer networks, and other media to circulate their ideas and counterknowledge on a variety of issues, contesting the mainstream and offering alternative views and politics. Likewise, activists in communities of color—like Oakland, Harlem, and Los Angeles—are setting up community computer and media centers to teach the skills necessary to survive the onslaught of the mediazation of culture and computerization of society to people in their communities.

Obviously, right-wing and reactionary groups have used the Internet to promote their political agendas. One can easily access an exotic witch's brew of ultraright web sites maintained by many Ku Klux Klans, myriad neo-Nazi groups including Aryan Nations, and various militia groups. Internet discussion lists also promote these views and the ultraright is extremely active on many computer forums, as well as their radio programs and stations, public access television programs, fax campaigns, video, and even rock music production. These groups are hardly harmless, having promoted terrorism of various sorts ranging from church burnings to the bombings of public buildings. Adopting quasi-Leninist discourse and tactics for ultraright causes, these groups have been successful in recruiting working-class members devastated by the developments of global capitalism, which have resulted in widespread unemployment for traditional forms of industrial, agricultural, and unskilled labor.

The Internet is thus a contested terrain, used by Left, Right, and Center to promote their own agendas and interests. The political battles of the future may well be fought in the streets, factories, parliaments, and other sites of past conflict, but politics today is already mediated by media, computer, and information technologies and will increasingly be so in the future. Those interested in the politics and culture of the future should therefore be clear on the important role of the new public spheres and intervene accordingly.

Now I would argue that these new modes of technological experience are at least positive supplements to interactions with nature, objects, and human beings and should not be posited as antithetical and simply dismissed as hyperreal, hyperactive, or some such negative valuative term. Such a dismissal would prematurely close off potentially exciting and life-enhancing new realms of experience and expansions of "reality," as I would put it. And so I would question Borgmann's real/hyperreal distinction and negative valorizations of technologically mediated activity, arguing that the new realms of cyberspace have positive potentials, as well as dangers and limitations. The challenge, then, is to use technology in life-enhancing, fulfilling, and socially progressive ways, to create a better society and mode of life. And I believe that for better or worse technology is our fate, that it is an inexorable force that is dramatically changing every aspect of our lives, and that it challenges us to devise ways to make it more life enhancing.

Borgmann, however, sometimes totalizes technology in negative terms, as when he writes, "If we agree to call this distinctive approach to the reordering of the world modern technology, we should put the challenge to postmodernism by asking whether postmodernism will be more than technology by other means" (1992, 80). What does this mean? Are there overtones here of Ellul's technique that takes over, becoming totalitarian, of autonomous technology coming to totally dominate us? Is technology merely a negative mode of domination, of an oppressive ordering that is counterposed against a "good" realm of nature, community, and reality?

I have argued for a deconstruction of the opposition real/hyperreal with the latter stigmatized tout court as inferior, deficient, and even harmful—though in some cases it might be. In the spirit of pragmatism, I have been attempting to distinguish between life-enhancing and -diminishing examples of uses of technology, as well as progressive and reactionary uses, and provided some examples of what I considered positive uses of new technologies.

Borgmann, however, might very well dismiss my examples and arguments as cases of what he calls "hyperintelligence," which he claims "is obviously growing and thickening, suffocating reality and rendering humanity less mindful and intelligent" (1992, 108). On the other hand, Borgmann positively valorizes an "active intelligence" in intimate contact and interaction with reality and I am arguing that cyberspace interaction could be interpreted as such an example of "active intelligence" that I would consider just another mode of reality, another realm of experience, and not an inferior or debased form of hyperreality.

We need to articulate a standpoint of critique, from which one can make distinctions between positive and negative uses of technology. I would suggest that those forms and uses of technology that enhance positive values such as democracy, community, freedom, self-development, and the like can be deemed meritorious, while those forms and uses of technology that promote domination and oppression, or subvert democracy, community, freedom, creativity, and other positive values, can be considered blameworthy. Of course, often one cannot make such a clear distinction, there can be unintended consequences of introducing technologies, and technologies are often highly ambivalent. Yet, it is a mistake, I believe, to dismiss technology per se as dehumanizing or life negating and to valorize only nontechnological activities and interaction as genuinely focal things and practices.

In conclusion, I want to return to the question of the postmodern divide and the issue of whether we have or have not crossed over such a divide.

Between the Modern and the Postmodern

And so we come to the question of whether we have crossed a postmodern divide, or, as I would argue, we are currently dwelling in a liminal stage between the modern and the postmodern. Postmodern theorists, like Baudrillard (1993), postulate that technology has propelled us into a brave new world of hyperreality, simulation, and the ecstasy of communication that constitutes the catastrophe of modernity and an entirely new postmodern technoscape, which he and his followers describe in hyperbolic terms. Borgmann is more modest in his concept of a postmodern divide that he mainly sketches out in terms of a paradigm shift most visible in philosophy. My own position is that there is a postmodern turn in theory and culture under way and something like a new postmodern society in the making, but I would argue that we haven't yet crossed over to the other side, and are existing between the modern and the postmodern (see Best and Kellner 1997 and forthcoming).

The radicality of postmodern discourse is the claim that we have entered a dramatically novel era that requires entirely new theories and politics. These postmodern theories posit an extreme break and rupture with modern culture and society, assuming that the historical epoch of modernity is over and that we are living in an entirely new social order, a postmodernity. The claim is analogous to the earlier modern arguments that the Enlightenment broke with the unreflective, childlike past (Kant), or that the French and industrial revolutions produced an altogether new modern society radically different from traditional society. Modern social theory arose as an attempt to describe the emergent modern societies. Classical modern theorists like Marx pointed out that for the first time in history, life was organized around the production (and later consumption) of commodities. Industrial men, women, and children were becoming species of a new type of *homo faber* (literally toiling animals), who were condemned to novel modern forms of misery and servitude. Nietzsche described in turn the innovative social forms of modern society and the rise of a modern state and mass society, while classical social theorists like Durkheim and Weber described the forms of social differentiation, rationalization, and secularization characteristic of modern societies (see Antonio and Kellner 1992).

Some postmodern theorists claim that a rupture has taken place in history every bit as great as the divide between modern and premodern societies. The break is described by postmodernists as the transition from modernity to postmodernity, by Marxists as the restructuring of global capitalism and the emergence of a new regime of post-Fordist accumulation and transnational capitalism, and by certain sociological theorists as the move to a "postindustrial" or "information society." A wide range of theorists interpret the contemporary moment in terms of specific mutations of the economy, polity, society, and culture, described in a variety of competing vocabularies. There are thus a proliferation of new discourses that have attempted to capture the novelty of the present moment, of which postmodern theories are the most prominent.

Although it is prudent to be skeptical of extreme postmodern claims that would render obsolete the assumptions, values, categories, culture, and politics of the modern era, it must be admitted that significant changes are taking place and that many of the old modern theories and categories can no longer adequately describe our contemporary culture, politics, and society. Whereas the modern era swept in unprecedented forces of secularization, rationalization, commodification, individualization, urbanization, nationalism, bureaucratization, and massification, the last three decades have seen the decline of the nation-state, decolonialization, explosions of ethnicity and fundamentalism, the rise to unparalleled power of media culture, the revolutionizing force of computers, and the spread of new forms of virtual reality and hyperreality described in new postmodern theories and literature like cyberpunk.

And yet the extreme claims for a postmodern break and rupture do violence to our sense of enduring continuities with the past and the fact that many ideas and phenomena that are claimed to be "postmodern" have their origins or analogues precisely in the modern era. Consequently, I would argue that we are living between a now aging modern era and an emerging postmodern era that remains to be adequately conceptualized, charted, and mapped. In Bernstein's (1991) appropriation of Benjamin and Adorno, we are living in a "new constellation" of changing elements that are irreducible to a common denominator, in a force field of dynamic interplay of the old and new.

Historical epochs do not rise and fall in neat patterns or at precise chronological moments. Perhaps our current situation is parallel in some ways to the Renaissance, which constituted a long period of transition between the end of premodern societies and the emergence of modern ones. Such periods are characterized by conflicts between the old and the new and the birth pangs associated with the eruption of a new era. Indeed, change between one era and another is always protracted and contradictory and usually painful. But the sense of "betweenness," or transition, requires that one grasp the continuities with the past as well as the novelties of the present and future. Thus, it is also important to capture the continuities with the modern, as well as discontinuities, in order to make sense of our current predicament.

Living in a borderland between the old and the new creates tension, insecurity, and even panic, as well as excitement and exhilaration, thus producing a cultural and social environment of shifting moods and an open but often troubling future. The discourse of the postmodern is therefore deeply implicated in the hopes and fears of the present and is an important component of our current situation. Thus, the ubiquity of the term "postmodern," its constant proliferation, its refusal to fade away, and its seeming longevity—several decades is a long time for a mere fad in our rapidly changing world—suggest that it is addressing contemporary concerns in a useful way, that it illuminates certain contemporary realities, that it resonates to experience, and that it is an important part of the contemporary critical lexicon that one has to come to terms with one way or another.

Consequently, it would be a mistake merely to dismiss the discourse of the postmodern out of hand as a mere fad or ephemeral fashion. Although many proclaim that the phenomenon is over, there continue to be waves of books, articles, conferences, and proliferation of postmodern discourse. People continue to feel passionately about the postmodern, and the discourse obviously speaks to important changes in our culture and society and by now has acquired a certain weight. Postmodern theory has penetrated almost all academic disciplines, producing critiques of modern theory and alternative postmodern theoretical practices in philosophy, social theory, politics, economics, anthropology, geography, science, and just about every academic field (see, for instance, the survey of postmodern modes of inquiry in Dickens and Fontana 1994 and the discussion of the postmodern paradigm shift in Best and Kellner 1997, chap. 6). Groups and individuals marginalized in the society, culture, and university have taken the term as their own and use it to oppose the established order of things. Since many of these individuals are younger, one expects that the discourse will continue to be used for some time to come.

In addition, the discourse is remarkably flexible and open, and individuals can use it to promote a lot of different agendas, as well as a lot of babble and gibberish. Although some discourse of the postmodern (e.g., that of Baudrillard and his followers) is exceedingly cynical and ultraskeptical, those who wish to promote religion have also used the discourse to attack modernity and Enlightenment rationalism (e.g., Smith 1982), as has Borgmann. Individuals have used it to promote a tremendous variety of theoretical and political agendas, producing a bewildering cacophony of postmodern discourse. Many have invested cultural capital in promoting the turn, while others are invested in attacking it. The discourse of the postmodern is thus an integral part of the contemporary scene and will, I believe, be with us for a long time to come. Therefore, we must take it seriously and engage it critically.

It is the merit of Borgmann's recent work to engage seriously and critically the postmodern turn and to construct a positive normative ethical and political version of postmodern theory that provides provocative perspectives on the contemporary era. He carries out a sustained and radical critique of modernity and projects positive postmodern alternatives that challenge us to focus on what is most important and fundamental to our lives. Thus Borgmann requires us to question anew the role of technology in the present moment and to develop appropriate ethical and political responses to its presence and demands.

REFERENCES

Antonio, Robert J., and Douglas Kellner. 1992. "Metatheorizing Historical Rupture: Classical Theory and Modernity." In *Metatheorizing*. Ed. George Ritzer. New York: Sage.

Baudrillard, Jean. 1993. *Symbolic Exchange and Death*. London: Sage.

Berman, Marshall. 1982. *All That Is Solid Melts in the Air*. New York: Simon and Schuster.

Bernstein, Richard. 1991. *The New Constellation*. Cambridge: Polity Press.

Best, Steven, and Douglas Kellner. 1991. *Postmodern Theory: Critical Interrogations*. London: Macmillan; New York: Guilford Press.

———. 1997. *The Postmodern Turn*. New York: Guilford Press.

———. forthcoming. *The Postmodern Adventure*. New York: Guilford Press.

Borgmann, Albert. 1984. *Technology and the Character of Contemporary Life: A Philosophical Inquiry*. Chicago: University of Chicago Press.

———. 1992. *Crossing the Postmodern Divide*. Chicago: University of Chicago Press.

Brecher, Jeremy, and Steven Costello. 1994. *Global Village or Global Pillage: Economic Reconstruction from Bottom Up*. Boston: South End Press.

Bronner, Stephen Eric, and Douglas Kellner, eds. 1989. *Critical Theory and Society: A Reader*. New York: Routledge.

Cooley, Mike. 1987. *Architect or Bee? The Human Price of Technology*. London: Hogarth.

Dickens, David R., and Andrea Fontana. 1994. *Postmodernism and Social Inquiry*. New York: Guilford Press.

Fiske, John. 1994. *Media Matters*. Minneapolis: University of Minnesota Press.

Horkheimer, Max, and Theodor Adorno. 1972. *Dialectic of Enlightenment*. New York: Continuum.

Jameson, Fredric. 1984. "Postmodernism: The Cultural Logic of Late Capitalism." *New Left Review* 146:

———. 1991. *Postmodernism, or The Cultural Logic of Late Capitalism*. Durham: Duke University Press.

Kellner, Douglas. 1984. *Herbert Marcuse and the Crisis of Marxism*. London: Macmillan; Berkeley and Los Angeles: University of California Press.

———. 1989a. *Critical Theory, Marxism and Modernity*. Cambridge: Polity Press; Baltimore: Johns Hopkins University Press.

———. 1989b. *Jean Baudrillard: From Marxism to Postmodernism and Beyond*. Cambridge: Polity Press; Palo Alto: Stanford University Press.

———, ed. 1989c. *Postmodernism/Jameson/Critique*. Washington: Maisonneuve Press.

———, ed. 1994. *Jean Baudrillard: A Critical Reader*. Oxford: Basil Blackwell.

———. 1995. *Media Culture*. London: Routledge.

———. forthcoming. "Rethinking Literacy in a Multicultural Society." *Educational Theory*.

Light, Andrew. 1995. "Three Questions on Hyperreality." *Research in Philosophy and Technology* 15:211–22.

Moody, Kim. 1988. *An Injury to One London*. London: Verso.

Smith, Huston. 1982. *Beyond the Post-modern Mind*. New York: Crossroad.

Stoll, Clifford. 1995. *Silicon Snake Oil: Second Thoughts on the Information Highway.* New York: Doubleday.

Waterman, Peter. 1990. "Communicating Labor Internationalism: A Review of Relevant Literature and Resources." *Communications: European Journal of Communications* 15, no. 1/2: 85–103.

————. 1992. *International Labour Communication by Computer: The Fifth International?* Working Paper Series 129. The Hague: Institute of Social Studies.

Zuboff, Shoshana. 1988. *In the Age of the Smart Machine: The Future of Work and Power.* New York: Basic Books.

FOURTEEN

Technology and Temporal Ambiguity

Mora Campbell

A BEGINNING

Temporal ambiguity is experienced when, in answer to the question of a given moment, "What time is it?" a plurality of synchronous answers can be given, each as compelling as the next. *Linguistic ambiguity* refers to the simple fact that many words and phrases admit of several possible meanings. Ambiguous speech—speech that is linguistically ambiguous—is often negatively associated with equivocation or the avoidance of commitment through using terms that are open to many interpretations. Generally, the temporal and spatial context for the use of a phrase such as "I will care for you," whether said to a spouse on a wedding day or to a stranger in a car accident, resolves ambiguity.

Temporal ambiguity and concomitant questions of commitment are more difficult to resolve. Given interpretations or answers to the question "What time is it?" are the very contexts for clarifying one's intent. For instance, on any given day, I wake up and ask myself, what time is it? The answer may be, it's 7:00 A.M., November 11, 1998. Time to get up, take a shower, and go to work. I may also answer that it is the first anniversary of my grandmother's death; a day to remember and celebrate her life with other family members. Or I may answer that it is the day of the rally downtown to protest government changes to the educational system.

In the case of linguistic ambiguity, choosing one interpretation of a word over another generally pushes other possible interpretations aside. In cases of temporal ambiguity, choosing one description does not rule out the others. If I choose to go to work, the ambiguity doesn't get resolved. It will still be compellingly true that today is the anniversary of my grandmother's death and a day of protest, and that I honor my intentions and commitments in relation to both of these.

For Albert Borgmann and Donna Haraway, theorists of technology, ambiguity is a fundamental characteristic of life in technological society.

Borgmann describes technology as the "characteristic and constraining pattern to the entire fabric of our lives" (Borgmann 1984, 3). He calls this pattern the "device paradigm" and argues that one of its defining features is the erasure of the contexts out of which the prevalent things in our lives, such as food or heat, arise. Haraway argues that *we are* technology. She uses the metaphor of the cyborg to illustrate how we are, in our collective fictional imaginations and lived experiences, "shot through" with technology.

Borgmann and Haraway differ quite radically, however, with regard to their views on the moral efficacy of ambiguity. Borgmann argues that if we are to meet our commitments to such things as social justice and the ending of environmental degradation, then the ambiguities in our experience must be resolved through the creation of contexts of significance and meaning. Haraway argues that the physical and existential ambiguity of the cyborg allows us to meet these very same commitments through the blurring of culturally guarded boundaries between humans and animals, organisms and machines, men and women, and so on, which underpin social injustices and the unjust treatment of nonhuman beings and places. This paper brings Borgmann and Haraway into conversation to examine the relationship between technology and temporal ambiguity and, ultimately, the question of the moral efficacy of ambiguity. Ultimately, their accounts serve as correctives to one another, but both would be strengthened by taking questions of individual and collective experiences of temporality more seriously.

TENSE AND CONTRADICTORY CONDITIONS

After giving birth to my second son, Timothy, I shared a hospital room with another woman and baby, her first child. Her mother was there often, giving advice on how to hold, burp, change, and feed the baby; "Feed him every three hours; you'll never be able to lead a normal life if you don't get him on a schedule." Sharing this room was difficult. The baby cried a lot. I just wanted to be alone and quiet with Tim.

That was seven years ago. Earlier this week Tim had the flu. Neither he nor my other son, Chad, had been sick for over a year. I'd forgotten about "the life of a fever." It started Monday afternoon after he got home from school. Uncharacteristically, he lay down on the couch and just wanted to read stories until even that seemed like too much. I looked at him closely and realized that his eyes were glassy and cheeks flushed. I took his temperature. It was 102. I felt bad. I'd been so impatient with him the past few days when he'd just played with his food at the supper table and was generally grumpy. If I'd been paying attention, I'd have noticed he was starting to not feel well. But I was so busy, focused on keeping up with work and home tasks.

With Tim sick I had to slow down, march to a different drum: the rhythm of a fever. Each morning, around five, he'd wake up, his temperature high. By seven or eight, after Tylenol, juice, and rest, his temperature was normal, and all Tim wanted to do was play. By two in the afternoon he'd be bouncing off the couch or practicing his karate moves. At these times, all the work I had to do would come crashing in as I wondered if he wasn't well enough, after all, to be in school. But sure enough, late afternoon came and the fever would return. I'd console myself, rather guiltily, that I'd get some work done in the evening because, at least, sick kids fall asleep easily and early. With undone work piling around me like heaps of laundry, Thursday was the hardest. Tim didn't have a fever, but I didn't send him to school, recalling my mother's rule: "You don't go to school till you haven't had a fever for a full day"; while often predictable, fevers can shift course.

Tim went to school today. But he was pale and his eyes looked so dark. I asked his teacher to call if he seemed like he was fading. Hating to admit it, I hoped she wouldn't call. Back at work, I was right into it; calling people when I knew they weren't there so I could leave quick messages; firing off e-mails with a minimum of punctuation; asking students to leave assignments under my door because I'd be sure to see them when I stepped on them. Home late and tired, I took Chad and Tim to McDonald's. Tim likes the Happy Meal. I have to stay focused, keep afloat in the other and "larger" rhythms of which I'm a part: tenure schedule, lecture and exam schedules; these tied to other work cycles. Professors are often heard saying that if their students can't meet paper and exam deadlines, they'll never cope in the real world of holding down a job.

Two weeks ago I gave a midterm in my large undergraduate environmental ethics class. I was almost late getting there. We had a snowstorm that produced, as it turns out, the largest amount of snow on this date for sixty years. But nothing seems to stop this city. Students were both kindly and selfishly passing throat lozenges around. It was distracting with so many people coughing. One student, wearing a tank top because she was so hot, actually moved out into the hall to finish writing because she couldn't stop hacking. I told her she could write the make-up test next week, but she declined: "If I stop now, the work will pile up, and I'll be even worse off later."

It is at times like these that I experience temporal ambiguity. Like an ambiguous word or expression, each moment admits of several possible interpretations. *What time is it?* Midwinter, late 1990s. A time of illness. Time of my small child just becoming ill. Midterm writing and marking time. These times are synchronous, yet not at ease with one another. Each time calls forth its own particular context and set of commitments. Is keeping all interpretations alive at once hedging commitment to any one of them?

In her *The Ethics of Ambiguity*, Simone de Beauvoir argues that life is ambiguous. We live in a tense and contradictory condition as beings who are both subject and object, consciousness and body. Further, ambiguity suggests that there can never be one rationally fixed meaning or ethical code for events. Instead, in life there is a profusion, even an excess, of meaning. Moral limits can never be fixed with complete seriousness. Seriousness, she says, is the most terrible affliction of the oppressed wherein worldly conditions appear natural and inevitable. We can't know, and that's the point of ambiguity. All anyone can do is try to be honest with themselves and others about what their intentions are (de Beauvoir 1948, 27).

REFORMING PERVASIVE AMBIGUITY?

In the first part of *Technology and the Character of Contemporary Life*, Borgmann reflects on ambiguity in outlining his approach to examining technology. He writes that he is doing *philosophy* of technology in a traditional way. This is a philosophy close to the Aristotelian sense of "*theoria,* the calm and resourceful vision of the world." Philosophy has not been widely practiced in this way for some time; rather,

> [t]heoria was eclipsed with the rise of the modern period, and ambiguity befell all eminently theoretical endeavors. Language is negatively ambiguous if it exhibits a disorienting or debilitating plurality of senses. In the everyday world a pervasive negative ambiguity makes itself felt in the suspicion and diffidence with which ambitious questions and assertions are met. Words of beauty are suspected of naiveté, words of salvation are thought to conceal egotism, words of profoundness are charged with obscurantism. The mere plurality of senses that attaches to every word is a prosaic matter, apparent in dictionaries, and normally counterbalanced by the resolving force of the context of discourse. But no such context seems to be at hand when weighty matters are at issue. Instead more and more claims pour forth, eroding and submerging all points of orientation. (Borgmann 1984, 6)

The response to this pervasive ambiguity, particularly among philosophers, has not been a concern with theory, "in the sense of a steady view of the world." Rather, it has been characterized by a seemingly endless and inconclusive epistemological turn toward questions of whether and how we can form theories at all. "It is for now," writes Borgmann, "simply a fact that the predominant response to ambiguity is not a desire to be open for what speaks with simple and salutary authority but the desire to gain authority over ambiguity by getting hold of its controlling conditions" (Borgmann 1984, 7).

Borgmann reflects further on ambiguity in its current manifestations in new communications technologies and the ways in which they blur distinctions between the actual and virtual or physical and textual. Speaking about computer-simulated intelligence, and the selves that people present and encounter on communication networks like the Internet, he writes:

> The confusion between textual and personal intelligence is more than a conceit of workers in artificial intelligence. It is becoming a force in *the* real world. Because no textual intelligence worthy of confusion with personal intelligence exists, people reduce themselves to textual intelligence and offer each other such reduced and ambiguous intelligence on communication networks. Far from finding the inconclusiveness of textual intelligence annoying, people take delight in pouring their aspirations and desires into the margins of discretion that a text allows for and in fact requires. It is as though people want to withdraw from the reach and presence of personal intelligence to indulge themselves with the pliability and ambiguity of textual intelligence. (Borgmann 1994, 282)

Writers such as Howard Rheingold argue that the new communications technologies are radically innovative and pathbreaking insofar as they democratize access to information and communication, and allow for the creation of new kinds of community (Rheingold 1991). Borgmann regards them as just another instance of the device paradigm (Borgmann 1984, 40–48); and the "radical" move we are making in embracing these technologies is simply to become ever more entrenched in this governing pattern. What permits the blurring of distinction between the virtual and the real, the textual and the physical on communication networks is lack of context, such as could be afforded by a face-to-face conversation. Negation of context is the defining feature of the device paradigm.

To highlight the device paradigm and its pervasiveness, Borgmann contrasts a *device* with a *thing*. Devices make things such as light, heat, and water available to us without the burdens of complex social relations, the need to understand where things come from and how they work, hard physical labor, or real dangers like a well freezing or wood house catching fire, they also deskill us. The fewer skills one possesses, the narrower their understanding of and involvement in the natural and social world. For Borgmann, the social and physical engagement required to learn and practice a skill like making a canoe or preparing a holiday meal opens us up to the cultural, historical, and natural dimensions of the world of canoeing or a celebration. This is the realm of focal things and practices.

Focal things and practices are, however, scattered and extremely fragile in a world of consumption and disengagement as ordered by the device paradigm. Many or most of the activities in our lives are like the procurement of heat, whereby a concealed machinery procures commodities. We see this in fairly small things such as televisions, watches, and radios, which procure things like entertainment, information, and time. We do not have to understand either the devices themselves or all the lives and labor that went into making what they generate, such as "the evening news." Borgmann argues that even larger things like government and the insurance industry are governed by the device paradigm insofar as they too have become a concealed machinery that, without engaging us, delivers things like policies or security in the form of an insurance check.

Borgmann wonders how superficial or "thin and disembodied" commodities can "become before the tie is ruptured that connects them with the things from which they are derived and from which their significance continues to draw nourishment" (Borgmann 1984, 55). In the case of new communications technologies like the web and hyperrealities, the answer is obvious as more and more people spend copious amounts of time interacting with textual, and often scripted, other selves on listserves, or play virtual golf, pool, football, or what have you. As we strive to create ever more brilliant, encyclopedically complete, and pliable virtual realities where we can go anywhere and know anything with the touch of a few buttons, computers are becoming "world-devices." They lie in the hidden machinery that procures "life and all it has to offer," with only a small range of skills and a small amount of strength and attention required (Borgmann 1992, 87–88).

The reform of technology, of the increasingly "thin and disembodied" character of our lives, lies in appreciating all those things that, like the mountain in his distinction between backcountry skiing and the experience of the indoor simulated Skiorama, possess "powerful presence and a vigorous continuity with the world at large" (Borgmann 1995, 42). A cathedral, a clay pot, or garden can be such a thing. Regular practices like hiking in the forest or gardening preserve focal things and order our lives in the worlds they articulate. A tomato plant brings us into its world of responsiveness to the seasons, to the sun, the rain, the characteristics of the soil.

Moreover, in a regular practice like hiking, our lives can cease to mirror the device paradigm. Rather than hiking being seen as a means to an end like health, in the context of something we enjoy, like nature, the forest becomes an orienting force, the place from which significance radiates into the rest of our lives. Here technology ceases to be the informing pattern of our lives and instead (in the form of such things as hiking boots and gortex jackets), technology is put in the service of our commitments. A larger societal clearing of significance in focal things can be found as more and more people engage in focal practices and come together in appreciation of the differences and commonalties of commitment expressed in the contexts of these practices.

There is considerable elegance and hope in the simplicity of Borgmann's proposals for the reform of technology through focal things and practices. There are, however, in temporal, gender, and cross-cultural terms, several difficulties. Time is, in many societies worldwide, generally structured by the divide between public work and school time on the one hand and private home and leisure time on the other. Further, this temporal pattern is, in most Western countries, set within the pattern of the Gregorian calendar, which remains one of the dominant arbiters of when there is time off from work and school. All of the focal practices that Borgmann names in his various writings, from running, family meals, gardening, music making, and hiking to baseball and religious celebration take place within the private side of this social divide (Zerubavel 1981, 1985).

Insofar as focal practices take place within private or leisure time, they are what you squeeze into your lunch hour or do when you have time left over from work. Public work and school time remains the dominant societal schedule to which other rhythms must accommodate themselves. The larger rhythms of significance in which focal practices can situate our lives, such as the seasons, are overshadowed by this dominant schedule. Moreover, the dominant societal schedule mirrors the device paradigm. Work is a means to the end of leisure, while leisure can easily become a means to the end of working more effectively. Unless focal practices serve to shift this overall pattern, they cannot, in temporal terms, significantly reorient the context of our lives. How little seasonal patterns, for instance, orient our lives is evident in the fact that few cities and workplaces shut down except in the case of storms that are literally life threatening. Moreover, in the summertime, while city dwellers pray for continuous days of sunshine, the farmers who provide their food may be praying for rain to end a drought (Campbell 1990, 1992).

Furthermore, a cathedral can be a site of orientation and significance in a large Western city because, under the rhythm of the Gregorian calendar, its presence is made commanding through the cycle of recurring events such as Easter and Christmas that "light it up," so to speak. A Buddhist or Hindu temple, for instance, will rarely have the same commanding presence insofar as their events are not "lit up" in our calendar. Temporal ambiguity must, at times, be fierce for those in Western societies whose work, school, and religious cycles continually overlap and conflict. Respect for difference will truly be realized only when, in actual practice, multiple temporal orientations (and the concomitant commitments they call forth) are acknowledged in the ways in which time is structured in our societies.

Furthermore, while Borgmann clearly shows the pattern of technology in the split between machinery and commodity, and the corresponding division of our lives into foreground and background, he does not deal with other divisions that also lie at the heart of patterns of domination, such as gendered ones or those between humans and animals, society and nature. For instance, in speaking of "communities of celebration," Borgmann writes, "I want to use [a notion of] community . . . that is close to Robert Bellah's notion of a community of memory and of practices of commitment, and which refers to a group of people who are in one another's bodily presence and engaged in a common enterprise" (Borgmann 1990, 320). The examples of communities of celebration he provides include the enactment of familial traditions, athletics, and religious practice. However, in his accounts of how these traditions can act as sites of orientation, gender is never problematized. In sports and familial and religious practices, women have either been excluded in significant ways or subordinated to roles defined for them by "their nature." Borgmann would wholeheartedly agree that gender roles must be brought into consideration; but he does not explain how traditions can, at one and the same time, serve as sites of significant orientation and also of vehement negotiation.

Finally, while in focal practices we are led to appreciate natural things as other, commanding of our respect, and as that which can focus our commitments in the continuity of rhythms like the seasons, distinctions between ourselves and animals, for instance, are kept intact. We are not necessarily led to an appreciation of our continuity with nature.

EMBRACING AMBIGUITY?

Haraway's account of technology addresses more adequately questions of difference among people and between people and nonhuman beings than does Borgmann's. Her "cyborg manifesto" begins with the assertion that it is "an effort to build an ironic political myth faithful to feminism, socialism, and materialism." Irony, she writes, "is about contradictions that do not resolve into larger wholes, even dialectically, about the tension of holding incompatible things together because both or all are necessary and true" (Haraway 1991, 149).

The cyborg is Haraway's ironic political myth. Through this myth she illustrates that technologies are not something that we can simply choose to place in the service of our commitments to such things as ending environmental degradation or promoting equality among people. Our technologies are us. We are them. Whether we are in labor and hooked up to a fetal monitor, playing with our children and their action figure toys that are part machine, part animal, and part human, indulging in "X-Files" fearful fantasies of aliens probing our bodies with instruments, driving a car, surgically altering our bodies, sending a fax, taking hormone therapy, or surfing the net, we are cyborgs. And yet, she states in an interview, "[t]he cyborg is a reminder, and it's not a friendly reminder, that there wasn't much choice about this. The argument was—and it's an argument I still hold onto—that historically, discursively, physically . . . [w]e are embodied in locations that have nothing to do with choice" (Darnovsky 1991, 68).

The cyborg is the very offspring of militarism and patriarchal capitalism. While we did not choose to be cyborgs, Haraway argues that those committed to equality and respect between humans and the rest of nature, and women, in particular, must choose to name themselves as such. The cyborg is a metaphor for the ambiguous "disassembled and reassembled, postmodern collective and personal self [that] feminists must code" (Haraway 1991, 205). The ambiguity results from the fact that the cyborg is a split being: human and animal, organism and machine. Yet this very ontological ambiguity serves to call into question the dualisms of civilization and nature, mind and body, men and women, human and animal, public and private, and organism and machine (Haraway 1991, 163).

In particular, the cyborg cuts through the "Gordian Knot tying women together with nature" and with it, hierarchical and dualistic assumptions that render that which is natural as inferior and subordinate to that which is considered rational, civilized, noninstinctive. Beings who experience, "intimately," the boundaries between human and animal, organism and machine, guarded boundaries that preserve anthropocentrism, sexism, racism, cyborg tongues can speak of other ways of living (Haraway 1991, 181).

The cyborg is the very embodiment of Haraway's epistemology of situated knowledges. She develops this notion out of two concerns. The first is to escape the controlling dyad of universalism and relativism that characterizes knowledge production in our time. Universality is the god-trick, as she describes it, of searching for the ideas, precise and encapsulating, that are true for all people at all times. Relativism is the god-trick of walking on water; any thing that anyone says can be walked on because it's all relative. The second concern is to rethink feminism and environmental politics through overcoming, in particular, philosophies and politics of feminism that search for places of innocence from which to speak: as laborer, mother, colonized person, organic farmer—places that are often touted as "closer to nature" and demonize sites of technology and science. Our positions and identities are clearly much more complicated than this: "Splitting, not being, is the privileged image for feminist epistemologies" (Haraway 1991, 193). The cyborg, in its very fluidity, tending in multiple directions and depths, creates the "possibility of webs of connection called solidarity in politics and shared conversations in epistemology" (Haraway 1991, 91), including the possibility of shared conversations with other beings to whom we are kin.

Haraway is, however, no simple optimist. Just as we are, in our being, "shot through" with technology, so too then are we riddled with militarism and patriarchal capitalism. With threats of a Stars Wars apocalypse, and increasing genetic modifications of living beings, cyborgs are implicated in prevailing social and political practice and are deadly serious. But as de Beauvoir said, seriousness, particularly as passive acceptance, is the most terrible affliction of the oppressed, wherein all worldly conditions appear natural and inevitable. For Haraway, the cure lies in the capacity to imagine the world as otherwise. As trickster beings, ones who can shape-shift into various gendered configurations of machine, animal, human, and organism, cyborgs are powerfully playful. What preserves this capacity is the postmodern virtue of ambiguity. A term that once evoked images of chaos, contracts you'd never sign, and the analytic sin of equivocation, ambiguity is now a term rich in its description of the fundaments of existence and a prescription for a way of being and deciding. In their very being and alliances, cyborgs confound classification. They are deeply ambiguous.

Katherine Hayles speculates that approximately ten percent of the U.S. population are, in the "technical sense," cyborgs. The percentage includes those with prosthetic limbs, artificial joints, hearing aids, drug implants, and faces modified by plastic surgery. However, she writes that the modifications that are really at issue are in the cybernetic reconstitution of the human body. These are in the move from "biomorphism to technomorphism," in the "psychic/sensory" reorganization occurring in the conjoining of the human sensorium with computer memories in complex feedback loops. Here the computer or video screen as boundary between user and data breaks down to create a new place, "cyberspace" (Hayles 1994, 178).

In cyberspaces or virtual realities, the cyborg is not just a being with attachable and detachable chemical and mechanical parts, but "body-plus-equipment-plus-computer-plus-simulation." Hayles sees warning signs in the fact that virtual reality technologies will likely be available only to a few. In the collusion of genetic/bioengineering and cybernetics, "Having an unmodified body will be like having a working-class accent; it will mark you as cannon fodder for the system" (Hayles 1994, 182).

And yet Hayles argues, like Haraway, that virtual reality both reinforces patterns of commodification and domination characteristic of late capitalism, while also subverting them through inaugurating new links between people in cyber-communities and new ways of seeing the world that can foster the deep appreciation of difference. We might, she writes, turning to the example of a "data puppet," commonsensically suppose that the subject in the cybernetic "body-plus-equipment-plus-computer-plus-simulation" is our body-self, while the object is the puppet. Yet information flow is a feedback loop; data puppet is subject too. Hayles calls this ambiguity one of cyberspace's most "disturbing and arresting features." The usual body boundaries that serve to mark *self* and *other* begin to dissolve, and also perhaps some of the damaging differences between our selves and other beings that so marks daily life. She writes:

> One can imagine scenarios in which the Other is accepted as both different *and* enriching, valued precisely because it represents what cannot be controlled and predicted. The puppet then stands for the release of spontaneity and alterity within the feedback loops that connect the subject with the world, as well as with those aspects of sentience that the self cannot recognize as originating from within itself. At this point the puppet has the potential to become more than a puppet, representing instead a zone of interaction that opens the subject to the exhilarating realization of Otherness valued as such. (Hayles 1994, 188)

Hayles concludes her article by stating that the "positive seduction of cyberspace leads us to an appreciation of the larger ecosystems of which we are a part, connected through feedback loops that entangle our destiny with their fates." We will "remember what cannot be replaced" (Hayles 1994, 188). Borgmann argues, as we saw, that a computer-generated experience such as the Skiorama "provides a disposable experience that is discontinuous with its environment," while a mountain calls forth our respect in possessing "a commanding presence and a telling continuity with the surrounding world" (Borgmann 1995, 38). Hayles is arguing that what has "commanding presence" are the very *processes* of feedback afforded in virtual realities. It is these interactive processes that will lead us to appreciate the complex and irreplaceable ecosystemic feedback processes in which our lives are embedded.

Haraway's metaphor of the cyborg is both an analysis of technology and a proposal for social change. While her analysis deals with dualisms and divisions that Borgmann's does not, it fails, in emphasizing the fluidity of the cyborg, to recognize what Borgmann sees; namely, that there is a pattern to this fluidity. Reflecting on deconstructionist postmodernism and its metaphors of trickster and shape shifter used to describe self and knowing and to displace the god-trick, Susan Bordo asks, "What sort of body is it that is free to change its shape and location at will, that can become anyone and travel anywhere?" She argues that the "epistemological fantasy of *becoming* multiplicity" defeats itself. With no limits to where self can travel, "the postmodern body is no body at all" (Bordo 1990, 45). The fantasy of becoming multiplicity, in Haraway's sense, defeats itself because, as cyborgs, we in fact instantiate the device paradigm.

Furthermore, while Haraway and Hayles acknowledge the deadly serious side of our cyborgness, they argue that the risks are worth it for the benefits that can arise in *reimagining our world*. We must choose to name and enact ourselves as cyborgs so that we can realize our commitments to end social injustice and environmental degradation. In so arguing they are, however, again mirroring the very means-end device paradigm that Borgmann identifies as the ruling pattern of our lives. As cyborgs we become devices, political devices, to achieve the ends to which we are committed.

Finally, in *temporal terms*, I do not agree with Haraway that the way to be faithful to our commitments is by choosing to name and embrace our being as cyborgs. I will illustrate my concerns by returning to Hayles's examples of virtual reality and by asking the question, What time is it for the cyborg?

AN ENDING

What time is it for Hayles's subject/object ambiguity-producing virtual reality data puppet? It's on time or off time depending on which way Hayles flicks the switch. It can be moving fast or slow time, depending on which skill level Hayles picks. Perhaps this is a smart program and the puppet can actually learn to surprise Hayles by changing its speed at random. The puppet may have learned to do this over time. But in what sense is it older? It may suddenly become an old-woman puppet, slow and frail; but in what recognizable sense is it aged? Key markers of difference among people and between people and nonhuman beings are temporal ones. Appreciating that others can be both "different and enriching," and that they cannot be "controlled and predicted," means knowing what time it is for them. What is their age and what cultural and personal significance is, for them, attached to being that age? What are their rhythms of eating, sleeping, dreaming, and sex? Are they in love? Did someone close to them just die? Do they have fever? Or are they in the middle of thinking about something that requires all of their attention? Asking sincerely the question "How are you?" is like asking "What time is it for you?" In temporal terms, I am not at all convinced that Hayles as cyborg is a body capable of becoming multiplicity and appreciating difference through the fuzzing of her own body boundaries. On the contrary, it appears like a form of cyborg narcissism, a virtual mirroring of herself in interaction with a being who is really on her time, one she can turn on or off.

It is entirely unclear then, how the virtual reality feedback loops that Hayles describes will lead us to an appreciation of the larger feedback loops or ecosystems of which we are a part. Ambiguity, argue Hayles and Haraway, preserves our capacity to imagine the world as otherwise and to appreciate our kinship with other beings; yet there is no ambiguity in cybertime. It is universal and relative, not situational. The net and web are always on. They purport to accommodate all the rhythms/times of beings on-line who can come together in new communities. But in encompassing all equally, important differences disappear. There are no night and day, no moon and sun cycles, no seasons, no tides in cybertime. There are no rabbits, no raccoons, no snowstorms on-line. Unless a trickster speaks as a lightening bolt to our power lines, we won't hear her rhythmic language.

In regards to promoting the appreciation of difference, I am much more hopeful about Borgmann's focal practices. Meditation is an example of focal practice. Learning to meditate is a matter of learning to observe and to cease controlling one of the things we most take for granted, our breath. At a poetry workshop I once attended, we were led through a guided meditation. For fifteen minutes we simply observed our breath, letting thoughts go by like clouds, not attaching to them, returning the breath, going in and out. We were then asked to imagine our breath floating under the door, down the sidewalk, lifted by an updraft into the lungs of a bird, exhaled in a raindrop, falling in a stream, carried as a bubble down to the ocean. "Poetry," said the workshop leader, "opens us to the larger cosmos," or feedback loops you could say, "of which we are a part." Meditation on the most unambiguous aspect of our being; the breath moving in and out can preserve our capacity to *imagine the world as otherwise*. With Haraway, we could even imagine that our breath, exhaled from an ocean wave, is inhaled in the hands of passing a sailor's watch, slowing time in its pause to breathe . . . in.

I often experience temporal ambiguity as a tug-of-war between the schedules, which frame my life like forward-moving cages, and the spiraling births, illnesses, changes, and endings that go on in and around them. At other times I feel the very embodiment of a forward-moving cage, the living manifestation of an agenda book! The first is a Borgmann-like account; the second a Haraway. I experience both as true. Schedules cut deep divisions into my life, and moments of relaxation or meditation can all too easily become the means to the end of working more efficiently (with a clearer mind), while work becomes a means to afford the end of relaxation and holiday time. At the same time, in my very being, I become entrained to the mechanical rhythm of the clock, eating, sleeping, even going to bathroom, when my schedule allows for it.

Being a cyborg *in the temporal sense* of embodying the rhythms of a clock, schedule, speed of a video game, or the all-now-time of the web doesn't necessarily, I have argued, lead to the appreciation of difference or subversion of patterns of domination. On the contrary, in embodying these machinelike rhythms we may enact patterns of domination in our own selves, working when our bodies are telling us "they" need to sleep or eat; doing this to ourselves, we do it to others too.

Meditation is a focal practice that allows me to slow down and gently listen to my own self; to notice that my back aches, that I'm tired; that although I didn't realize it, my heart is troubled by an argument I had the other day. In meditation, I can ask of myself, what time is it? And when I engage in it regularly, I am better able to ask the same question of others and to hear their answers.

But all of Borgmann's examples of focal practices—running, gardening, making music, or preparing and sharing a meal with family and friends— take place within the confines of scheduled time and, more than that, they all take place in our so-called free or leisure time. Borgmann's answer to the reform of technology does not resolve the temporal ambiguity and question of commitments of which I spoke earlier. Snowstorms, student illnesses, midterms, an irritable child becoming ill and in need of rest all exist, but each moment calls forth its own context and set of commitments, and it's just not true that you can keep all meanings alive at once without hedging on engagement to at least one of them. As I have argued, temporal ambiguity is heightened in a multicultural society. Except for the extremely wealthy or those with flexible employers and/or jobs, it is difficult for those who practice, for instance, Muslim or Native North American traditions, to enact their commitments while living in a society governed by the Gregorian calendar. A society that honors the rhythms embodied in the traditions and beliefs of all its peoples would go a long way in the appreciation of difference.

In sum, by failing to deal with the temporal parameters and implications of their proposals, neither Borgmann nor Haraway goes far enough in imagining a world that can be otherwise.

References

Bordo, Susan. 1990. "Feminism, Postmodernism, and Gender-Scepticism." In *Feminism/Postmodernism*. Ed. Linda J. Nicholson. New York: Routledge.

Borgmann, Albert. 1984. *Technology and the Character of Contemporary Life: A Philosophical Inquiry.* Chicago: University of Chicago Press.

———. 1990. "Communities of Celebration: Technology and Public Life" *Research in Philosophy and Technology* 10:315–45.

———. 1992. *Crossing the Postmodern Divide*. Chicago: University of Chicago Press.

———. 1994. "Artificial Intelligence and Human Personality." *Research in Philosophy and Technology* 14:271–83.

———. 1995. "The Nature of Reality and the Reality of Nature." In *Reinventing Nature: Responses to Postmodern Deconstruction*. Ed. Michael E. Soulé and Gary Lease. Washington: Island Press.

Campbell, Mora. 1990. "The Temporal Structure of Technological Society." Doctoral dissertation, University of Waterloo.

———. 1992. "Time Waits for No Beast: Temporality and the Dis-integration of Nature." *Alternatives* 19:22–27.

Darnovsky, Marcia. 1991. "Overhauling the Meaning Machines: an interview with Donna Haraway." *Socialist Review*, April–June, 65–84.

de Beauvoir, Simone. 1948. *The Ethics of Ambiguity*. Trans. Bernard Frechtman. New York: Philosophical Library.

Haraway, Donna. 1991. *Simians, Cyborgs, and Women*. New York: Routledge.

Hayles, Katherine N. 1994. "The Seductions of Cyberspace." In *Re-thinking Technology*. Ed. Vernon A. Conley. Minneapolis: University of Minnesota Press.

Rheingold, Howard. 1991. *Virtual Reality*. New York: Summit Books.

Zerubavel, Eviatar. 1981. *Hidden Rhythms: Schedules and Calendars in Social Life*. Chicago: University of Chicago Press.

———. 1985. *The Seven Day Circle*. New York: Macmillan.

FIFTEEN

Trapped in Consumption: Modern Social Structure and the Entrenchment of the Device

Thomas Michael Power

Albert Borgmann has detailed the failure of modern technology to deliver on its promise of a richer, fuller life (Borgmann 1984). Instead of living "the good life" of creative endeavor, active citizenship, and ennobling physical adventures, we have become passive consumers, disengaged from and disburdened of those connections with the physical and social world that give substance and meaning to life. In that sense, technology has impoverished us rather than enriching us.

Although he has discussed the many ways in which social structures, including political, legal, and economic institutions, not only have come to support that technology but have also often taken on the form of a "device" themselves (Borgmann 1984, 75, 105), he has not explored the role of these institutions in helping to mold and entrench technology. As a result, it is never made clear why technology developed as it did; instead, we are provided only with suggestions as to how we arrived in our current technological predicament. One suggestion is that "we" trusted the promises made on behalf of technology and accepted that technology expecting those promises to be fulfilled, and we have simply been disappointed. Or, alternatively, the negative descriptions of the consequences of "our" choices might be read as implying that we individually and collectively simply did not have the moral insight and courage to resist the cheap promises of technological consumption. One possible variation on these suggestions would focus upon the transformation of individually rational decisions into a collective error, the aggregate outcome being a quite different reality from the one expected—an "invisible foot" effect as opposed to Adam Smith's "invisible hand."

To the social scientist, all of these explanations are less than satisfactory because they ignore the role of social institutions in constraining and guiding the choices individuals make. For instance, when technology was disburdening us, why did it focus upon production of consumer goods rather than the production of more leisure or a higher quality of work experiences? Why did technology distort our lives and society in the particular ways that it did? This chapter seeks to examine the analysis of technology from a social science perspective: in particular, from the perspective of economics. In that sense, it seeks to develop what Andrew Feenberg in chapter 16 labels the "sociohistorical dimensions of technical action." If these dimensions are ignored, the understanding of both modern technology and the sources of its destructiveness is incomplete and distorted in ways that mitigate against effective reform.[1]

The Social Institutions Committing Us to Consumption

The choices we make individually and collectively are heavily influenced by the way existing social institutions constrain the range of choices or set the implicit costs associated with the choices available. These constraints are important. At the extreme, where violence or the threat of violence is being used to impose one group's will on another, we do not focus exclusively on the choices made by those being coerced. We accept the relevance of the threats and coercion that may be guiding the observed choices. Similarly, social institutions can play a direct role in guiding the choices that individuals "freely" make. If we ignore these institutions, we are likely to misjudge the values and motivation of the individuals observed. In addition, we are unlikely to map out a useful strategy for change.

What are the primary social institutions that support and guide the commitment to consumption found in most of the modern world? I find that in a capitalist economy insecurity is intentionally manufactured and maintained to discipline the workforce. Our ideology and social institutions also insist on an "individualistic dance" in an overwhelmingly social setting that also generates social insecurity. Combined, these two sources of very real material insecurity assure that, no matter what the level of collective affluence attained by the overall economy, we individually experience material deprivation that makes further growth in our income a top priority.

1. This social science analysis involves stepping outside of the formal philosophic context into a rather eclectic economic context. My intent is not to provide an economic critique of the philosophy of technology, but rather a commentary on that analysis based in economics. That imposition of the concepts from one field of inquiry onto another is bound to mangle the subtleties of one or the other or both fields. For that, I offer apologies at the outset.

Also, although I consider myself an economist working within the professional mainstream when it comes to economic analysis, I am a rather harsh critic of the profession when it comes to the ideological baggage it often uncritically imports into its analysis. Because that very ideological baggage is the subject of this paper, it is important to warn the noneconomist reader that my characterization of both economics and the market economy would be rejected heatedly by many mainstream economists.

In addition to this sense of deprivation, the availability of two of the primary sources of meaning and satisfaction in life, productive work and community involvement, are severely limited by our social institutions, while, simultaneously, a relatively minor contributor to our sense of well-being, material consumption, is made readily available in a myriad of dazzling choices. The emphasis on consumption in this setting can be seen either as a rational adjustment to the limited choices available or as the perverse result of severe deprivation. Woven through this analysis will be the role played by one of our dominant social institutions, the market economy.

Socially Maintaining Economic Insecurity

Over the last sixty years, after accounting for inflation, the U.S. economy has tripled the value of the output available to the average citizen (U.S. Department of Commerce 1994, table 691).[2] With such a massive increase in material productivity, it is clear that we as a society could have eliminated the worst aspects of poverty, hunger, inadequate health care, homelessness, etc. Similarly, we could have constructed a social "safety net" for all citizens so that they would not have had to fear going without basic necessities because of individual or collective economic hard times. Although we took some steps in this direction, by the 1990s, despite rising poverty, hunger, homelessness, and unemployment among our impoverished minorities, we were busily working on dismantling the modest social safety net we had previously put in place.

This continued coexistence of economic insecurity with affluence is not an accident. Within the dominant economic ideology, the threat of going without basic necessities is considered a crucial motivating force. Providing income to the unemployed is seen as creating an unproductive dependence that undermines the individual's willingness to contribute their labor effort to the overall economy. That is, we collectively believe that in order to motivate productive and disciplined involvement in the economy, serious and real economic insecurity has to be constantly at the door. Without that, adults get surly and lazy on the job, they use collective bargaining to demand profit-reducing or inflation-causing wage increases or even drop out of the workforce altogether. These effects could drastically reduce the overall productivity of the economy. The conclusion drawn from these assumptions is that in order to enjoy the affluence that we have been able to produce, we must, as a matter of public policy, maintain a significant

2. It does not matter whether personal consumption expenditures, gross domestic product, or gross national product is used to make this calculation. All increased about threefold from 1930 to 1990.

level of economic insecurity. That insecurity is part of the basic motivational structure that makes the economy productive.

In this context, high levels of "natural" unemployment, weakening of labor unions, dismantling of our meager "welfare state," undermining a minimum wage by inflation, and recent very high levels of immigration all make sense. They create and maintain economic insecurity while our economy continues to generate "affluence." When all of us, upper middle class to minimum-wage day laborer, are constantly reminded that we could be out of work tomorrow and unable to take care of ourselves or our families, it is not surprising that we strongly support as rapid a rate of economic growth as possible, even if that produces a consumption stream that is not directly a high priority. The high priority is the ready availability of employment opportunities, making consumption itself a by-product of a quite different pursuit.

An Individualist Dance in a Social Setting

Although our modern society creates almost complete mutual interdependency, that society tells its members that they are "on their own." Complex economic arrangements are used to guide goods and services from around the world to individual households. Those individuals live on top of one another in high-density settlements where almost every production and consumption activity has significant impacts on others. The cultural institutions of that society, especially the mass media, tell us how we are to live if we want to be full members of the society upon which we so extensively depend. As a result of these interdependencies, our individual well-being is closely linked to the actions and decisions of others. That, however, is not the economic story we are told. We are taught that we are individuals, entirely responsible for our own well-being and with rights to behave as we individually determine is appropriate. "Society," we are told, is nothing but the aggregation of sovereign individuals. Any other social formulation is a liberty-threatening collectivist concept.

This particular combination of social reality and social ideology assures ongoing social insecurity no matter what level of material affluence is achieved. We judge our individual well-being, like all previous human societies, by our status relative to others. That is, our individual well-being, is socially determined. But we are taught the opposite, that we individually determine it through the success of our own actions. One obvious aspect of this is the "keeping up with the Joneses" treadmill. As we all seek to protect or enhance our relative social status, we implicitly struggle against each other in a competition that cannot boost our aggregate well-being because the goal, relative status, is a zero-sum game. We all pay a price for this competition, but the aggregate benefit is zero. We are collectively worse off, impoverished. The solution is obvious: a collective decision not to behave in this costly manner. But in our individualist setting, such a social decision is excluded. To individually leave the competition is to threaten oneself or one's family with social marginalization. The individually rational decision in this prisoner's dilemma setting is to continue the costly individualistic fantasies that consume our lives and the planet.

The combination of the very real threat of economic insecurity and the ongoing and escalating competition to simply maintain one's relative social status assures that households experience material insecurity no matter what the level of affluence achieved. Expanded consumption is experienced as a necessity even when from an objective point of view the contribution that consumption is making to satisfying "needs" is trivial.

Deprivation of Fundamental Needs: Useful, Satisfying Work

Critics of modern technology correctly emphasize the ways in which the degradation of work has impoverished our lives (Borgmann 1984, 114–24). It is through socially meaningful work that most adults define themselves. The quality of that work is one of the dominant determinants of satisfaction with life (Argyle 1989, 33–37). Despite that fact, the quality of work and job content has been systematically degraded as integrated crafts production has been replaced by mechanized, assembly line production requiring very little skill on the part of the workforce (Braverman 1974).

Even though work quality is a central component of human well-being, it is important to note that there is little or no "market" available where work quality can be pursued. Because labor markets are regularly disciplined by recessions, depressions, and technology-induced industrial declines, employers tend to control the terms upon which employment is taken. Unemployed workers or those threatened with unemployment are not in much of a position to drive a hard bargain. When labor markets are regularly in a labor-surplus condition, workers' preferences for certain types of work environments can be largely ignored. The balance of power lies with the employer in defining the character of the work opportunities.

Of course, workers did not accept the degradation of work without significant struggle. As Borgmann puts it, "It took a long, arduous, and sometimes violent process to discipline people from pretechnological work to divided labor, and the directors of this forcible development were the early entrepreneurs and capitalists" (1984, 120). This struggle was true not only in capitalist world but also in the "socialist" world. Whenever a dominant class seeks to confiscate part of the surplus being produced over and above subsistence, that class, whether it be Stalin's commissars, slave owners, military rulers, or capitalists, will choose technologies that increase their ability to extract that surplus and protect their dominant position.[3] Those technologies will be chosen and imposed, even when violence is required and overall productivity is reduced. This can be seen in the adoption of most of the centralized technologies from the medieval grain mill (versus the use of hand mills), to the early factories and assembly lines that made use of the same power sources and machines as previous production methods but organized the workforce differently, to Stalin's collective, mechanized farms (Marglin 1974, 1975; Braverman 1974).

So despite its importance in the determination of overall well-being, work quality is not among the choices readily available to the typical worker. Individual workers do not freely choose the technologies with which they will work. Those technologies are largely imposed upon them through the disciplining forces of the labor market. Those who resist pay a very high monetary price or face the police powers of the state. Powerful class relationships come to be embodied in the technologies that are used, making their replacement very difficult. It also means that work often is anything but satisfying. It builds serious emotional and social deprivation into workers' lives and impacts heavily on how those workers then proceed to live those lives.

Deprivation of Fundamental Needs: The Experience of Community

We are social animals with a strong desire to be part of something larger then ourselves or our nuclear families. Membership in a larger community contributes to our sense of identity, security, and satisfaction. Market economies, however, are not very supportive of stable communities. In a market economy, the character and timing of the use of the resources available to a community to support itself are decided not by the community but by the private individuals who own those resources. The community has no control over its economic base. Partly as a consequence of this, people are encouraged to be highly mobile, constrained by as few community "roots" as

3. Borgmann seems to suggest that concepts such as "extraction of surplus," and "class-based exploitation" are of little use in explaining the degradation of work (e.g., 1984, 82–85, 120–21). I do not find this convincing. Technologies tend to empower some and reduce the power of others. It seems highly unlikely that this impact is ignored by those affected by technology. Once one realizes that the past and present communist regimes were relatively rigid class societies heavily focused on collecting and using the surplus generated by the economy, the similarities between "capitalist" and "socialist" development in terms of choice of technologies is readily understandable. See Albert and Hahnel 1981; Betterheim 1976, 1979; Harrington 1976; and Sherman 1988. One does not have to postulate either technology as an autonomous force or a new class of "technocrats."

possible. That is necessary so that the workforce can shift geographically as economic activity shifts with changes in national and international markets.

The ideal market setting is one in which resources can quickly shift from one type of economic activity to another. That is true of workers as well as capital. In this context, "community" is a premodern, precapitalist concept that inhibits the efficient use of resources (including human labor effort). This is made explicit in our economics textbooks, where "society" is defined as simply the numerical aggregation of the relevant individuals and their preferences. In this setting, "community" is a dangerous "collectivist" concept that is likely to inhibit the "free" use of property, a concept, therefore, that has to be opposed.

We have fashioned our larger social and economic institutions to facilitate rootlessness" Our insurance policies, unemployment support, low-income support, retirement income, etc. are all transferable from one community to another. We no longer depend upon the residents living at any one particular location for support. This rootlessness undermines individuals' commitment to place as well as any place's commitment to individuals.

Hence, another human need of central importance is systematically frustrated by our social institutions. Those seeking to maintain a commitment to place and people are forced to pay a high price. Often, only the elderly can afford to maintain that commitment. The shrinking and aging populations in our agricultural towns and rusting manufacturing cities demonstrate the impact. For those who feel they cannot afford to pay the price associated with maintaining a commitment to place, an overall pattern of rootlessness has to simply be accepted as a condition of life, leaving another important source of human satisfaction frustrated. That frustration also has consequences in other areas of human activity.

Rational Adaptation to Deprivation

As outlined above, our social institutions, especially the operation of the market economy as guided by the dominant interests, seriously constrain the satisfaction of certain human needs, especially the need for productive, satisfying work and the need to belong to a stable, supportive community. While constraining these choices, that very market economy is simultaneously generating a broad range of choices in the realm of consumption. High costs are placed on the pursuit of certain fundamental needs while low costs are placed on the pursuit of consumption preferences. Within this divided structure of choices, it is not surprising that most people choose to pursue consumption opportunities while abandoning the pursuit of quality, craftslike work or involvement in a stable community. To the structured set of choices presented by our social institutions the rational adjustment is to focus one's energy on consumption.

Such nearly exclusive emphasis on consumption in an already affluent society can be characterized as perverse from a larger perspective. But the perversity lies not in the choices made by individuals but in the social structure of those choices. This is important in weighing solutions or corrections. Preaching to those trapped by an artificially constrained set of choices, urging them to ignore the high costs associated with certain choices, is unlikely to be as productive as seeking to remove those constraints on choice. Fewer saints are needed if one can focus on the root of the problem, the social institutions that are distorting people's lives.

The Market as the Institution of Central Concern

In the discussions above, unregulated markets were repeatedly the center of concern. Their operation in the context of the particular pattern of owner-ship and economic power that has developed over the last several centuries has contributed significantly to the commitment to material consumption and the degradation of contemporary life. Stating that, however, offers little in the way of confidence that there are solutions available. In the mid-1990s, the market ideology has reached a level of global hegemony unrivaled by any previous social worldview. In the nineteenth century, traditional social structures dominated most peoples' lives despite the success of the European powers in imposing their colonial structure on most of the world. For most of the twentieth century, socialist thought and institutions often did successful battle with capitalism and market institutions. With the collapse of the Soviet Union and the abandonment of "communism," there appears to be no serious opposition to the extension of market institutions to most of the world's peoples.

These criticisms of the market may be read by some as a disgruntled "socialist" attack on this victory of market institutions. This is partly correct but also seriously in error. If by "socialist" one means belief in the reality of social interests and institutions and the social character of human beings, then I am clearly a socialist. As already stated, it is clear to me that our individual well-being is largely socially determined. Given that, it is important to recognized that social context and craft it in a way that enhances the human experience rather than degrading it. At the same time, it is important to recognize the market economy as a social institution with significant positive potential. It does certain things better than any alternative social institution currently available. My criticism of the way markets have constrained our choices and degraded our lives should not be read as an attack on all uses of the market. The criticism is much more specific. It focuses on the market operating within the context of certain patterns of the distribution of wealth and power and the market extended to arenas where it performs very badly from a social perspective.

Carving out room for a limited critique of our "actual existing" market institutions will not be easy in these days of global ideological hegemony. But without a critique of those market institutions and effective modification of them, most suggestions for change primarily involve the search for the few saints among us who will ignore the very high costs associated with "right behavior." That is unlikely to be very productive. Borgmann, for instance, in discussing work and labor, points out that "[t]echnology has given us enormous power, and it is a legitimate question whether we, as citizens of the technological society, are *able or willing* to employ that power on behalf of good work"(1984, 115, emphasis added). He then focuses heavily on the "willing" but says almost nothing about the "able." Our social institutions play an important role in determining the latter.

ECONOMICS AND THE GLORIFICATION OF THE DEVICE

In many ways, the market economy and economics embody the "device paradigm." Economists have made the close association of the economy and something like the device clear for almost a century as they have sought to describe rational economic activity in terms of the "market mechanism." Even the very first economic writings indicate a device of sorts. Initially the circulatory model of the human body was used; later, the clocklike mechanics of Newtonian physics. Nineteenth-century economics texts with titles such as *Mathematical Psycics* (Edgeworth [1881] 1932) and *Mathematical Principles of the Theory of Wealth* (Cournot [1838] 1897) made clear what the intellectual goal was. The initial conceptual steps taken in economic analysis set the stage: First, the economy and economic activity were envisioned as separate realms of human activity that could be studied outside of their social and political contexts and had an existence separate from the rest of people's lives. The second conceptual step is to appeal to a mechanical model to explain the economy. Below we explore the embedding of the device paradigm into economics and its implications for how we conceptualize the economy.

Economics and the Device Paradigm

Borgmann characterizes the "device paradigm" as (1) a radical separation of means and ends with the intent of (2) disburdening or disengaging by making certain (aspects of) things that are (3) easily and widely available while (4) hiding the full human context of the thing or activity (1984,

40–48). Unfortunately, economics explicitly advocates exactly these steps as appropriate and praises the market economy because it allows the accomplishment of all of these. It is in that sense that both economics and the market economy propagate the device paradigm.

Contemporary neoclassical economics begins its analysis by distinguishing between resources (means) and people's preferences (ends). Those resources that are treated as mere means include human labor effort (labor) and the natural environment (land and natural resources). The objective is to manipulate the means in order to reach the highest possible level of satisfaction of the population's preferences. The individual's preferences are taken as given (and sovereign) and are not explored. Human beings are reduced to the status of individual optimizers focused upon squeezing the most they can out of the limited resources at their disposal. The "economic person" is isolated and atomistic. They are acquisitive and calculating. Social institutions, cultural values, moral constraints, and all other social context are largely stripped away.

The degree of reductionism here is made clear by the treatment of work as a disutility to be avoided.[4] As pointed out above, economics assumes that society is nothing but the aggregation of these atomistic individuals, so there are no social objectives to the individual's decisions either.[5] The assumption is that people want to be disburdened of work, of social constraints, and of the complex traditional contexts in which people used to live their lives. This reduces the determinants of the individual's well-being largely to the level of personal consumption achieved. Individuals are explicitly labeled consumers rather than producers or citizens. Constraints of any sort are assumed to reduce the range of choice and therefore the level of satisfaction attainable. In that sense constraints reduce well-being and are irrational if self-imposed or not opposed.

The Economy as a Device

One of the important accomplishments of economics has been to first distinguish the economy as a separate realm of human activity and then to

4. Economics is perfectly capable of looking at work as an activity that has both positive and negative features. Work can be modeled as an activity of positive value that simply has a time cost associated with it. That, however, is not the way it is typically modeled.

5. Economics *can* model objectives that involve interdependency among individuals' preferences. My well-being can be modeled as depending, at least partially, upon your well-being. Similarly, my level of satisfaction can be modeled as depending not only upon my consumption goods available to me but also upon my level of consumption relative to yours. Such interdependencies are not typically modeled by economists; when they are, some of the familiar conclusions of microeconomics about the optimality of unregulated market results cannot be derived.

see it as managed by an automatic mechanism that is both impersonal and socially rational. Economics teaches that the market mechanism is both self-adjusting and rational even though no rational social thought is involved in its operation. Conscious direction of the economy is not only not necessary but inappropriate and destructive. Through the competition among self-interested parties, the narrow self-seeking that motivates those individuals is canceled out and an outcome intended by none of the participants emerges. That outcome, economics insists, is socially rational in the sense of minimizing costs and using scarce resources efficiently in satisfying the aggregate preferences of the population.

The market mechanism allows the social economy to operate invisibly (the "invisible hand") in the background, resolving automatically for us a broad range of social issues: the distribution of income, the character of work, the use of natural resources, the rate and character of economic growth, the level of unemployment, the mix of goods and services, etc. All we focus upon in the foreground is the level of total production, the gross domestic product, and other aggregate outcomes such as the level of employment and average incomes. We do not need to know how these outcomes emerge since an automatic mechanism is always invisibly at work producing the best outcome possible. In fact knowing too much about the operation of the market economy can be dangerous since it may encourage the heretical idea that conscious social intervention could improve economic outcomes.

The market mechanism is praised for being impersonal. It disburdens us of having to be concerned about medieval issues such as just prices, usurious interest rates, fair wages, or profiteering. We no longer need to know or care about who it is we have come to depend upon and who it is that depends upon us. The market-mediated transactions with their emphasis only on price and quantity are all we have to concern ourselves with. The social character of production and exchange is purposely hidden in favor of the impersonal mechanism. We strip economic activity of its social character and present it instead as the inevitable and indisputable result of an objective mechanism. Social disengagement is not an unintended consequence of this reliance upon the market, it is one of the objectives. Because human beings are assumed to be incapable of making rational social decisions, those social decisions are purposely left to a nonhuman mechanism, and human economic decisions are limited to the individual private realm. We purposely choose to be ruled by a device.

This is the device writ large, national and global in scale, reaching into almost all aspects of our productive activities.

SEEDS OF HOPE: SUBVERSIVE IDEAS, CHOICES, AND PRACTICES

If one believes that social institutions have a strong and persistent impact on the range and character of the choices that confront us, the centrality of market-related institutions in the explanation of technology's character is disconcerting if not demoralizing. Given our strong ideological and institutional commitments to market institutions, commitments that only seem to be getting stronger as we enter a new century, there would not appear much possibility for significant reform of technology. We are committed to one of the most pervasive manifestations of the device paradigm, it would seem. Given that commitment to the device on the grandest of scales, it is unclear that we can resist the device paradigm elsewhere in our individual and social lives.

Despite confident assertions that "free-market capitalism" has once and for all defeated "collectivism" in its many forms, the hegemony of market ideology is nowhere near complete. We turn now to some of the openings where the market device does not dominate.

The Emergence of New Nonmarket Realms

The free-market ideologues' depiction of the steady expansion of market relations to a larger and larger sphere of human activity is not altogether accurate. Rather, what has been taking place is an ongoing adjustment of the line that separates the realm in which we rely primarily upon commercial markets and the realm where we rely upon noncommercial institutions. In addition to centralized and decentralized government institutions, the noncommercial realm includes private nonprofit organizations (nongovernmental organizations or NGOs) as well as informal communities and families.

That this boundary does not shift relentlessly one direction can be seen by considering two areas of contemporary concern: the environment and medical care. Over the last several decades, environmental resources that previously had been freely available to commercial firms were placed under the direct control of government agencies. Those environmental resources, including air and water quality, wildlife habitat, and many natural resources, were literally worth hundreds of billions of dollars to commercial firms. Over the last several decades our "market society," however, has systematically socialized them. It should not be surprising that there continue to be heated political battles over environmental policy, given the value of the resources that the public has sought to shift from the private realm to the social realm. That battle, of course, is not yet over. The point here is simply that environmental policy represented a major expansion of nonmarket institutions at the expense of the commercial realm.[6]

Something similar *almost* happened to medical care in the United States. Significant parts of health care have always lain outside of the commercial realm. Most large hospitals and much medical research are lodged in the nonprofit sectors and always have been. With the establishment of Medicare for the elderly and Medicaid for the poor, the United States shifted at least the financing of health care for these groups to the public realm. Americans continue to debate the merits of having the government play a much larger role in managing the entire health care system. For the time being, commercial businesses seem likely to continue to play a significant role in this field, but the line between the commercial and noncommercial has been shifting at the expense of commercial markets. What is unclear is how far it will continue to shift that direction.

The point here is that the line between the realm of commercial markets and the noncommercial sectors is not set rigidly and is not shifting uniformly in one direction. There is room for significant institutional change.

The widespread concern about environmental degradation and support for social measures to contain and remediate it legitimizes collective, nonmarket actions. The environmental arena provides dramatic examples of "market failure" that capture the public's attention. The environment is one of the areas where collective action still appears legitimate. That is one of the reasons that promarket forces have focused so much attention on the "irrationality" of typical "command and control" environmental regulation. Environmentalism is seen as one of the last significant refuges of "socialists." Environmentalists are accused of being watermelons: green on the outside but red on the inside. This simply underlines the importance of environmental issues in helping craft an opening in the dominant ideology and institutions for noncommercial, nonmarket concerns.

In addition, as Borgmann has pointed out (1984, 234), the public's concern with environmental issues has helped them make the distinction between money income and the quality of life. This awareness that well-being is importantly tied to noncommercial, nonmarket aspects of their communities and landscapes is exactly the type of intellectual and institutional opening that is required if we are to step away from the device

6. There is increasing interest in rationalizing environmental regulation and management by making more use of market mimicking institutions and incentives. Some of these efforts are more controversial than others. Container (and appliance) deposit laws are widely accepted. Marketable air pollution permits are still heatedly debated. The introduction of marketlike institutions within environmental management does not necessarily represent a retreat from social regulation of environmental quality. Rather, it may involve primarily a search for the most effective and efficient ways of implementing that social regulation.

paradigm. It is important to realize that this concern with the quality of the social, human-built, and natural environments in which we live is not only a political or philosophic concern. People are acting on those concerns. On a regular basis they have to confront the cost to themselves in terms of sacrificed money income and other lost commercial economic potentials associated with choosing to live in one location rather than another. People regularly make significant sacrifices in the pursuit of higher-quality living environments.[7] Unfortunately, people are not always entirely conscious of the meaning and implications of these choices. Often the very people who have made the largest sacrifices in the pursuit of noncommercial, nonmarket values are the most rigid defenders of the dominance of the market. The social character of what they seek and the need for collective action to protect it remain hidden from their view. It is crucial to the reform of technology and the market that these very real nonmarket values be brought into the foreground, where they can influence political and social decisions.

Within economics there has been an ongoing struggle to come to grips adequately with the full dimensions of environmental concerns. The long-term (over many generations) and irreversible nature of some environmental impacts have prompted at least some economists to review how well commercial market institutions actually deal with environmental degradation and resource depletion. These analyses tend to demonstrate the relatively short-run focus of commercial markets compared to the much more permanent time dimensions associated with many important environmental problems.[8]

This new environmental economics has also sought to distinguish typical commodities sold in commercial markets from many of the environmental values being sought by people. Consider one example of a focal concern: the experience of pristine nature (Borgmann 1984, chap. 22). In comparing preservation of pristine natural landscapes with their development for commodity potential, economists have made the following types of economic distinctions (Krutilla and Fisher 1975, Krutilla 1967).

---

7. People move to areas with fewer commercial opportunities, that pay lower wages, that have higher unemployment rates, that have less developed commercial and public infrastructures, that have higher housing costs, etc. Analysis of these location decisions and the sacrifices made in the pursuit of preferred living situations allows economists to estimate the money value of cleaner air, lower crime rates, reduced congestion, scenic beauty, outdoor recreation opportunities, etc. See Power 1996b, chap. 6.

8. For a contemporary view of these issues see Norton 1995. For the original development of these issues by economists see Krutilla et al. 1975. For a textbook presentation see Randall 1981.

Pristine Nature	*Commodity*
• unique, nonreproducible, gift of nature	• mass-manufactured commercial product, reproducible
• no close substitutes	• substitutes readily available
• context and origin important	• context and origin unimportant
• experienced directly; not an intermediate good	• an intermediate input in the pursuit of an end
• valued for itself; integrity and original character important	• valued as a means to an end; alternative means are acceptable
• potential for irreversible change	• replacement and repair possible and practical
• nonconsuming experience is the objective	• consumption is the objective

These distinctions have been shown to have important economic impli-cations, tending to increase significantly the economic value of preservation compared to the value of commodity development. What is as interesting is that these are also some of the characteristics that Borgmann and other philosophers have pointed to in explaining the ways in which the experience of pristine nature can be a focal concern that helps us escape the device paradigm (Borgmann 1984, 182–96). The recognition *within economics* of values that are not commodity- and market-related is a crucial initial step if that discipline is going to distance itself from the commercial market as *the* overreaching social device.

Changing the Nature of Work

The degraded nature of much of modern work is not uniformly applauded by either the business community or the economics profession. Two oppos-ing trends appear within the economy. One seeks to systematically reduce the commitments that employers make to their employees by deskilling jobs so that almost anybody can do the job, by hiring part-time and temporary help, and by contracting for services rather than hiring personnel. These efforts allow businesses to cut both wages and benefits. Commitments are limited and jobs are defined so that uncommitted workers can be used without causing problems. The other trend has sought to enrich jobs and to get workers more involved in managerial decisions. Self-directed teams, sharing a variety of tasks, are the focus of the workplace. This approach sees worker commitment to the firm and worker involvement in decisions as crucial in raising the productivity of the operation. Workers often are not only involved in firm decision making but are also given an ownership stake in the company (Appelbaum and Batt 1994). The first of these trends appears focused on repeating the patterns of nineteenth-century industrialization. The second appears to recognize the human capacity and desire for productive and responsible social management of production.

Economists have also become increasingly aware of the unexpected expansion of work time and decline in leisure in the American economy. Workers have actually threatened strikes during the 1990s in order to escape the burden of mandatory overtime. The average work week, which has not shortened over the last forty years and has been growing over the last two decades, apparently is not entirely voluntary on the part of workers (Schor 1991, 1). In addition, the growth of voluntary part-time work to match nonmarket interests and obligations has become increasingly important to a significant part of the population (Kahne 1985). These struggles over how the work schedule is defined indicate an increased awareness of the importance of nonmarket activities in determining individual and household well-being. This offers another opening for emphasizing the importance individually and socially of stepping back from total reliance on the market device.

The dominant trend, however, is headed in the opposite direction and provides another example of how social institutions bias the "choices" we make. Workers tend to face a very limited choice: they can work full-time plus as specified by the employers or not work at all. Instead of people choosing the mix of hours worked in the commercial sector, hours committed to home-based production, and hours for leisure activities that best match their values, people are told that if they want relatively well-paid employment, they *must* work longer than they would prefer and accept higher levels of income and consumption in trade for the lost time at home or in leisure. Economic insecurity in the form of unemployment or part-time employment in the low-wage sector operates to encourage a consumption-oriented "choice."

The Shift from Goods to Services

Although much has been written decrying the transformation of the economy away from goods production, most of the fears that have been expressed are misconceived or factually in error (Power 1996a, chap 3). There are widely shared cultural values that picture goods production as more real and reliable because it produces a tangible product that can be stored, shipped long distances, and easily exchanged impersonally on the market. Goods are more fungible. Services, on the other hand, by definition involve nonstorable values that have to be produced at the time they are demanded. They often are pictured as insubstantial and of fleeting value. As widely shared as these cultural perceptions are, they clearly are little more than conservative expressions of a preference for a past reality. They are built around a largely masculine cultural preference for economic activities that combine muscle, sweat, and raw materials in the production of a material, quantifiable output. It is not at all clear that such work is morally or economically superior to that which combines people's minds and hearts in rendering valuable services to their neighbors. Medical, engineering, educational, business, and other professional services are hard to dismiss as insubstantial in their contribution to both the economy and our well-being. The shift to services is at least partially an economic reflection of the fact that it is not by bread alone that we live.

Diane P. Michelfelder in chapter 12 discusses the moral importance of the experiences and perspectives that are formed within the context of familial and friendly relations. She specifically focuses on the moral significance of caregiving. Care, and the trust and responsibility associated with it, can be a way of countering the moral impoverishment of a society in which narrow, calculating self-interest is offered as a behavioral norm. The realm of caregiving and nurturing, of course, has been traditionally assigned to women. This has allowed it to be largely ignored or dismissed in both economic and philosophic discourse. The current economic concern that the economy is being undermined by the shift to services is simply a reflection of the fear that the center of gravity of the economy is shifting toward "women's work." As discussed below, rather than fearing this shift, there are reasons to cautiously welcome it on both economic and moral grounds.

This economic transformation that is already well under way has the potential for changing how it is that we think about the economy and economic activity. In the past the language of economics has emphasized the material, the quantitative, and the necessary. The metaphors for "economic issues" makes this clear ("putting food on the table," "bread-and-butter issues," "bringing home the bacon," "putting a roof over our heads"). This allowed economic discussions to focus on the production and consumption of commodities. The shift to services has the potential of allowing us to emphasize the fact that economic activity primarily involves scratching each other's backs and taking in each other's wash. That is, economic activity increasingly involves the exchange of services among skilled craftspeople, people skillfully caring for each other. That activity has an important human dimension that when degraded degrades the economic value of the service. For that reason, that human dimension may continue to be protected even by the market. In addition, services tend to be locally produced rather than produced at distant locations around the world. Finally, mass production technology has had difficulty adapting to services. Labor productivity has grown relatively slowly in services compared to goods production.

All of these characteristics are attractive from the point of view of putting an integrated human dimension back into our economic activity. There is at least the potential that skilled, human-oriented, community-based economic activity that is socially meaningful and satisfying can develop around these service activities. Of course, no one should underestimate the power of corporate-dominated markets to transform this potential into the same dead-end to which it carried most goods production.

Paul Thompson in "Farming as Focal Practice" (chapter 9) seems to implicitly confirm the moral inferiority of a services-oriented economy compared to "primary" production such as farming, fishing, or other natural resource–based economic activities. Thompson claims a primacy for such traditional subsistence activities as "quintessential focal practices." The basis of this claim appears to be associated with the following characteristics of farming:

a. individuals and social groups evolved a way of being around it;

b. that way of being is highly attuned to the place in which the participants are situated; and

c. the activity is productive of the sustenance of the participants' lives.

There is no disputing the fact that this combination of characteristics ensures a high level of integration and connectedness in the lives of farm families and communities. As Thompson says, "farming unifies 'achievement and enjoyment of mind, body, and the world' " (168). There is, however, an element of what I would call *economic fundamentalism* associated with this position. Variants of this economic fundamentalism currently confuse and distort public environmental and economic development policies.[9] Note that Thompson's "quintessential focal practices" include only traditional subsistence activities: farming, fishing, and hunting and gathering. The contemporary antienvironmental lobby would add logging and mining to this list. In my reading, Thompson would accept these candidates only in the past when the technology used was nonindustrial.

This economic fundamentalism sees activities that directly support human biological survival (food, protection from the elements, etc.) as both economically and morally superior to those activities that are not directly necessary for survival. Thompson incorporates this element by emphasizing

9. Economic fundamentalism holds that all wealth springs from the earth. For that reason, economic activities that develop natural resources are held to be more central to human well-being than other types of economic activities. All other economic activities are seen as largely parasitic on "primary" activities such as agriculture, mining, logging, fishing, etc. For a critique of this position see Power 1996a, 173–75; 1996b, chaps. 2 and 3; 1996c.

farming's role in providing "sustenance" and "life" and making "the material production of thought or experience" possible. This emphasis on the role of physical necessity in making human experience somehow more authentic and valuable is puzzling. Much of human culture and social organization aims at keeping pure physical need at bay because, at its worst, it brings on a morally and socially destructive struggle for survival. Human beings are capable of far more than physical survival and it is within that margin of safety where we are protected against imminent loss of life that our art, thought, play, love, and hope evolve into human cultures. Although our pursuit of sustenance may provide the material for a culture that is integrated into our daily lives, other central aspects of that culture will not be directly tied to physical survival: beauty, love, the pursuit of truth, the worship of our gods, etc. Often these will intentionally draw us away from the demands of physical survival.

Thompson's criticism of a pluralism that sees many things that might serve as focal objects and many different focal practices that "might, for different groups and individuals, [be] edifying, unifying, and salvific" (166) and his claims for "primary" economic activities are not convincing. In fact, they are dangerous given the way they play into the atavistic antienvironmental backlash influencing public policy since the mid-1990s. One can focus on the importance of being connected to place, of respecting and understanding the natural processes that allow us to live in a particular place, and of integrating our productive work into our lives and that of our community without idealizing subsistence activities. It is *not* only the farmer who understands the earth. It is *not* only the irrigator or dam builder or commercial angler who understands the river. It is *not* only the logger who understands the forest or the miner who understands the mountain. To me it is questionable to suggest that it is only those who have a commercial or subsistence survival connection with a place who can appreciate and care for that place and integrate it into their lives. As human beings we can relate to the natural world of which we are a part in many, many more dimensions than this economic fundamentalism suggests. The focus on physical survival impoverishes us as human beings in many different ways.

The Market Economy as a Conscious Social Construct

Although the popular depiction of the market economy as an objective and socially rational device that springs spontaneously from simply protecting our property and liberty is built primarily upon ideology and religious-like faith rather than upon empirical social science, a direct attack on markets is unlikely to be consequential in this period of ideological hegemony for the free marketeers. An indirect approach is required. We need to bring to the foreground and out of hiding the actual social context required for markets to operate.

A market economy is not a "mechanism" that emerges automatically from "a system of natural liberty." A market economy can operate with some degree of social rationality only when an appropriate context has been socially crafted. It is laws, social institutions, values, and ethics as well as conscious adaptations to historical experience that *allows* a market economy to operate productively. Remove any significant part of these and a market economy degenerates into bandit capitalism of the sort seen in Somalia before and after American intervention or in the drug dealers' turf in our central cities or in ex-Soviet countries dominated by the Russian Mafia. That is, the market can contribute to socially rational and productive outcomes only when it is embedded in an appropriate social structure. Crafting the social structure that allows markets to perform productively is a central project of any society that wishes to rely heavily upon a market economy.

Similarly, we need to bring into the visible foreground the human-crafted context that also limits the market. Every human society carefully delimits where market transactions are productive and appropriate and where they are not. Marketlike transactions do not always facilitate the pursuit of human objectives. That is why we make extensive use of nonprofit organizations, both private and governmental. It is also why we seek to shield broad areas of human activity and experience from commercial forces. The family and household are one obvious example. But one could include the justice system, most of education, a good part of medical care, scientific research, spirituality and religion, etc. We need to emphasize what we already know: in some realms market transactions damage the values we are pursuing. Paying a spouse for a delicious meal or particularly sensitive and pleasurable lovemaking degrades the experience rather than enhancing or facilitating it. Buying and selling votes, judicial decisions, or government rulings does not make the social system more efficient. The ban on voluntary slavery (indentured servitude) is not an irrational interference in the economy, and so with many other examples. The more important and central a concern is to our well-being, the more likely it is that we will collectively set it off-limits to commercial transactions.

Yet this conscious crafting of the social context in which we trust the market to operate is systematically hidden from us in the rhetoric and ideology of the impersonal market mechanism. Rather than attacking the role of commercial markets, we need to work toward illuminating both the legal, cultural, and ethical structure we have crafted that allows the market to be productive and rational and those broad regions of social activity we have set off-limits to commercial calculation and the reasons we have done so. If illuminating the socially crafted character of the market economy is successful, the market will be seen in its rich social context, cease to be an overreaching device, and become trusted as an effective instrument in certain realms. When the market is constrained appropriately, we may also enjoy being disburdened from broad ranges of economic decisions (e.g., how many red patent leather shoes the economy should manufacture).

Finally, by focusing on the social context that allows the market to perform in a socially rational way, we can begin to explore the ways in which commercial markets operate to limit rather than enhance the choices we make. This disclosure is a crucial step in reclaiming the human promise originally seen in technology.

This task will not be easy. The economy as device, with its socially crafted nature hidden in the background, serves powerful interests. Current ideology conveniently hides the massively unequal distribution of income, wealth, and economic power. Claims about the market's automatic, objective social rationality render proposals for conscious democratic social intervention illegitimate, leaving privately organized social institutions, such as our multinational corporations, unconstrained. Antigovernment rhetoric draws on a theme that has been central to American history since the first European settlers arrived: A primitive Daniel Boone fantasy of complete independence from any larger social unit than the nuclear family continues to have enormous emotional appeal. We are atomistic individuals whose liberty is threatened even by the presence of neighbors who might have expectations of us. Social settings, like democratic government, are a danger, not a potential. We cannot and should not act together. One can reject these ideological elements as primitive slogans not worthy of a response, but they have the power to paralyze political dialogue and suffocate social imagination. That, of course, is their function—to protect the status quo. The challenge is to navigate around them and find the openings that allow us to speak directly to people's unambiguous experience in ways that allow them to see the positive potential of change.

Borgmann, in the conclusion to *Crossing the Postmodern Divide,* might be read as urging something similar: a return to communal celebration built around focal things and practices. "We must talk in the public forum about the things that finally matter and about common measures that will give these things a secure and prominent place in our midst" (Borgmann 1992, 116). There is a difference, however. For Borgmann this is largely a matter of people individually and collectively choosing to commit themselves appropriately. "In a finite world, devotion to one thing will curb indulgence in another" (1992, 116). Phrasing it as moral choice can be "de-moralizing" when social institutions structure the choices so that only the truly heroic and saintly can afford to make the right choices. The public dialogue in which we engage must not only lay out the proper alternative for our personal and collective behavior; it must also help people to understand that it is not primarily personal moral failure that makes moving in that direction difficult. Understanding the ways in which social institutions structure our choices so as to encourage the wrong choices is crucial if we are going to maintain the social energy that supports the type of "patient vigor" for which Borgmann calls. Ignoring the role of social institutions encourages us in our moral failures to lash out at ourselves and our neighbors as we sink into the cynicism and sullenness Borgmann rightly decries.

REFERENCES

Albert, Michael, and Robin Hahnel. 1981. *Socialism Today and Tomorrow.* Boston: South End Press.

Appelbaum, Eileen, and Rosemary Batt. 1994. *The New American Workplace: Transforming Work Systems in the United States.* Ithaca, N.Y.: ILR Press.

Argyle, Michael. 1989. *The Psychology of Happiness.* London: Routledge.

Bettelheim, Charles. 1976. *Class Struggles in the USSR.* Vol. 1. New York: Monthly Review Press.

————. 1979. *Class Struggles in the USSR.* Vol. 2. New York: Monthly Review Press.

Borgmann, Albert. 1984. *Technology and the Character of Contemporary Life.* Chicago: University of Chicago Press.

————. 1992. *Crossing the Postmodern Divide.* Chicago: University of Chicago Press.

Braverman, Harry. 1974. *Labor and Monopoly Capital: The Degradation of Work in the Twentieth Century.* New York: Monthly Review Press.

Cournot, Augustin. [1838] 1897. *Researches into the Mathematical Principles of the Theory of Wealth.* Trans. Nathaniel T. Bacon. New York: Macmillan.

Edgeworth, F. Y. [1881] 1932. *Mathematical Psycics.* Reprint, London: London School of Economics Reprints.

Harrington, Michael. 1976. *The Twilight of Capitalism.* New York: Simon and Schuster.

Kahne, Hilda. 1985. *Reconceiving Part-Time Work: New Perspectives for Older Workers and Women.* New York: Rowman & Allaheld.

Krutilla, J. V. 1967. "Conservation Reconsidered." *American Economic Review* 57:777–86.

Krutilla, J. V., and Anthony C. Fisher. 1975. *The Economics of Natural Environments: Studies in the Valuation of Commodity and Amenity Resources. Resources for the Future.* Baltimore: Johns Hopkins University Press.

Marglin, Stephen. 1974. "What Do Bosses Do? The Origins and Functions of Hierarchy in Capitalist Production." Pt. 1. *Review of Radical Political Economics* 6, no. 2: 60–112.

————. 1975. "What Do Bosses Do? The Origins and Functions of Hierarchy in Capitalist Production." Pt. 2. *Review of Radical Political Economics* 7, no. 1: 20–37.

Norton, Bryan G. 1995. "Ascertaining Public Values Affecting Ecological Risk Assessment." School of Public Policy, Georgia Institute of Technology, Atlanta (issue paper prepared for Risk Assessment Forum, U.S. Environmental Protection Agency).

Power, Thomas M. 1996a. *Lost Landscapes and Failed Economies: The Search for a Value of Place.* Washington: Island Press.

———. 1996b. *Environmental Protection and Economic Well-Being: The Economic Pursuit of Quality.* Armonk, N.Y.: M. E. Sharpe.

———. 1996c. "Custom and Culture's Worst Enemy Speaks." *High County News,* October 14, 17.

Randall, Alan. 1981. *Resource Economics.* Columbus: Grid Publishing, 1981.

Schor, Juliet B. 1991. *The Overworked American: The Unexpected Decline in Leisure.* New York: Basic Books.

Sherman, Howard J. 1988. *Foundations of Radical Political Economy.* New York: M. E. Sharpe.

U.S. Department of Commerce. 1994. *Statistical Abstract of the United States.* Washington: Government Printing Office.

SIXTEEN

From Essentialism to Constructivism: Philosophy of Technology at the Crossroads

Andrew Feenberg

What Heidegger called "the question of technology" has a peculiar status in the academy today. After World War II, the humanities and social sciences were swept by a wave of technological determinism. If technology was not praised for modernizing us, it was blamed for the crisis of our culture. Whether interpreted in optimistic or pessimistic terms, determinism appeared to offer a fundamental account of modernity as a unified phenomenon. This approach has now been largely abandoned for a view that admits the possibility of significant "difference," i.e., cultural variety in the reception and appropriation of modernity. Yet the breakdown of determinism has not led to quite the flowering of research in philosophy of technology one might hope for.

On the one hand, mainstream philosophy, which was never happy with the intrusion of technological themes, sticks happily to its traditional indifference to the material world. Where the old determinism overestimated the independent impact of artifactual on social reality, the new social-scientific approaches appear to have so disaggregated the question of technology as to deprive it of philosophical significance. It has become matter for specialized research.[1] And for this very reason, most professional philosophers now feel safe in ignoring technology altogether, except of course when they turn the key in the ignition.

On the other hand, those few philosophers, notably Albert Borgmann, who continue the earlier interrogation of technology have hesitated to assimilate the advances of the new technology studies. They remain faithful to the determinist premises of an earlier generation of founders of the field, such as Ellul, Heidegger, and the Frankfurt School. For these thinkers

---

This chapter is adapted from my book *Questioning Technology* (London: Routledge, 1999), 183–236. Reprinted by permission of Taylor and Francis. The following, however, is not simply a reprint of that material, but has been reworked for this occasion.

1. See, for examples, Pinch, Hughes, and Bijker 1989.

modernity continues to be characterized by a unique form of technical action and thought that threatens nontechnical values as it extends itself ever deeper into social life. They argue that technology is not neutral. The tools we use shape our way of life in modern societies where technique has become all-pervasive. The results of this process are disastrous: the triumph of technological thinking, the domination of nature, and the shattering of community. On this account, modernity is fundamentally flawed.

While the problems identified in this tradition are undoubtedly real, these theories fail to discriminate different realizations of technical principles relevant to the alternatives we confront. As a result, technology rigidifies into destiny and the prospects for reform are narrowed to adjustments on the boundaries of the technical sphere. It is precisely this essentialist reading of the nature of technology that recent social-scientific investigations refute without, however, relating their nonessentialist conception of technology to the original problematic of modernity that preoccupies the philosophers.[2] Here I attempt to preserve the philosophers' advance toward the integration of technical themes to a theory of modernity without losing the conceptual space opened by social science for imagining a radically different technological future.

I now begin to present my argument with a brief reminder of Heidegger's approach.

HEIDEGGER

Heidegger is no doubt the most influential philosopher of technology in this century. Of course he is many other things besides, but it is undeniable that his history of being culminates in the technological enframing. His ambition was to explain the modern world philosophically, to renew the power of reflection for our time. This project was worked out in the midst of the vast technological revolution that transformed the old European civilization, with its rural and religious roots, into a mass urban industrial order based on science and technology. Heidegger was acutely aware of this transformation, which was the theme of intense philosophical and political discussion in the Germany of the 1920s and 1930s (Sluga 1993). At first he sought the political significance of "the encounter between global technology and modern man." The results were disastrous and he went on to purely philosophical reflection on the question of technology (Heidegger 1959, 166).

Heidegger claims that technology is relentlessly overtaking us (Heidegger 1977a). It is transforming the earth into mere raw materials, which

2. For an exception, see Latour 1993.

he calls "standing reserves." We ourselves are now incorporated into the mechanism, mobilized as objects of technique. Modern technology is based on methodical planning that itself presupposes the "enframing" of being, its conceptual and experiential reduction to a manipulable vestige of itself. He illustrates his theory with the contrast between a silver chalice made by a Greek craftsman and a modern dam on the Rhine (Heidegger 1977a). The craftsman gathers the elements—form, matter, finality—and thereby brings out the "truth" of his materials. Modern technology "de-worlds" its materials and "summons" nature to submit to extrinsic demands. Technology thus violates both humanity and nature at a far deeper level than war and environmental destruction. Instead of a world of authentic things capable of gathering a rich variety of contexts and meanings, we are left with an "objectless" heap of functions.

Translated out of Heidegger's ontological language, this seems to mean that technology is a cultural form through which everything in the modern world becomes available for control. This form leaves nothing untouched: even the homes of Heidegger's beloved Black Forest peasants are equipped with TV antennas. The functionalization of man and society is thus a destiny from which there is no escape. Heidegger calls for resignation and passivity rather than an active program of reform that, in his view would simply constitute a further extension of modern technology. As Heidegger explained in his last interview, "Only a god can save us" from the juggernaut of progress (Heidegger 1977b).

Although Heidegger means his critique to cut deeper than any social or historical fact about our times, it is by no means irrelevant to a modern world armed with nuclear weapons and controlled by vast technology-based organizations. These latter in particular illustrate the basic concepts of the critique with striking clarity. Alain Gras explores the inexorable growth of such macrosystems as the electric power and airline industries (Gras 1993). As they apply ever more powerful technologies, gain control over more and more of their environment, and plan ever further into the future, they effectively escape human control and indeed human purpose. Macrosystems take on what Thomas Hughes calls momentum, a quasi-deterministic power to perpetuate themselves and to force other institutions to conform to their requirements (Hughes 1989).

Heidegger's basic claim that we are caught in the grip of our own techniques is thus all too believable. Increasingly, we lose sight of what is sacrificed in the mobilization of human beings and resources for goals that remain ultimately obscure. So far so good. But there are significant ambiguities in Heidegger's approach. He warns us that the essence of technology is nothing technological; that is to say, technology cannot be understood through its usefulness, but only through our specifically technological engagement with the world. But is that engagement merely an attitude or is it embedded in the actual design of modern technological devices? In the former case, we could achieve the "free relation" to technology that Heidegger demands without changing technology itself. But that is an idealistic solution in the bad sense, and one that a generation of environmental action would seem decisively to refute.

Heidegger's defenders point out that his critique of technology is not concerned merely with human attitudes but also with the way being reveals itself. Again roughly translated out of Heidegger's language, this means that the modern world has a technological form in something like the way in which, for example, the medieval world had a religious form. Form in this sense is no mere question of attitude but takes on a material life of its own: power plants are the gothic cathedrals of our time. But this interpretation of Heidegger's thought raises the expectation that criteria for a reform of technology qua device might be found in his critique. For example, his analysis of the tendency of modern technology to accumulate and store up nature's powers suggests the superiority of another technology that would not challenge nature in Promethean fashion.

Unfortunately, Heidegger's argument is developed at such a high level of abstraction he literally cannot discriminate between electricity and atom bombs, agricultural techniques and the Holocaust.[3] All are merely different expressions of the identical enframing, which we are called to transcend through the recovery of a deeper relation to being. And since he rejects technical regression while leaving no room for a better technological future, it is difficult to see in what that relation would consist beyond a mere change of attitude. Surely these ambiguities indicate problems in his approach.[4]

3. In a 1949 lecture, Heidegger explained: "Agriculture is now the mechanized food industry, in essence the same as the manufacturing of corpses in gas chambers and extermination camps, the same as the blockade and starvation of nations, the same as the production of hydrogen bombs" (quoted in Rockmore 1992, 241).

4. I would of course be willing to revise this view if shown how Heidegger actually envisages technological change. What I have heard from his defenders is principally waffling on the attitude/device ambiguity described here. Yes, Heidegger envisages change in "technological thinking," but how is this change supposed to affect the design of actual devices? The lack of an answer to this question leaves me in some doubt as to the supposed relevance of Heidegger's work to ecology. One enthusiastic defender informed me that art and technique would merge anew in a Heideggerian future, but was unable to cite a text. That would indeed historicize Heidegger's theory, but in a way resembling Marcuse's position in *An Essay on Liberation* (1969) with its eschatological concept of an aesthetic revolution in technology. It is not clear how the case for Heidegger is fundamentally improved by this shift, which would not make much difference to the substantive arguments presented here. For an interesting defense of Heidegger's theory of technology that eschews mystification, see Dreyfus 1995.

A Contemporary Critique
Technology and Meaning

Heidegger holds that the restructuring of social reality by technical action is inimical to a life rich in meaning. The Heideggerian relation to being is incompatible with the overextension of technological thinking. It seems, therefore, that identification of the structural features of enframing can found a critique of modernity. I intend to test this approach through an evaluation of some key arguments in the work of Albert Borgmann, the leading American representative of philosophy of technology in the essentialist vein.[5]

Borgmann's social critique is based on the concept of the "device paradigm" as the formative principle of a technological society that aims above all at efficiency. In conformity with this paradigm, modern technology separates off the good or commodity it delivers from the contexts and means of delivery. Thus the heat of the modern furnace appears miraculously from discreet sources in contrast with the old wood stove that stands in the center of the room and is supplied by regular trips to the woodpile. The microwaved meal emerges effortlessly and instantly from its plastic wrapping at the individual's command in contrast with the laborious operations of a traditional kitchen serving the needs of a whole family.

The device paradigm offers gains in efficiency, but at the cost of distancing us from reality. Let us consider the substitution of fast food for the traditional family dinner. To common sense, well-prepared fast food appears to supply nourishment without needless social complications. Functionally considered, eating is a technical operation that may be carried out more or less efficiently. It is a matter of ingesting calories, a means to an end, while all the ritualistic aspects of food consumption are secondary to biological need. But what Borgmann calls "focal things" that gather people in meaningful activities that have value for their own sake cannot survive this functionalizing attitude.

The unity of the family, ritually reaffirmed each evening, no longer has a comparable locus of expression. One need not claim that the rise of fast food causes the decline of the traditional family to believe that there is a significant connection. Simplifying personal access to food scatters people

5. For another interesting contemporary approach that complements Borgmann's, see Simpson 1995. Simpson denies that he is essentializing technology, and yet he works throughout his book with a minimum set of invariant characteristics of technology as though they constituted a "thing" he could talk about independent of the sociohistorical context (Simpson 1995, 15–16, 182). That context is then consigned to a merely contingent level of influences, conditions, or consequences rather than being integrated to the conception of technology itself.

who need no longer construct the rituals of everyday interaction around the necessities of daily living. Focal things require a certain effort, it is true, but without that effort, the rewards of a meaningful life are lost in the vapid disengagement of the operator of a smoothly functioning machinery (Borgmann 1984, 204 ff.).

Borgmann would willingly concede the usefulness of many devices, but the generalization of the device paradigm, its substitution for simpler ways in every context of daily life, has a deadening effect. Where means and ends, contexts and commodities are strictly separated, life is drained of meaning. Individual involvement with nature and other human beings is reduced to a bare minimum, and possession and control become the highest values.

Borgmann's critique of technological society usefully concretizes themes in Heidegger. His dualism of device and meaning is also structurally similar to Habermas's distinction of work and interaction (Habermas 1970). This dualism always seems to appear where the essence of technology is in question.[6] It offers a way of theorizing the larger philosophical significance of the modernization process, and it reminds us of the existence of dimensions of human experience that are suppressed by facile scientism and the uncritical celebration of technology. Borgmann's contrast between the decontextualization of the device and the essentially contextual focal thing reprises Heidegger's distinction between modern technological enframing and the "gathering" power of traditional craft production that draws people and nature together around a materialized site of encounter. Borgmann's solution, bounding the technical sphere to restore the centrality of meaning, is reminiscent of Habermas's strategy (although apparently not due to his influence). It offers a more understandable response to invasive technology than anything in Heidegger.

However, Borgmann's approach suffers from both the ambiguity of Heidegger's original theory and the limitations of Habermas's. We cannot tell for sure if he is merely denouncing the modern attitude toward technology or technological design, and in the latter case, his critique is so broad it offers no criteria for the constructive reform of technology itself. He would probably agree with Habermas's critique of the colonization of the lifeworld, although he improves on that account by discussing the all-important role of technology in modern social pathologies. But like Habermas, he lacks a concrete sense of the intricate connections of technology and culture beyond the few essential attributes on which his critique focuses. Since those attributes have largely negative consequences,

---

6. In the next part of this paper I will attempt to resituate this dualism within technology itself, to avoid the ontologized distinctions characteristic of essentialism.

we get no sense from the critique of the many ways in which the pursuit of meaning is intertwined with technology. And as a result, Borgmann imagines no significant restructuring of modern society around culturally distinctive technical alternatives that might preserve and enhance meaning.

But how persuasive is this objection to Borgmann's approach? After all, neither Russian nor Chinese communism, neither Islamic fundamentalism nor so-called Asian values have inspired a fundamentally distinctive stock of devices. Why *not* just reify the concept of technology and treat it as a singular essence? The problem with that is the existence of smaller but still significant differences that may become more important in the future rather than less so as essentialists assume. What is more, those differences often concern precisely the issues identified by Borgmann as central to a humane life. They determine the nature of community, education, medical care, work, our relation to the natural environment, the functions of devices such as computers and automobiles, in ways either favorable or unfavorable to the preservation of meaning and focal things. Any theory of the essence of technology that forecloses the future therefore begs the question of difference in the technical sphere.

Interpreting the Computer

I would like to pursue this contention further with a specific example that illustrates concretely my reasons for objecting to Borgmann's approach. The example I have chosen, human communication by computer, is one on which Borgmann has commented fairly extensively. While not everyone who shares the essentialist view will agree with his very negative conclusions, his position adequately represents that style of technology critique and is therefore worth evaluating here at some length.[7]

Borgmann introduces the term "hyperintelligence" to refer to such developments as electronic mail and the Internet (Borgmann 1992, 102 ff.). Hyperintelligent communication offers unprecedented opportunities for people to interact across space and time, but paradoxically it also distances those it links. No longer are the individuals "commanding presences" for each other; they have become disposable experiences that can be turned on and off like water from a faucet. The person as a focal thing has become a commodity delivered by a device. This new way of relating has weakened connection and involvement while extending its range. What happens to the users of the new technology as they turn away from face-to-face contact?

> Plugged into the network of communications and computers, they seem to enjoy omniscience and omnipotence; severed from their

7. For another critique of the computer similar to Borgmann's, see Slouka 1995.

network, they turn out to be insubstantial and disoriented. They no longer command the world as persons in their own right. Their conversation is without depth and wit; their attention is roving and vacuous; their sense of place is uncertain and fickle. (Borgmann 1992, 108)

This negative evaluation of the computer can be extended to earlier forms of mediated communication. In fact Borgmann does not hesitate to denounce the telephone as a hyperintelligent substitute for more deeply reflective written correspondence (Borgmann 1992, 105).

There is an element of truth in this critique. On the networks, the pragmatics of personal encounter are radically simplified, reduced to the protocols of technical connection. It is easy to pass from one social contact to another, again following the logic of the technical network that supports ever more rapid commutation. However, Borgmann's conclusions are too hastily drawn and simply ignore the role of social contextualizations in the appropriation of technology. A look, first at the history of computer communication and second at its innovative applications today refutes his overly negative evaluation. We will see that the real struggle is not between the computer and low-tech alternatives, but within the realm of possibilities opened by the computer itself.

In the first place, the computer was not destined by some inner techno-logic to serve as a communications medium. The major networks, such as the French Teletel or the Internet were originally conceived by technocrats and engineers as instruments for the distribution of data. What actually happened in the course of the implementation of these networks? Users appropriated them for unintended purposes and converted them into communications media. Soon they were flooded with messages that were considered trivial or offensive by their creators. Teletel quickly became the world's first and largest electronic singles bar (Feenberg 1995, chap. 7). The Internet is overloaded with political debates dismissed as "trash" by unsympathetic critics. Less visible, at least to journalists, but more signifi-cant, there gradually appeared all sorts of other applications of computers to human communication, from business meetings to education, from discussions among medical patients, literary critics, and political activists to online journals and conferences.

How does Borgmann's critique fare in the light of this history? It seems to me there is an element of ingratitude in it. Because Borgmann takes it for granted that the computer is useful for human communication, he appreciates neither the process of making it so nor the hermeneutic transformation it underwent in that process. He therefore also overlooks the political implications of the history sketched above. Today the networks constitute a fundamental scene of human activity. To impose a narrow regimen of data transmission, to the exclusion of all human contact, would surely be perceived as totalitarian in any ordinary institution. Why is it not a liberation to break such limitations in the virtual world that now surrounds us?

In the second place, Borgmann's critique ignores the variety of communicative interactions mediated by the networks. No doubt he is right that human experience is not enriched by much of what goes on there. But a full record of the face-to-face interactions occurring in the hall rooms of his university would likely be no more uplifting. The problem here is that we tend to judge the face-to-face at its memorable best and the computer-mediated equivalent at its transcribed worst. Borgmann simply ignores more interesting uses of computers, such as the original research applications of the Internet and teaching applications that show great promise (Harasim et al. 1995). It might surprise Borgmann to find the art of reflective letter writing reviving in these contexts.

Consider for example the discussion group on the Prodigy Medical Support Bulletin Board devoted to ALS (amyotrophic lateral sclerosis or Lou Gehrig's disease). In 1995, when I studied it, there were about five hundred patients and caregivers reading exchanges in which some dozens of participants were actively engaged (Feenberg et al. 1996). Much of the conversation consisted of exchanges of feelings about dependency, illness, and dying. There was a long running discussion of problems of sexuality. Patients and caregivers wrote in both general and personal terms about the persistence of desire and the obstacles to satisfaction. The frankness of this discussion may owe something to the anonymity of the online environment, appropriated here for very different purposes than those Borgmann criticizes. Here the very limitations of the medium open doors that might have remained closed in a face-to-face setting.

These online patient meetings have the potential for changing the accessibility, the scale, and the speed of interaction of patient groups. Face-to-face self-help groups are small and localized. With the exception of AIDS patients they have wielded no political power. If AIDS patients have been the exception, it is not because of the originality of their demands: patients with incurable illnesses have been complaining bitterly for years about the indifference of physicians and the obstacles to experimental treatments. What made the difference was that AIDS patients were networked politically by the gay rights movement even before they were caught up in a network of contagion (Epstein 1996, 229). Online networks may similarly empower other patient groups. In fact, Prodigy discussion participants established a list of priorities they presented to the ALS Society of America.

Computer networking may thus feed into the rising demand by patients for more control over their own medical care. In that case, subversive rationalization of the computer would enable a parallel transformation of medicine.

It is difficult to see any connection between these applications of the computer and Borgmann's critique of hyperintelligence. Is this technologically mediated process by which dying people come together despite paralyzing illness to discuss and mitigate their plight a mere instance of "technological thinking"? Certainly not. But then how would Heidegger incorporate an understanding of it into his theory, with its reproachful attitude toward modern technology in general? The ambiguities of the computer are far from unique. In fact they are typical of most technologies, especially in the early phases of their development. Recognizing this malleability of technology, we can no longer rest content with globally negative theories that offer only condemnation of the present and no guidance for the future.

Borgmann's critique of technology pursues the larger connections and social implications masked by the device paradigm. To this extent it is genuinely dereifying. But insofar as it fails to incorporate these hidden social dimensions into the concept of technology itself, it remains still partially caught in the very way of thinking it criticizes. His theory hovers uncertainly between a description of how we encounter technology and how it is designed. Technology, i.e., the real-world objects so designated, both is and is not the problem, depending on whether the emphasis is on its fetish form as pure device or our subjective acceptance of that form. In neither case can we change technology in itself. At best, we can hope to overcome our attitude toward it through a spiritual movement of some sort.[8]

I propose a very different conceptualization that includes the integration of technologies to larger technical systems and nature, and to the symbolic orders of ethics and aesthetics, as well as their relation to the life and learning processes of workers and users and the social organization of work and use. On the essentialist account, one could still admit the existence of these aspects of technical life, but they would be extrinsic social influences

---

8. Andrew Light has argued that I underestimate the significance of Borgmann's distinction between device and thing for an understanding of the aesthetics of everyday life. The distinction is useful for developing a critique of mass culture and could provide criteria for subversive rationalizations of the commodified environment. The story of the ALS patients told here could be interpreted in this light as an example of the creation of a meaningful community through the creative appropriation of the hyperreal technological universe Borgmann describes (Light 1996, chap. 9). I am in general agreement with this revision of Borgmann's position, but in some doubt as to whether Borgmann himself would be open to it.

or consequences. Essentialism proposes to treat all these dimensions of technology as merely contingent and to hand them over to sociology while retaining the unchanging essence for philosophy. A certain conception of philosophy is implied in this approach.

INSTRUMENTALIZATION THEORY
The Irony of Parmenides

Heidegger and Borgmann have undoubtedly put their fingers on significant aspects of the technical phenomenon, but have they identified its "essence"? They seem to believe that technical action has a kind of unity that defies the complexity and diversity, the profound sociocultural embeddedness that twenty years of increasingly critical history and sociology of technology have discovered in it. Yet to dissolve the technical realm into the variety of its manifestations, as constructivists sometimes demand, would effectively block philosophical reflection on modernity. The problem is to find a way of incorporating these recent advances in technology studies into a conception of technology's essence rather than dismissing them, as philosophers tend to do, as social influences on a reified technology "in itself" conceived apart from society.[9] The solution to this problem is a radical redefinition of technology that crosses the usual line between artifacts and social relations assumed by common sense and philosophers alike.

The chief obstacle to this solution is the unhistorical understanding of essence to which most philosophers are committed. I propose, therefore, a kind of compromise between the philosophical and the social-scientific perspective. In what follows, I will attempt to provide a *systematic* locus in the concept of essence for the sociocultural variables that diversify technology's historical realizations. On these terms, the "essence" of technology is not simply those few distinguishing features shared by all types of technical practice that are identified in Heidegger and Borgmann. Those constant determinations are not a technological a priori, but are partial moments abstracted from the various concrete stages of a process of development.

I now attempt to work out this historical concept of essence as it applies to technology. Is the result still sufficiently "philosophical" to qualify as philosophy? In claiming that it is, I realize that I am challenging a certain prejudice against the concrete that is an occupational hazard of philosophy. Plato is usually blamed for this, but in a late dialogue Parmenides mocks the young Socrates' reluctance to admit that there are ideal forms of "hair or mud or dirt or any other trivial and undignified objects" (Cornford

---

9. Like the turtles in Feynman's famous story, the hermeneutics of technology "goes all the way down."

1957, 130C–E).[10] Surely the time has come to let the social dimension of technology into the charmed circle of philosophical reflection. Let me now offer, if only schematically, a way of achieving this.

Primary Instrumentalization: Functionalization

Substantivist philosophies of technology drew attention away from the practical question of what technology *does* to the hermeneutic question of what it *means*.[11] The question of meaning has become defining for philosophy of technology as a distinct branch of humanistic reflection. More recently, constructivism has sharpened reflection on a third range of questions concerning who makes technology, why, and how. My strategy here will consist in incorporating answers to the substantivist and constructivist questions into a single framework with two levels. The first of these levels corresponds more or less to the philosophical definition of the essence of technology, the second to the concerns of social sciences. However, merging them in the framework of a two-level critical theory transforms both.

This approach marks a break with essentialism, which privileges one attribute of technical artifacts—function—over all the others. This choice appears obvious because of the tacit identification of the functional and physical properties of the artifacts. Whereas social attributes such as the place of technologies in vocations are relational and seem therefore not to belong to technical artifacts proper, function looks like a nonrelational property of technology in itself. But in reality function is just as social as the rest. For example, the sharpness of a knife is indeed a measurable physical property, but sharpness is only a function rather than a hazard or a matter of pure indifference, through a social construction. All the properties of technologies are relational insofar as we recognize their technological character. As mere physical objects abstracted from all relations, these artifacts have no function and hence no properly *technological* character at all.[12] But if function is a social property of technological artifacts, then it should not be privileged over other equally important social dimensions.

On this account, the essence of technology has not one but two aspects,

10. Compare Latour's account of a similar episode involving Heraclitus (Latour 1993, 65–66).

11. Many of the ideas in this section and the next were first presented in an earlier version in Feenberg (1991, chap. 8).

12. Thus considered as just a thing, an automobile is no better parked with its wheels on the ground than in the air. It is only insofar as it is *assigned* a function that it must be considered as a technical device and placed squarely right side up. The spontaneous confusion between these two levels is no doubt less likely in non-Western societies. One who lives in a Japanese home with both tatami mat and wooden floors is well aware that what's underfoot is not just a thing on which to walk but also a whole national tradition.

an aspect that explains the *functional constitution* of technical objects and subjects, which I call the "primary instrumentalization," and another aspect, the "secondary instrumentalization," focused on the *realization* of the constituted objects and subjects in actual technical networks and devices. Essentialism offers insight only into the primary instrumentalization by which functions are separated from the continuum of everyday life. Primary instrumentalization characterizes technical relations in every society, although its emphasis, range of application, and significance vary greatly. Technique includes those constant features in historically evolving combinations with a secondary instrumentalization that includes many other aspects of technology. The characteristic distinctions between different eras in the history of technology result not only from new inventions, but also from different structurings of these various moments.

The primary instrumentalization consists in four reifying moments of technical practice: decontextualization, reductionism, autonomization, and positioning.

Decontextualization. To reconstitute natural objects as technical objects, they must be de-worlded, artificially separated from the context in which they are originally found so as to be integrated to a technical system. The isolation of the object exposes it to a utilitarian evaluation. The tree conceived as lumber and eventually cut down, stripped of bark, and chopped into boards is encountered through its usefulness rather than in all its manifold interconnections with its environment and the other species with which it coexists. The isolated object reveals itself as containing technical schemas, potentials in human action systems, which are made available by decontextualization. Thus inventions such as the knife or the wheel take qualities such as the sharpness or roundness of some natural thing, a rock or tree trunk, for example, and release them as technical properties. The role these qualities may have played in nature is obliterated in the process. Nature is fragmented into usable bits and pieces that appear as technically useful after being abstracted from all specific contexts.

Reductionism. Reductionism refers to the process in which the de-worlded things are simplified, stripped of technically useless qualities, and reduced to those aspects through which they can be enrolled in a technical network. These are the qualities of primary importance to the technical subject, the qualities perceived as essential to the accomplishment of a technical program. I will therefore call them "primary qualities," it being understood that their primacy is relative to the subject's program. Quantification is the most complete reduction to primary qualities. "Secondary qualities" are what remains, including those dimensions of the object that may have been most significant in the course of its pretechnical history. The secondary qualities of the object contain its potential for self-development. The tree trunk, reduced to its primary quality of roundness in becoming a wheel, loses its secondary qualities as a habitat, a source of shade, and a living, growing member of its species. The Heideggerian enframing is the reduction of all of reality to such primary qualities.

Autonomization. The subject of technical action isolates itself as much as possible from the effects of its action on its objects. Metaphorically speaking, it thus violates Newton's third law, according to which "for every action there is an equal and opposite reaction." The actor and the object in mechanics belong to the same system, hence the reciprocity of their relations. This is not a bad description of ordinary human interactions. A friendly remark is likely to elicit a friendly reply, a rude one, a correspondingly unpleasant response. By contrast, technical action "autonomizes" the subject. This is accomplished by interrupting the feedback between the object and the actor. In an apparent exception to Newton's law, the technical subject has a big impact on the world, but the world has only a very small return impact on the subject. The hunter experiences a slight pressure on his shoulder as the bullet from his gun strikes the rabbit; the driver hears a faint rustling in the wind as he hurtles a ton of steel down the highway. Administrative action too, as a technical relationship between human beings, presupposes the autonomization of the manager as subject.

Positioning. Technical action controls its objects through their laws. There is thus a moment of passivity with respect to those laws in even the most violent technological intervention. The technical conforms with Francis Bacon's dictum "Nature to be commanded must be obeyed." The laws of combustion rule over the automobile's engine as the laws of the market govern the investor on the stock market. In each case, the subject's action consists not in modifying the law of its objects, but in using that law to advantage. Of course there are considerable differences between these two examples; for one thing the engine is an artifact designed in conformity with natural law whereas the investor can only adopt a strategic position with respect to the objective process of the market. Location, as they say in real estate, is everything: fortunes are made by being in the right place at the right time. By positioning itself strategically with respect to its objects, the subject turns their inherent properties to account. The management of labor and the control of the consumer through product design have a similar situational character. There are no natural laws of worker and consumer behavior that would allow one to design them as one would a machine, but one can position oneself so as to induce them to fulfill preexisting programs they would not otherwise have chosen. In these social domains, Baconian obedience is a kind of navigation in the turbulent waters of interests, expectations, and fantasies that cannot be controlled, only anticipated and used.

Secondary Instrumentalization: Integration

The primary instrumentalization lays out in skeletal fashion the basic technical relations. Far more is necessary for those relations to yield an actual system or device: technique must be *integrated* with the natural, technical, and social environments that support its functioning. The process of integration compensates for some of the reifying effects of the primary instrumentalization. Here technical action turns back on itself and its actors as it is realized concretely. In the process, it reappropriates some of the dimensions of contextual relatedness and self-development from which abstraction was originally made in establishing the technical relation. The underdetermination of technological development leaves room for social interests and values to participate in the process of realization. As decontextualized elements are combined, these interests and values assign functions, orient choices, and ensure congruence between technology and society at the technical level itself.

On the basis of this concept of integration, I argue that the essence of technique must include a secondary instrumentalization that works with dimensions of reality from which abstraction is made at the primary level. This level of includes four moments: systematization, mediation, vocation, and initiative.

Systematization. To function as an actual device, isolated, decontextualized technical objects must be combined with other technical objects and reembedded in the natural environment. Systematization is the process of making these combinations and connections, in Latour's terms, of "enrolling" objects in a network (Latour 1992). Thus individual technical objects—wheels, a handle, a container—are brought together to form a device such as a wheelbarrow. Add paint to protect the wheelbarrow from rust and the device has been embedded in its natural environment as well. The process of technical systematization is central to designing the extremely long and tightly coupled networks of modern technological societies but plays a lesser role in traditional societies where technologies may be more loosely related to each other functionally, but correspondingly better adapted to the natural and social environment.

Mediation. In all societies, ethical and aesthetic mediations supply the simplified technical object with secondary qualities that seamlessly embed it into its new social context. The ornamentation of artifacts and their investment with ethical meaning are integral to production in all traditional cultures. The choice of a type of stone or feather in the making of an arrow may be motivated not only by sharpness and size, but also by various ritual considerations that yield an aesthetically and ethically expressive object. Heidegger's chalice exemplifies such expressive design. By contrast, production and aesthetics are differentiated in modern industrial societies. The goods are produced first, and then superficially styled and packaged for distribution. The social insertion of the industrial object appears as an afterthought. From this results the unfortunate separation of technique and aesthetics characteristic of our societies; unfortunate, I would argue, because no one denies the prevailing ugliness of so much of our work and urban environment. Ethical limits too are overthrown in the breakdown of religious and craft traditions. Recently, medical advances and environmental crises have inspired new interest in the ethical limitation of technical power. These limitations are eventually embodied in modified designs that condense considerations of efficiency with ethical values. A similar condensation appears in the aesthetics of good industrial design. Thus mediations remain an essential aspect of the technical process even in modern societies.

Vocation. The technical subject appears autonomous only when its actions are isolated from its life process. Taken as a whole, the succession of its acts adds up to a craft, a vocation, a way of life. The subject is just as deeply engaged as the object—Newton is vindicated—but in a different register. The doer is transformed by its acts: the individual of our earlier example, who fires a rifle at a rabbit, will become a hunter with the corresponding attitudes and dispositions should he pursue such activities professionally. Similarly, the chopper of wood becomes a carpenter, the typer at the keyboard a writer, and so on. These human attributes of the technical subject define it at the deepest levels, physically, as a person, and as a member of a community of people engaged in similar activities. "Vocation" is the best term we have for this reverse impact on users of their involvement with the tools of their trade. In traditional cultures and even in some modern ones, such as the Japanese, the concept of vocation or "way" is not associated with any particular kind of work, but in most industrial societies it is reserved for medicine, law, teaching, and similar professions. Perhaps this is an effect of wage labor, which substitutes temporary employment under administrative control for the lifelong craft of the independent producer, thereby reducing both the impact of any particular skill on the worker and the individual responsibility for quality implied in vocation.

Initiative. Finally, strategic control of the worker and consumer through positioning is to some extent compensated by various forms of tactical initiative on the part of the individuals submitted to technical control. Before the rise of capitalist management, cooperation was often regulated by tradition or paternal authority, and the uses of the few available devices so loosely prescribed that the line between producer programs and user appropriations was often blurred. It is capitalism that has led to the sharp split between positioning and initiative, and the marginalization of the latter. Nevertheless, a certain margin of maneuver belongs to subordinated positions in the capitalist technical hierarchy. That margin can support conscious cooperation in the coordination of effort and creative user appropriation of devices and systems.

We have examples of alternatives to bureaucratic control in the collegial organization of certain professionals such as teachers and doctors. Refined and generalized, collegiality might be able to reduce the operational autonomy of management, substituting complex self-organization for control from above.[13] In the sphere of consumption, we have numerous examples, such as the computer, where creative appropriations by users result in significant design changes. As noted above, this is how human communication became a standard functionality of a technology that was originally conceived by computer professionals as a device for calculating and storing data.

The secondary instrumentalization constitutes a *reflexive metatechnical practice* that supports the reintegration of object with context, primary with secondary qualities, subject with object, and leadership with group. It treats functionality as raw material for higher-level forms of technical action. There is of course something paradoxical about this association of reflexivity with technology; in the substantivist framework technical rationality is supposed to be blind to itself. Reflection is reserved for another type of thought competent to deal with such important matters as aesthetics and ethics. We have here the familiar split between nature and *Geist* and their corresponding sciences.

CAPITALISM AND SUBSTANTIVE THEORY OF TECHNOLOGY

Substantivism identifies technology in general with modern Western technology. There are undoubtedly universal achievements underlying that technology, many of them borrowed from other civilizations in the first place. However, the particular form in which these achievements are realized in the West incorporates values that are not at all universal but belong to a definite culture and economic system. Modern Western technology is uniquely rooted in capitalist enterprise. As such it privileges the narrow goals of production and profit. The enterprise organizes the technical control of

13. For a discussion of this theme in the context of modern production, see Hirschhorn 1984.

its workers and dispenses with the traditional responsibilities for persons and places that accompanied technical power in the past. It is this peculiar indifference of modern capitalism to its social and natural environment that frees the entrepreneur to extend technical control to the labor force, the organization of work, and aspects of the natural environment that were formerly protected from interference by custom and tradition.[14] To define technology as such on these terms is ethnocentric.

What does a broader historical picture show? Contrary to Heideggerian substantivism, there is nothing unprecedented about our technology. Its chief features, such as the reduction of objects to raw materials, the use of precise measurement and plans, the management of some human beings by others, large scales of operation, are commonplace throughout history. The same could be said of Borgmann's device paradigm. It is the exorbitant role of these features that is new, and of course the consequences of that are truly without precedent.

Those consequences include obstacles to secondary instrumentalization wherever integrative technical change would threaten the maximum exploitation of human and natural resources. These obstacles are not merely ideological but are incorporated into technological designs. Only a critique of those designs is adequate to the problems, and only such a critique can uncover the technical potential available to solve them. If we define technology exclusively in terms of the dimensions privileged by modern capitalism, we ignore many currently marginalized practices that belonged to it in the past and may prove central to its future development. For example, before Taylor, technical experience was essentially vocational experience. Using technology was associated with a way of life; it was a matter not just of productivity but also of character development. This link was broken when capitalist deskilling transformed workers into mere objects of technique, no different from raw materials or machines. Here, not in some mysterious dispensation of being, lies the source of the "total mobilization" of modern times.

Similarly, the old craft guilds with their collegial forms of organization have been replaced by capitalist management. Collegiality, like vocational investment in work, survives only in a few specialized and archaic settings such as universities. Not the essence of technology but the requirements of

14. It is important to resist the temptation to dismiss capitalism as a factor on the grounds that Soviet communism and its imitators did no different and no better. These regimes never constituted an alternative; they followed the capitalist example in essential respects, importing technology and management methods, and in some cases, such as protection of the environment, carrying its irresponsibility even further. I have discussed this problem in more detail in Feenberg 1991, chap. 6.

capitalist economics explain this outcome (Braverman 1974; Noble 1984). A different social system that restored the role of the secondary instrumentalizations would determine a different type of technical development in which it would be possible to recover these traditional technical values and organizational forms in new ways. Thus reform of this society would involve not merely limiting the reach of the technical, but building on its intrinsic democratic potential.

Because its hegemony rests on extending technical control beyond traditional boundaries to embrace the labor force, capitalism tends to identify technique as a whole with the instrumentalizations through which that control is secured. Meanwhile, other aspects of technique are forgotten or treated as nontechnical. It is this capitalist technical rationality that is reflected in the essentialism of Heidegger and Borgmann. Because they characterize technology by the privileged instrumentalizations of capitalist modernity, they are unable to develop a socially and historically concrete conception of it. They take their own labor of abstraction, by which they eliminate the sociohistorical dimensions of technical action, for evidence of the nonsocial nature of technology.

Conclusion: The Gathering

In conclusion I would like to return briefly to Heidegger's critical account of our times to see how it stands up to the theory I have presented. For Heidegger modern technology is stripped of meaning by contrast with the meaningful tradition we have lost. Even the old technical devices of the past shared in this lost meaning. For example, Heidegger shows us a jug "gathering" the contexts in which it was created and functions (Heidegger 1971). The concept of gathering resembles Borgmann's notion of the "focal thing." These concepts dereify the thing and activate its intrinsic value and manifold connections with the human world and nature. Heidegger wants to show us the way back to another mode of perception that belongs to the lost past or perhaps to a future we can only dimly imagine. In that mode we share the earth with things rather than reducing them to mere resources. Perhaps a redeemed *techne* will someday disclose the potentiality of what is rather than attempting to remake the world in the human image.

The undeniable insight here is that every making must also include a letting be, an active connection to what remains untransformed by that making. This is Heidegger's concept of the "earth" as a reservoir of possibilities beyond human intentions. In denying that connection the technocratic conception of technology defies human finitude. The earth, nature, can never become a human deed because all deeds presuppose it (Feenberg 1986, chap. 8). Yet I would like to share David Rothenberg's interpretation, according to which Heidegger would also want us to recognize that our contact with the earth is technically mediated: what comes into focus as nature is not the pure immediate but what lies at the limit of *techne* (Rothenberg 1993, 195 ff.). Despite occasional lapses into romanticism, this is after all the philosopher who placed readiness-to-hand at the center of *Dasein's* world.

The cogency of Heidegger's critique thus ultimately comes down to whether technology is *fundamentally* Promethean. Only then would it make sense to demand liberation from it rather than reform of it. It is true that the dominant ideology, based on a narrow functionalism, leaves little room for respect for limits of any kind. But we must look beyond that ideology to the realities of modern technology and the society that depends on it. The failure of Heidegger and other thinkers in the humanistic tradition to engage with actual technology is not to their credit but reveals the boundaries of a certain cultural tradition.[15]

Beyond those boundaries we discover that technology also "gathers" its many contexts through secondary instrumentalizations that integrate it to the world around it. Naturally, the results are quite different from the craft tradition Heidegger idealizes, but nostalgia is not a good guide to understanding technology. When modern technical processes are brought into compliance with the requirements of nature or human health, they incorporate their contexts into their very structure, as truly as the jug, chalice, or bridge that Heidegger holds out as models of authenticity. Our models should be such things as reskilled work, medical practices that respect the person, architectural and urban designs that create humane living spaces, computer designs that mediate new social forms. These promising innovations all suggest the possibility of a general reconstruction of modern technology so that it gathers a world to itself rather than reducing its natural, human, and social environment to mere resources. It is now the task of philosophy of technology to recognize that possibility and to criticize the present in the light of it.

REFERENCES

Borgmann, Albert. 1984. *Technology and the Character of Contemporary Life: A Philosophical Inquiry.* Chicago: University of Chicago Press.
———. 1992. *Crossing the Postmodern Divide.* Chicago: University of Chicago Press.
Braverman, Harry. 1974. *Labor and Monopoly Capital.* New York: Monthly Review.
Cornford, Francis. 1957. *Plato and Parmenides.* New York: Liberal Arts Press.

15. For a discussion of that tradition as it shapes philosophy of technology, see Mitcham 1994.

Dreyfus, Hubert. 1995. "Heidegger on Gaining a Free Relation to Technology." In *Technology and the Politics of Knowledge*. Ed. Andrew Feenberg and Alastair Hannay. Bloomington: Indiana University Press.

Epstein, Steven. 1996. *Impure Science: AIDS, Activism, and the Politics of Knowledge*. Berkeley and Los Angeles: University of California Press.

Feenberg, Andrew. 1986. *Lukács, Marx, and the Sources of Critical Theory.* New York: Oxford University Press.

———. 1991. *Critical Theory of Technology.* New York: Oxford University Press.

———. 1995. *Alternative Modernity: The Technical Turn in Philosophy and Social Theory.* Berkeley and Los Angeles: University of California Press.

Feenberg, Andrew, J. Licht, K. Kane, K. Moran, and R. Smith. 1996 "The Online Patient Meeting." *Journal of the Neurological Sciences* 139:129–31.

Gras, Alain. 1994. *Grandeur et dépendence: Sociologie des macro-systèmes techniques.* Paris: Presse Universitaire de France.

Habermas, Jürgen. 1970. "Technology and Science as Ideology." In *Toward a Rational Society.* Trans. Jeremy Shapiro. Boston: Beacon Press.

Harasim, Linda, Star Roxanne Hiltz, Lucio Teles, and Murray Turoff. 1995. *Learning Networks: A Field Guide to Teaching and Learning Online.* Cambridge: MIT Press.

Heidegger, Martin. 1959. *An Introduction to Metaphysics.* New York: Doubleday Anchor.

———. 1971. "The Thing." In *Poetry, Language, and Thought.* Trans. Albert Hofstadter. New York: Harper and Row.

———. 1977a. *The Question Concerning Technology.* Trans. William Lovitt. New York: Harper and Row.

———. 1977b. "Only a God Can Save Us Now." Trans. D. Schendler. *Graduate Faculty Philosophy Journal* 6:5–27.

Hirschhorn, Larry. 1984. *Beyond Mechanization: Work and Technology in a Postindustrial Age.* Cambridge: MIT Press.

Hughes, Thomas. 1989. "The Evolution of Large Technological Systems." In *The Social Construction of Technological Systems.* Ed. Trevor Pinch, Thomas Hughes, and Wiebe Bijker. Cambridge: MIT Press.

Latour, Bruno. 1992. "Where Are the Missing Masses? The Sociology of a Few Mundane Artifacts." In *Shaping Technology/Building Society: Studies in Sociotechnical Change.* Ed. Wiebe Bijker and John Law. Cambridge: MIT Press.

———. 1993. *We Have Never Been Modern.* Trans. C. Porter. Cambridge: Harvard University Press.

Light, Andrew. 1996. "Nature, Class, and the Built World: Philosophical Essays between Political Ecology and Critical Technology." Ph.D. diss., University of California, Riverside.

Marcuse, Herbert. 1969. *An Essay on Liberation.* Boston: Beacon.

Mitcham, Carl. 1994. *Thinking through Technology: The Path between Engineering and Philosophy.* Chicago: University of Chicago Press.

Noble, David. 1984. *Forces of Production.* New York: Oxford University Press.

Pinch, Trevor, Thomas Hughes, and Wiebe Bijker, eds. 1989. *The Social Construction of Technological Systems.* Cambridge: MIT Press.

Rockmore, Tom. 1992. *On Heidegger's Nazism and Philosophy.* Berkeley and Los Angeles: University of California Press.

Rothenberg, David. 1993. *Hand's End: Technology and the Limits of Nature.* Berkeley and Los Angeles: University of California Press.

Simpson, Lorenzo. 1995. *Technology, Time, and the Conversations of Modernity.* New York: Routledge.

Slouka, Mark. 1995. *War of the Worlds.* New York: Basic Books.

Sluga, Hans. 1993. *Heidegger's Crisis: Philosophy and Politics in Nazi Germany.* Cambridge: Harvard University Press.

SEVENTEEN

Philosophy in the Service of Things

David Strong

If we assume that Albert Borgmann is more or less correct about the illusions of technology and postmodernism and about his hopeful alternatives of focal things and postmodern realism, where do we go from here philosophically? How should we do philosophy? What will the character of philosophy be? What kinds of tasks will it yet face? Here I summarize and characterize some of the types of arguments Borgmann employs in order to present a broad and suggestive outline of what I take to be his new way of doing philosophy.

Borgmann has written his "technology book" *(Technology and the Character of Contemporary Life,* or *TCCL)* and "postmodernism book" *(Crossing the Postmodern Divide,* or *CPD),* but he has not yet written his "philosophy book." When readers of his two books want to know what his general philosophy is, they must do some work to unify and make explicit what is mostly implicit in these two books. On the one hand, he seems to reject much of traditional philosophy. For instance, in *CPD,* he writes,

> The conventional norms of ethical theory have as much bearing on hyperreality as digestion has on sucrose polyester, a hyperreal fat. . . . Nor does hyperreality as such appear to be in any sense a moral problem. Just as olestra is "digestive inert," hyperreality appears to be morally inert. . . . Traditional theories of reality, what philosophers call ontologies, are as powerless to explicate the difference between the real and the hyperreal as are conventional theories of morality. Hyperreality is ontological inert, one might say. (Borgmann 1992, 94–95)

Yet, on the other hand, Borgmann maintains that to evaluate the difference between the real and hyperreal "is a task that is at once ontological, moral, aesthetic, theological, and political" (Borgmann 1992, 96). The challenge is to do philosophy in a radically new way. As opposed to traditional philosophy, Borgmann's technology and postmodernism books exemplify what I will call "philosophy in the service of things." What are the character and potential of a philosophy in the service of things? I will consider this question in three ways. First, what is the need for a philosophy in the service of things? Second, how do Borgmann's two books exemplify this new way of doing philosophy? Third, what philosophical challenges does this philosophy of things face, and what is its unique achievement?

The Need for a Philosophy in the Service of Things

Ultimately a philosophy of things is needed because traditional philosophy fails to challenge technology at a radical level *and* to account for things in their own right. More generally, philosophical thinking has been taken as fully adequate for dealing with substantive concerns, and, in Borgmann's view, it is not equal to that task. He constructs arguments for these points throughout his two books.

Philosophy is limited first in what it can really do. Here Borgmann distinguishes between substantive or significant truth and trivial or more formal truth (deictic vs. apodeictic discourses). Tautologies are at the extreme end of what is perhaps a continuum while mere empirical facts alone are not much more significant. Significant truths have the power to move us, to orient us in the world. Significant truths have to do with the things that bear on the very meaning of our lives: the existence of God, another's love, or a centering thing. Insofar as empirical truths or even logical truths serve these substantive concerns, these less significant truths take on significance as well, albeit derivatively. A weather forecast of twenty-eight degrees becomes very significant if you are a farmer or gardener and it's late spring. The reasoning of philosophy and the empirical sciences generally are quite adequate and appropriate for dealing with less than substantive truths—with one proviso: these truths somehow make reference, as above, to more substantive concerns. Here, with this less than substantive truth, agreement can be compelled (Borgmann 1984, 22–31, 179–80). However, and this is the point he is most concerned to make, reason and philosophy do not have the power to demonstrate cogently significant truths. Borgmann points this shortcoming out in arguments of Anselm of Canterbury, Blaise Pascal, John Stuart Mill, and John Rawls, concluding:

> The attempt to begin with little and end with much has a long ancestry. It has become the dominant move of moral discourse in our era because it seems so adequate to the modern temper. But the critics of these various moves have invariably shown that, if one assumes little, one can conclude but little. If a strong conclusion is arrived at, then strong assumptions have joined the argument on its way from the initial assumptions; or the latter turn out,

>on closer inspection, to have been stronger and hence less easily
>acceptable than initially thought. (Borgmann 1984, 175–76)

These significant truths, then, cannot be forced on another person appropriately (or without manipulation); rather they require a pointing out or a re-
minding, employing the discourse of testimony and appeal, not knockdown
arguments. Here is a limit of appropriate philosophy that Borgmann accepts
and argues that we should accept (Borgmann 1984, 176–82). Because a
firm line is drawn between demonstrable and testimonial truth here, I will
call this critique *Borgmann's knife.*

Philosophy comes up against two more limits as Borgmann confronts it
with the inescapable horns of a dilemma. Let's call this argument *Borgmann's
fork.* Metaphysical accounts of reality are either significant but at the price
of being impoverished accounts of reality, *or* they are equal to the richness
of reality but at the price of being so complex that they are unhelpful. Either
these accounts are harmfully reductive or they lack significance. It would
not help us to explore the postmodern divide square foot by square foot
(Borgmann 1992, 4). To understand social reality with the kind of detail
of the microontology of physics is impossible, of course, but even if it were
not, it would be unhelpful because we would not be able to distinguish
between the significant and trivial information (Borgmann 1984, 74–76).
Accordingly, for Borgmann, science yields too much truth. And so too
do the traditional ontologies with which we began this chapter. What we
want to know is what is significantly true. But when theory alone seems
to yield significant truth (or when information *as* reality supplants actual
reality, as he argues in the recent *Holding On to Reality*), it does so at the
price of impoverishment. We are really more interested in controlling than
knowing (Borgmann 1984, 70). So we are back again to the other horn of
the dilemma.

The remaining web of arguments against philosophy's adequacy may
be thought of as *Borgmann's spoon.* They all go to show that philosophy
never quite reaches the world, faces in the flesh, things in their full round,
or "contingency" (Borgmann 1999). Philosophical ideas are thought to
avoid the provincialism and prejudice of particularity because they are
vague and hence ambiguous. But Borgmann finds that what appears to
be openness is in reality a resolved ambiguity because such optimism
concerning ideas naively ignores the way our assumed and unchallenged
context typically and forcefully prejudices our interpretation of general
ideas, ideals, and values. Given the rule of technology, an ideal, such as
self-realization, is both fulfilled only technologically through consumption
and simultaneously thought by members of that society to be value-neutral

with regard to which "lifestyle" is chosen. Self-realization as an ideal is subverted by technology and that subversion is hidden from members of the consumer society. Hence, pretending to keep questions of the good life open while not challenging the basic framework of orientation involves liberal philosophers in an impossible contradiction.

Moreover, even if we were to pry an ideal such as self-realization away from the forces of the rule of technology, it would still be an unhelpful standard. Just as philosophy in the passage from the later book is unable to evaluate the difference between the real and hyperreal, so John Rawls's Aristotelian principle[1] *alone* is unable to rank fly-fishing over computer games. Similarly, values of complexity, diversity, beauty, integrity, and stability could be used to justify both the protection *and* the development of nature (Borgmann 1984, 186–87). Principles, values, and ideas are just too abstract to tell us much, without at least implicit orientation from either the rule of technology or, alternatively, from things. In short, traditional philosophy is unable to measure what Paul B. Thompson in chapter 9 calls "the ontological loss."

Finally, in addition to yielding an unworldly sense of existence, ideals in general make the means and ends split of the device seem only natural; we think we are merely actualizing this ideal by an alternative means. Here such ideals tempt technological subversion in a second way. The full depth of things yields to superficial aspects of them. Versions of this technological subversion critique are made repeatedly in *TCCL*, especially in part 2.

In response to these knife, fork, and spoon problems, contemporary philosophy may become an entirely self-critical, endlessly prefatory, and deconstructive enterprise. If it did so, it would suffer paralysis. Or philosophy can give up the notion of its assumed sufficiency and learn to serve things that matter.

PHILOSOPHY IN THE SERVICE OF THINGS ILLUSTRATED

In the most general terms the two main points of *TCCL* are:
1. When technology, namely devices and the commodities they procure, are substituted for focal things, disengagement ensues. Technology cannot, though it may seem to, procure a good life.
2. Technology in the form of the device is appropriate and beneficial when made to serve more important focal things.

I will argue that Borgmann thinks of philosophy in a parallel fashion.

1. "Other things equal, human beings enjoy the exercise of their realized capacities (their innate or trained abilities), and this enjoyment increases the more the capacity is realized, or the greater its complexity" (Rawls 1971, 426). For discussion, see Borgmann 1984, 213–16.

1. We have already seen that philosophy cannot produce significant truth, at least not without impoverishment. Nor can it be substituted for engagement with real things without troubling consequences.
2. Now we will see that philosophy can be beneficial and appropriate when carried out in the service of things.

More strongly, just as devices are indispensable to an appropriate technology where they serve things, so too is philosophy essential but in a likewise more humble role. *TCCL* and *CPD* exemplify philosophy in this fashion.

Philosophy as theory tethered to things. How does a philosophy in the service of things avoid the knife, fork, and spoon arguments outlined in the previous section? How, to begin with, does a philosophy in the service of things avoid the second argument, the fork of either impoverishing reductionism or unhelpful adequacy? Borgmann takes up this question most directly in the chapter on "paradigmatic explanation" in *TCCL*. There he argues for grounds on which one theory could and should be seen as superior to another even if both theories are equally consistent, precise, and applicable. Superiority is not gained through one theory being adequate to the complexity of social reality. That, Borgmann shows, is impossible, for "there are indefinitely many patterns that can be highlighted" (Borgmann 1984, 75). Even if it were possible to account for all the patterns, it would be like examining the postmodern divide square foot by square foot, and that, as we have seen, is unhelpful. (Nor, for the same reason, does he want his theory of information in *Holding On to Reality* to be mired in endless qualifications, such as with distinctions between signs and symbols, etc.) So a philosophy in the service of things is not going to outdo metaphysical accounts in terms of adequacy.

How can a theory be said to be superior if not by these standards of metaphysics? Borgmann's own device paradigm theory is only one among many theories of social reality. It is better than these others, he argues, only because it more adequately addresses an extratheoretical concern, that is, our bonds of engagement with things.

> In most concrete phenomena of the technological universe, the cut between commodity and machinery, foreground and background can be made in more than one way. What should guide the incisions is our concern to shed light on changes that imperil things, practices, and engaging human relations, and the desire to make room for such phenomena when they are struggling to assert themselves against the dominant pattern of availability. Such a guiding concern is a response to the claim of things in their own right. (Borgmann 1984, 76–77)

The device paradigm, here called the pattern of availability, does not avoid reductionism—all helpful theories fall prey to this—but the reduction is not an impoverishment because it serves centering or focal things, things threatened with replacement and displacement by devices. The paradigm— or any paradigm—is measured according to the way it meets our concerns with these things.

This is a significant and unique move for philosophy, one not recognized, much less appreciated, by current criticisms of paradigms and theory as such. Let us begin by noting some important features of it. First, philosophy as theory is affirmed not only as a worthy undertaking; it is indispensable for meeting our concerns with our relationship to things. Second, claims of universality for the theory are avoided by making its helpfulness and importance contingent upon an already existing concern with focal things and practices. For those who (even after they have been deictically informed) care little for these things, this theory will not be important. They may respond to it by saying, "So what?" Third, these same claims of universality are avoided by acknowledging that the paradigm's simplification does not capture all the important forces influencing our culture, as Thomas M. Power shows in chapter 15. Fourth, similar universal claims are avoided by testing other philosophical, social, and political theories against the device paradigm and showing that the latter is more incisive and comprehensive when measured by things. While concentrating on the machinery aspect of technology, most of these theories are blind to the split of means and ends with the device and do not comprehend the consequences for engagement that the "mere end" of a commodity entails (a subtle but principal argument made repeatedly in *TCCL* and especially clear in chapter 11, "Devices, Means, and Machines"). Fifth, Borgmann implicitly challenges us to develop a theory of technology that will be better, not by metaphysical standards, but because it is more comprehensive, incisive, and helpful for meeting these extratheoretical concerns with things. Borgmann thinks of his device paradigm as advancing over Heidegger's framework by these three criteria. The more recent hyperreality/focal reality distinction in *CPD* is a similar advance over the earlier device/thing distinction since this new distinction gets us to focus more on the presence of "mere ends," the place where Borgmann finds the deepest problems with technology.

Developing points one and two above, we can see that theory is always tethered to things for a philosophy in the service of things. Things reveal the need for philosophy. Conventionally, reason has been accused of being unfeeling toward things because it is used to clear-cut valleys, dam streams, pulverize mountains, or even view trout in terms of their cash value; but in this conventional sense, reason, or more generally philosophy, is made to serve neither things nor philosophy itself but ultimately the goal of consumption. A philosophy in the service of things attempts to do the opposite by reflecting in ways that remain continuous with our feelings and our concerns with things. Feeling prompts reflection and becomes clarified through philosophical articulation. So both *TCCL* and *CPD* (as well as *Holding On to Reality*) begin by asking us to consider our sentiment. Then we are made to see that we need to develop a language that gives voice to the concern we feel in our gut. Finally both books promise to provide us with such a language. In the philosophy of these two books the task is to move from a forefeeling, foreboding, or foresensing to fully articulate expression of what at bottom bothers us. The task of philosophy, at least in part, is to provide us with a language within which we can comprehend what troubles us (our bonds of engagement have become disrupted), see what is decisively at issue for us (technology as a way of life), and choose alternatives deliberately (hyperreality vs. focal reality).

If we do not feel troubled in some sense by technological change, much of Borgmann's philosophical theory will lose its point. It will be just another theory to set beside other theories, and reform will be uncalled for. Still, even though his works begin by asking us to consider our troubled feelings, a philosophy in the service of things avoids subjectivism by quickly locating our trouble with the things that concern us. Our bonds with things are severed because things themselves are injured and ignored. All our more profound involvements are taken over by television, prepared foods, cyberspace, and so on. The deep issue, then, is whether or not things will be given a place or say in our lives.

Grasping this deep issue is exactly the right place to begin, but our philosophical theory—the device paradigm—shows us that we will remain troubled in our relations with things unless we resolve the problem of technology in various and far-reaching ways. Thus theory is indispensable for showing us how to disentangle ourselves from the rule of technology in order to give a place and say to things.

This importance of philosophy as theory tethered to things can be appreciated by understanding its application in *CPD*. If the device paradigm is the theory that enables us to comprehend what is disturbing about trends in the late twentieth century, then we should expect to find a version of the paradigm at work in the critical and constructive parts of *CPD*. Yet the device paradigm is not even indexed and an explicit discussion of it rarely surfaces. Not withstanding these observations, the device paradigm, I will show, is everywhere present.

CPD begins by arguing that a new philosophy is needed that will enable us to comprehend and overcome the modern period and help us chart the landscape beyond, showing the decisive choices we face. Such a philosophy will discover landmarks and dwell on particulars. How is this the philosophy in the service of things found in the earlier book?

Just as social reality is too complex to be captured by theory, so too historical reality is seen as similarly complex. "Every historical account has a particular point of departure, a particular goal in mind, and is guided by particular considerations, acknowledged or not" (Borgmann 1992, 14). The historical account presented is, then, admittedly selective, choosing the landmarks of aggressive realism, universalism, and individualism. But why are these three landmarks chosen and not some others, and how does his historical account avoid the impoverishing reductionism of Borgmann's fork? The device paradigm avoids an impoverishing reduction because its value is derivative from our more fundamental concern with particular things and whether and to what extent it helps us with these things. Given that it is guided by these things, the paradigm itself has a derivative power to guide. Although he does not explicitly raise this issue, the device paradigm has guided Borgmann to select the triadic landmarks in history he does, and so too, it has guided him in making his most decisive critiques of both modernism and hypermodernism.

The domination of nature is a major theme of *TCCL*, but our attention gets mainly focused on a dyadic relationship: the machinery and commodity components of the device. As things get transformed into devices, the world itself becomes split into a universe of familiar surfaces resting on unfamiliar depths. However, a third component to this reduction really resides in the background of *TCCL*, that is, the commodities and machinery rest on a resource base of timber, coal, ore, water, oil, gas, etc. The third component appears in the triadic structure of *CPD*, where the reduction of the world of nature to mere resources is discussed under the landmark of *aggressive realism.*

The rise of *universalism* roughly corresponds to the machinery side of the device. With the device, the machinery is radically variable. These variations are carried out in light of Cartesian rules of abstraction, dissection, reconstruction, and control, the marks of what *CPD* calls methodical universalism. In *TCCL*, this process extends beyond physical machinery to include institutions such as the insurance industry and even to the planet itself, imaged as spaceship Earth. Here we find its closest relatives with *CPD*, for it is the corporation that typifies methodical universalism. For the conquest of nature to take place on a large scale, better integration and organization were needed and the modern response to this need was the corporation, a "rational, mechanical, and inclusive design" (Borgmann 1992, 35).

Individualism wears two faces in *CPD,* but its more consequential face corresponds to the commodity side of the device. As the author of enterprise, mythically imaged as a rugged individual, it corresponds better to the machinery of the device, but this rugged individualism has come under severe criticism by all postmodernism. Borgmann's unique and more important contribution is concerned with commodious individualism, the softer counterpart to rugged individualism. Commodious individualism— lives oriented around passive, private consumption—fully corresponds to the commodity. Passive consumption, essentially an activity enjoyed by oneself, constitutes 80 percent of the way we spend our leisure time (Borgmann 1992, 44).

Once we see the underlying theory, we can begin to see *CPD* in a different way too. To be sure, Borgmann has learned from the various postmodern critiques of modernism, but two fundamental points to which the device paradigm guides us remain pivotal. Theorists, whether modernist or postmodernist, usually miss the debilitating character of mere ends, of commodities. Second, so long as commodities and consumerism are not challenged in a radical way, the rule of technology will continue. Hence, the intellectual critiques of modernism do not understand that harm done by the modern project was "done by a collective productive effort. The actual identity of the individual is that of a consumer. If individualism is not recognized and restrained in consumption, it will continue to flourish" (Borgmann 1992, 80). The device paradigm, in service of things, directs our attention to these two weaknesses—overlooking the presence of mere ends and the corresponding commodious individualism—of postmodern critiques of modernism.

Seen from this standpoint of the earlier book's device paradigm (not Borgmann's explicit one), hypermodernism may be only a sham "new age": the mature outcome of modernism. For all its apparent differences, Borgmann's analysis finds that hypermodernism's alternative to modernism is in fundamental agreement with modernism because it has not yet disagreed with modernity over technology as a way of life. It is only "technology by other means" (Borgmann 1992, 82). For example, the postmodern economic critique of modernism is tied directly to the goal of greater consumption in the future. This fundamental agreement needs to be challenged before we can finally move beyond modernism in a radically new way that Borgmann calls postmodern realism.

To meet the challenges of hyperrefined virtual reality the device paradigm's terminology has had to undergo adjustments. The earlier book's "availability" (quick, easy, ubiquitous, safe) becomes disposability. The separations caused by devices are now generally discussed in terms of discontinuity (although this latter term is now said to be more immediately the result of the disposable nature of hyperreality rather than the means-and-end division of the device). The appealing but ephemeral quality of the hyperreal is called glamour or "experiences." As mentioned previously, because Borgmann now distinguishes between hyperreality and mechanical reality, his characterization of hyperreality itself generally emphasizes the commodity component of the device rather than the device as a whole, where machinery and commodity are on an equal footing (Borgmann 1992, 118–19). The critique of disengagement becomes expanded so that along with the earlier characterization of mature technology as disburdening, disengaging, diverting, distracting, and making us lonely, we can now add the terms disposability, discontinuity, disconnection, disorientation, and diffusion. We could go on with how philosophy as theory in the form of the device paradigm teaches us to critique hyperreality, how it shows us what it will take for focal realism to flourish, or how it informs *Holding On to Reality* in much the same manner, but for now, in order to fill out this picture of a philosophy in the service of things, we need to turn our attention to Borgmann's philosophical alternatives to the knife and spoon critiques respectively.

Philosophy in tandem with testimony. The passage below would seem to indicate that Borgmann thinks that his philosophy does not fall prey to the knife critique where philosophy cannot provide knockdown arguments for substantive truth.

> The real point of the technological paradigm is its critical office. It is exercised through the demonstration that, if we are concerned about the loss of engagement, the device paradigm reveals more clearly than any other just how and to what extent people move away from engagement. If that concern is granted, the demonstration can attain at least a measure of cogency. (Borgmann 1984, 77)

The conditional—"if we are concerned about the loss of engagement"—is key. The device paradigm does arm us with arguments that will be compelling for those people already fully and reflectively committed to focal things and practices. But what about people who are either on the fence or people who, while attracted to things, have been choosing commodities over things in a complicitous relationship with technology? These people too can be reached and brought to weigh technology, but it will not occur by strong-arming them with compelling arguments. Rather, the stronger case will be made by philosophical theory in tandem with testimony, for which the fine arts play the central role.

Both *TCCL* and *CPD* bring us not so much to a conclusion as to a basic choice. Borgmann's concern is not with first principles but first choices. The question here is: How can we really weigh the alternatives concerning technology and centering things? Because we can be ignorant and wrong about what matters, if we are going to make this choice intelligently, it will not do to be blind to what counts on either side of the alternative.

Approached in another way, Borgmann finds that social agreement is decisive for the distinguishing character of any age. The medieval period's social agreement gave way to the social agreement underlying the modern project. That agreement is now weakening, and we are on the threshold of a new social agreement concerning the postmodern period. But this new social agreement, Borgmann believes, has not been decided yet between postmodern hyperrealism and postmodern realism. What role can a philosophy in the service of things play in helping us to form this new agreement?

To be sure, even the device paradigm alone, that is, without an appeal to our bonds of engagement with things, makes us more reflective about choices we may be making with little or no reflection. Arguments alone also may help. Borgmann constructs one (and Gordon G. Brittan in chapter 4 evaluates it) showing that consumption falls short of the claims of traditional excellence. He constructs others that show that technology subverts the claims of social justice and our endeavor "to complete the social plank of the Enlightenment platform" (Borgmann 1992, 26). These arguments do change some people's minds, but what is unique to Borgmann's philosophy is that we can do better by supplementing argumentation with an appeal to things.

It is Borgmann's contention that people are not deeply and articulately reflective about the things of their lives. In order for the device paradigm to do its real work, people will have to become more thoughtful about these things. They will need to be reminded vividly of them, perhaps awakening to them for the first time, as it were. For this to happen, these things need to make their appearance in language. They do so in quiet, intimate conversations with each other, in stories, poems, plays (and works of art generally), in speeches, and even in classrooms. The languages that bring people to consider the things of their lives are not typically that of compelling arguments. Our words here are most instructive when they are spoken with the candor of testimony, when they are simply wrung out of us.

Philosophy in the service of things realizes that philosophical discourse, in its discursive forms, is incapable of showing things in the fullness of their considerability. It points to and relies on other, more appropriate, testimonial languages—narrative, poetry, and the like—that show things better. And as a philosophical task, it points beyond these languages to the things they evoke. Not only philosophy, but other languages too, need to serve things and need to be reminded of that from time to time. These disclosive or revelatory languages are better at getting us to consider things in the fullness of their considerability. In these languages that disclose, things are more nearly present in, if not their commanding presence, their binding presence. Philosophy must give way finally to the resolving powers not of arguments but of things. Rather than expelling art from its republic, philosophy must accept a more humble and subordinate position.

Philosophy in the light of things. Borgmann meets his objections to the spoon arguments by developing philosophy in the service of things as philosophy in the light of things. The spoon criticism, it will be recalled, charges that philosophical principles and ideals, such as the ideal of self-realization, never quite reach things in their full round. In a way, principles remain worldless. Thus their ambiguity is resolved in favor of technology (or, if the principles are disengaged from the rule of technology they remain unhelpful), and finally they make the means-and-ends split of technology seem only natural. In response to this problem, Borgmann attempts more or less successfully, to recall or evoke the presence of things and then to weigh devices, commodities, and hyperreality against them.[2] So in *TCCL* we *see* musicians playing classical instruments, the wood-burning stove, the wagons and the wheelwright, the bottle of wine, the focal meal, hiking in the wilderness, and running. Against all the divisions that devices entail, we see displayed the unities and unifying powers of, say, focal fly fishing. Through the exercise of "encompassing and discriminating knowledge" and "intricate bodily skills," this focal practice unifies achievement and enjoyment, competence and consummation, mind and body, person and world, individual and community, mortality and divinity. Similarly, whereas conventional theories of morality and of reality are "powerless to explicate the difference between the real and the hyperreal," Borgmann is able to evoke that difference by presenting us with a runner who sees a mountain lion take a snowshoe hare. We see the commanding presence, the continuity and centering power that is missing in hyperreality. We see these abstract characterizations, what should be thought of as experiential ideas, on the basis of the thing and practice conjured up.

Perhaps, on the contrary, we think of this runner as being only an instance of what is meant by ideals of "engagement," "continuity," and "centering." This particular idyllic story of a run up Missoula's Rattlesnake Creek, a made-up one I believe, does lend itself to that interpretation.

---

2. *Holding On to Reality* is far more successful than Borgmann's two earlier books in this regard.

But a philosophy in the light of things insists that *things* finally guide us. Borgmann makes this point in two ways. First, wherever we deeply reflect on the meaning of "engagement," "centering," "values," and "the good life," "what we have really done is to bring activities back to the things to which we respond in those activities" (Borgmann 1984, 217). Here, even as a challenge to the above story, we may offer firsthand accounts of times when we felt pulled most strongly by the things of our life. We feel that our interpretations and thinking move at deeper levels for having undergone such encounters. Along these lines, a philosophy of things thus attempts to push the conversation so that those involved in it must tap, behind our more abstract ideas and theories, the guiding experiential basis from which we gain our stance and bearings as human beings. I think we all know when we have heard a real story from this air-clearing, sometimes silencing, gut level of finality.

Second, things guide us in the sense they are essentially unpredictable. "The ultimate givenness of a focal thing [is] something that unforethinkably addresses us in its own right" (Borgmann 1984, 215). In some sense we all know about what Colin Fletcher calls an "expectancy barrier" that prevents us from really encountering things such as wilderness because we expect to see this or be impressed with that, often closing ourselves to nature on its own terms (Fletcher 1984, 51). As academics we know how cerebral and out of touch our contact with things can be. We know too how sensitively and subtly this theme of reality that overtakes us from behind can be developed in works of Meister Eckhardt and the Zen tradition. For Borgmann the appealing and commanding presence of things, the eloquence of reality, cannot be discovered by *coming at* things as if they were instances of guiding concepts; rather that presence requires that we *be with* them as they, in their own time and way, address us. It is when we take things at this level that they finally light up. Catching a steelhead trout can be revelatory in this way, as Henry Bugbee puts it: "If one eventually lands it, and kneels beside its silvery form at the water's edge, on the fringe of the gravel bar, if one receives this fish as purely as the river flows, everything is momently given, and the very trees become eloquent where they stand" (Bugbee 1976, 87). It is only when we are brought to see in this kind of light that we are able to intelligently evaluate the ontological loss between the real and the hyperreal, and so intelligently decide between the alternatives, between the fundamental material choices that Borgmann poses for us in *TCCL* and *CPD*.

The theoretical bent of mind dominant in professional philosophy, on the other hand, responds to evaluational difficulties by reaching for a new theory that supposedly answers those difficulties. Accordingly, Rawls looks forward to "a relatively precise theory and measure of complexity" to handle these evaluational difficulties arising from the Aristotelian principle. But for a philosophy in the service of things that is tantamount to buying a four-wheel drive as "the answering machine for the call of the wild," as one advertisement has it. It finds that theory as such is precisely the wrong kind of answer that can be given for what otherwise needs to be shown in the light of things. Even a philosophy as theory tethered to things, which I find best exemplified in the writings of Borgmann, needs to be counterbalanced with a philosophy not as theory but in the light of things, which I find ultimately better exemplified by Henry Bugbee's *The Inward Morning.*

A guiding idea? Once we have limited philosophy to being in the service of things—tethered to things, working in tandem with art, and carrying out evaluations in the light of things—is there room for any guiding ideas in this philosophy? Of course, the device paradigm is a guiding idea, but as we saw it is really tethered to things and perhaps more importantly depends upon a forethinkable, predictable uniformity in the development of technological devices. Things, if they are to retain their leadership in the guiding role, must be allowed to speak to us in unforethinkable ways. Still, given this limitation, are there roles for broader guiding ideas, covering both things and devices, in a philosophy in the service of things? On my reading, at least, I find that there exists such an idea in Borgmann's philosophy in the service of things. Pivotal for him is the idea that there exists "a symmetry between human life and its setting" (Borgmann 1992, 96). (This idea is at the bottom of Carl Mitcham's discussion of character in chapter 7.) Our very being is tied to things in this philosophy in the service of things.

Things and ourselves are codisclosed in this relationship. In the past, the correlative to human existence has been a world of things, as Borgmann evokes so well in his discussion of the wheelwright (Borgmann 1984, 44–47). As people act and develop in relation to things, the things themselves are also disclosed in their manifold depth. So the potential both of what people are capable of and what things are capable of are simultaneously realized in this relation. Since both human beings and things emerge into being at the same time in this codisclosive process, and since the two require each other for this coemergence, this symmetrical relationship can be called *correlational coexistence.* By responding to things in their full dimensions, I too emerge in the fullness of my dimensions. If I lack the power to be equal to them, neither do things emerge into the fullness of what they can be. If things are not allowed to be, neither am I allowed to be. If I sever my bonds with things by dominating them, I too am diminished (Strong 1995, 70).

Focal things provide a commanding presence and center to life. They gather and illuminate our world. Eloquent reality calls forth eloquence and eloquently lived lives. On the other hand, "[a] hyperreal setting fails to provide the tasks and blessings that call forth patience and vigor in people. Its insubstantial and disconnected glamour provokes disorientation and distraction" (Borgmann 1992, 96). The central heating dial does not evoke exertion and fidelity to daily tasks. The buttons of the stereo system do not evoke musicianship. McDonald's hamburger does not center family life. The television mostly evokes couch potatoes. The hyperreal run cannot gather to the center of a life. The self-realized hyperreal text does not require literacy, imagination, or resourcefulness on the part of the viewer. Because humans do not stretch to their fullness but nevertheless assume an overpowering stance in their dealings with devices, this asymmetrical relationship can be called one of *petty anthropocentrism* (Strong 1995, 71).

CPD shows that the human side of the symmetry of correlational coexistence requires more than individuals or families centering their lives around focal things and practices. To be really adequate to what in that book has now become focal reality, we need genuine communities. On the other hand, for the hyperreal to flourish all that is necessary to hold us together is an impersonal design.

In other words, this idea of symmetry goes right to the heart of a philosophy in the service of things. The most immediate source of this idea I take to be Heidegger. In "The Question Concerning Technology," Heidegger finds that even our relationship to objects has changed. As resource, nature has lost its capacity to be object, to stand against us; we simply overpower it with our massive machinery. Human-made objects, such as the jet, have lost this capacity as well. They are merely "on call" (Heidegger 1977, 17). Borgmann, who at times calls the device paradigm the "availability paradigm," follows this insight of Heidegger's: the device makes a commodity available. In *CPD,* disposability is at the center of hyperreality, causing, in contrast to commanding presence, discontinuity and all the other *dis-* words.

As we can see here, the idea of symmetry has come to light because of the troublesome transformation of things into available commodities in our time. But it seems to be more substantively significant than the more limited device paradigm. Epistemologically, for instance, Borgmann uses the idea of symmetry to address—not to resolve but to respond appropriately to—problems of subjectivism (Borgmann 1984, 181–82). More importantly, the idea seems to guide our way of seeing things: they command, order, challenge, gather, resist, and appeal from afar. From this standpoint, we can see that traditional philosophy has largely ignored things and that working out the implications of correlational coexistence poses a large philosophical task. Yet even if we accomplish this task, still the real work remains. What particular things, not just possible things, call us, as individuals and communities, as we move toward reforming technology and toward postmodern maturity? What things are entirely inappropriate or no longer appropriate for our context? That is, what things no longer call? These questions cannot be answered by appealing to the idea of symmetry; however guiding this idea may be, here the light of things plays the primary leadership role. As philosophers, we need to remember to listen, which is easier to say than to do.

Having said this about the idea of symmetry, I am not sure whether I am correct because so far Borgmann has not produced a text that directly attends to this idea, elaborating or justifying it; he has mostly used this idea of symmetry to critique technology and hypermodernism, and to point out alternatives. It may be an idea in the traditional metaphysical sense, and, at the very least, it needs to answer some of the questions of traditional metaphysics. The same can be said of Borgmann's version of a philosophy in the service of things. If not a text, at least some kind of explanation is in order.

WHERE FROM HERE?

Obviously a philosophy in the service of things should meet these concerns of the discipline. Along these same lines, Borgmann's particularist approach—or what one can make of it from *TCCL* and *CPD*—is unlike a traditional absolutism or a traditional relativism. It seems, however, that he is committed to a version of relativism: first principles obviously play no important role in this philosophy of things, and clearly Borgmann is at ease with the idea that each epoch has its own social agreement. Yet just as clearly, he is not in the camp of Richard Rorty's style of relativism, for Borgmann believes that the appealing powers of things can resolve disagreement and bring about social agreement in the coming postmodern era. Things for him serve as a kind of final court of appeal. Or at least the real decisive showdown, as he understands it, will be between eloquent focal reality and glamorous hyperreality as they appeal to us for the next fundamental agreement in the postmodern era. What kind of relativism is this then? It's time to make its character explicit in a more theoretical way.

The history of philosophy needs to be rethought in terms of focal things. For instance, most recently Borgmann has done this rethinking with Aristotle's separation of doing and making, of morality from production. Ever since the distinction was made, ethics has been concerned about conduct and not about what we make. On the other hand, a philosophy in the service of things finds that the "development and adoption of a device already constitutes a moral decision" (Borgmann 1992, 112). Typically, for all too many of us, once we invite the television into our house, our alternatives have reduced to what we are going to watch tonight. As our private and public lives become filled with devices, only the consumption of commodities is elicited from us and we become petty anthropocentric consumers. Technology is neither value neutral nor morally neutral. So fundamental material choices—what we make—are moral choices. From the standpoint of the idea of symmetry or correlational coexistence, we see that it is and was from the beginning a mistake to separate doing from making. This mistake is a basic reason why hyperreality appears morally inert to traditional philosophy.

So too, the history of philosophy from Thales through Heidegger can be reinterpreted in the light of the retrieval of things. Perhaps the polis, retrieved as the public thing around which Aristotle's ethical philosophy is centered, will yield new insights into virtue ethics. Perhaps the recollective powers of the divine in Augustine's *Confessions* can be reread in terms of the recollective powers of things. Perhaps Heidegger's epochs of Being need the retrieval of things around which fundamental epochal agreements are formed. Borgmann surely has more to say about the history of ethics and about the history of the disappearance of focal reality.

*

Borgmann's most important philosophical achievement is his advance beyond Heidegger at the most fundamental level. Here I do not mean that Heidegger's enframing becomes usefully specified and rigorously developed as the device paradigm. Nor do I mean that practices and embodiment become tied to things in ways Heidegger neglected. Neither is the most fundamental difference the way Borgmann makes ethical and political dimensions central to his philosophy and the reform of technology. Nor is it his willingness to engage with more sophistication the natural and human sciences. Nor that Borgmann actually provides in language a kind of house of Being within which we can come to terms with technology and come to dwell with it. All of these advances (or articulations) and many others step beyond Heidegger, certainly. Borgmann's most important philosophical achievement beyond and departure from Heidegger, for whom the essence of technology is nothing *technological,* is his physicalism: getting us to attend to the significance of our physical world and tangible things. Matter matters. I doubt that many of Borgmann's readers realize how much of new a philosophical beginning has been initiated here, especially given that many of his examples could be facilely dismissed as nostalgic, as Gordon G. Brittan and Lawrence Haworth point out, or as faddish and yuppyish, such as running or fly fishing, as Jesse S. Tatum indicates, or Catholic, such as wine, cathedrals, and the Eucharist. However, his thinking about physical things and materiality is radically new, refreshingly profound and sophisticated, and urgently called for to meet the real challenges we face.

On the one hand, Borgmann's philosophy in the service of things departs from and returns to particular material things like the hearth. "The focal significance of a mental activity should be judged, I believe, by the force and extent with which it gathers and illuminates the tangible world and our appropriation of it" (Borgmann 1984, 217). By this standard, his device paradigm and discussion of focal things and practices are outstanding. We have achieved genuine insight into both things and devices as we have never had before. On the other hand, Borgmann has a thingly, or at least physical, cast to his thinking when he addresses more global matters. For instance, speaking of physics, "Any credible view of reality must be consistent with the cosmological and microphysical conditions so far uncovered by physics. But the reality that finally matters lies between the physical microscale and macroscale. It must be granted its proper scale like a painting that would vanish as such if viewed through a microscope or from a satellite" (Borgmann 1992, 118). (These things that lie between require the kind of poetic disclosure of them that Phillip Fandozzi in chapter 8 discusses as the redemption of physical reality.) When Borgmann argues that labor has replaced work, he begins: "Technology did not enter an empty stage but a world that was filled with work and celebration, with hardship and joy. Human life is always full at any one time, and innovations can take place only by displacing some tradition" (Borgmann 1984, 116). This image of finitude and displacement is entirely physical; work, celebration, and other activities and events are always tied directly to a material setting and physical things. It is within this understanding of material finitude that our saying "yes" to focal things, focal reality, and their respective practices will simultaneously mean "no" to consumerism and hyperreality, bringing about fundamental cultural reform. It is also within a material framework that Borgmann wrestles with the philosophical problem of freedom and determinism, finding a resolution that at once honors human dignity and the dignity and eloquence of physical things (Borgmann 1979). Finally, when Borgmann makes a global appeal to become concerned about technology, it is an appeal to wonder about physical rearrangement. "What we in fact witness about us . . . is the most radical and forceful reshaping of the world ever. Something is going on here that needs to be illuminated and understood" (Borgmann 1984, 73). Technology has radically transformed the natural and built environments within which we live. We have but a weak understanding of the cumulative effects of this physical rearrangement upon Earth's ecological environment, and we have an even weaker understanding of what cumulatively this physical rearrangement means for us humans and the quality of our lives. What role do physical things play in the decisive issues of who we become and the quality of lives we lead? The way modern technology has problematized the physical has made possible genuine advances in our thinking about it.

Just how far can we take these things? One of these genuine advances has to do with religion, although this advance is confused in the writings of Borgmann. "People feel a deep desire for comprehensive and comprehending orientation. To be human is to have a capacity for the beginning and end of all things and for assuming a position among them" (Borgmann 1992, 144). Art and athletics have provided only partial illumination, Borgmann maintains, while religion offers a view of the world as whole, exemplified at the end of *CPD* by Bishop Paul Moore's farewell at the Episcopalian Cathedral of Saint John the Divine in upper Manhattan. Yet I wonder if this invocation of traditional religion is not at odds with the religious, or philosophical, vision of things underlying Borgmann's philosophy. I feel misled by my Methodist upbringing in a basic way that I find all traditional world religions misleading: they have not had the opportunity to develop under the problematization of the physical.[3] Borgmann's *deeper* concern, it seems to me, is "that we are in danger of losing our sense of reality" (Borgmann 1992, 12). We are in this danger because of the great transformation of the world from focal reality to hyperreality, from things to devices. If we lose this sense of reality, we have lost everything substantive, from our humanity to nature in its own right to a sense of the divine. Human existence will become predictable and trivialized.

So unlike Heidegger, Borgmann's philosophy shows that we are in this danger of losing our sense of reality because of the way we have rearranged and are continuing to reshape our material circumstances on Earth. In the process we are eliminating all meaning, divinity, and even our own freedom in its highest sense. This "end of all things" and "assuming our position among them" I do not find in Christianity or other religions. This is an entirely new version of the death of God, the death of anything divine, the loss of our souls, and the endangerment for all humanity to come. Nor do I find in these religions a good answer to "something is going on here [with the massive transformation of Earth] that needs to be

3. I am indebted here to Hubert Dreyfus's discussion of Michel Foucault's use of "problematize" with regard to practices at a National Endowment for the Humanities seminar at the University of California at Santa Cruz, summer 1997. Dreyfus stressed that problematizing in this sense is not something humans will to do. They only respond to its occurrence.

illuminated and understood." The closest I have come to understanding this transformation is through Borgmann's philosophy in the service of things. Too often religion has focused on good and evil actions, ignored the witness of things, and thus mislocated the real issue for us. If religion is illuminating, then it too, I believe, must be guided by things, be in a subordinate role to them, in other words, be a *religion in the service of things.* Otherwise invoking religion is more likely to confuse both the issue and the alternative. Although it may turn out that more profound articulations of their eloquence require the languages of religion, it is essential to remember that the things that finally matter are physical things.

Accordingly, Borgmann's works get us most concerned about preserving the eloquence and commanding presence of physical reality. "Rivers are muted when they are dammed; prairies are silenced when they are stripped for coal" (Borgmann 1992, 118). A device makes a commodity technologically available. Hyperreality is likewise disposable; it offers a *commanded* "presence." Even eloquent animals, like horses, can become "weak and faint," as "a debarked poodle forever confined to a condominium" (Borgmann 1992, 120). The chief danger here is not that we are losing this or that cherished thing, but *all* physical things are perishing, becoming choked out by physical hyperreality. The commanding presence or eloquent reality becomes silenced, muted, or marginalized. We are in danger of losing our sense of *this* reality.

This danger to physical reality is Borgmann's 'Dis,' which resembles Dante's Dis only metaphorically. The Dis of separation has its origin in the means-and-ends split of the device. This is a physical separation and discontinuity. The modern world presents us with a new kind of dualism, a *material dualism.* A new problem requires a new kind of resolution: "devicification," if you will, is the character of the reality of our built environment that needs to be modified. Transcendence of material dualism will turn upon material arrangements, requiring arrangements that bespeak continuity rather than division. Neither devices nor hyperreality will do; only eloquent physical things that bear a world offer the kind of continuity and unifying powers that enable transcendence of this material dualism.

Discontinuity is the price paid for having a world always at our disposal. That position of disposability makes us petty tyrants, imperiously manipulative, and self-centered. Everything else is beneath us; nothing is revered or respected in its own right. Divinity, in any sense, whether Christian or pagan, monotheistic or polytheistic, is entirely missing. We may attend Mass and speak of our religious beliefs, but if our paycheck, the shopping mall, the television and what is advertised on it, and net surfing occupy the time of our life, our life bespeaks a deeper atheism.

On the other hand, we all know people who profess to be conventional concerned atheists and who still respect, celebrate, and even revere the commanding presence of eloquent physical reality. If the consequences are not too harsh anyone can find a kind of catharsis, a welcome appreciation of discovered reality, even in being driven out of the mountains by a summer snowstorm or sandbagging a flooding river, although, for my part, I find the more consequential powers of commanding presence to be its attraction, its pulling powers rather than pushing powers. If this is a religious attitude, and if building one's life around things of commanding presence—things more and other than oneself rather than at one's disposal—is a religious life, and I kind of think it is, then Borgmann's vision here is deeply religious. Still I hesitate to call it that because it is so radically different from any received understanding of religion. What is genuinely enlightening is the qualitative difference, the ontological difference, between the disposability of hyperreality and the commanding presence of things in their own right. I prefer to leave it at that, and so does Borgmann at times, as in *TCCL*.[4] Speaking of wilderness as a sacred place, he writes,

> I do not propose that we transfer the traditional notions of divinity and worship from religion to nature. These concepts and their associated practices have generally become so desiccated that little would be gained in shifting them from one area to another. Rather, I think, it is now a matter of learning again from the ground up what it is to recognize something as other and greater than ourselves and to let something be in its own splendor rather than procuring it for our use. (Borgmann 1984, 190)

As with wilderness, so too for all other centering things. The early twentieth-century German poet Rainer Maria Rilke asks, in his "Ninth Elegy," "—oh why, *have* to be human?" Why? "because being here amounts to so much, because all this Here and Now, so fleeting, seems to require us and strangely concerns us" (Rilke 1939, 83). All that is here, vanishing so quickly, are things, he tells us later. Things need us and strangely concern us. They give us a reason to struggle with living lives of genuine destiny. Without the

---

4. Recently, he seems to depart from this position again. Engagement with things is not quite enough and is only "a proximate remedy" to the disquiet wrought by technology. "Many of us find it hard to face up and to be faithful to persons and things" (Borgmann 1999, 232–33). Beyond the resoluteness of a focal practice, it now seems to him that we need ("owe what fidelity . . . they possess to") a belief in something comparable to the Christian history of salvation, although there are still said to be manifold constructive responses. Naturally, I find this idea to be in tension with a philosophy in the service of things and just how far we can take these things.

presence of real physical things and a setting—a habitat as it were—that allows and encourages these things, the better part of ourselves may just lie unawakened. That is what we really have to work with. That is all we know and really need to know.

Borgmann's treatment of muted and silenced reality is thorough and compelling, and the logic, at least, of how disposability leads to debility and impoverishment is also clear. To speak of eloquent reality, the opposite of silenced reality, is no doubt more puzzling for us, however. What does it mean for reality to be eloquent? How do we listen to it? How does nature speak? What would it mean to give it a say? Here, in contrast to reality that lacks eloquence, we can begin to jot down a few simple notes but the symphony is not yet composed. What would it mean to really listen to the eloquent reality of the greater Yellowstone ecosystem, for instance? Can we discover and communicate eloquently new things and focal practices such as snowboarding, mountain biking, hang gliding, windsurfing, and scuba diving? Don't we feel challenged to make and to discover new things, and to find languages that distinguish them and their ontological gain from hyperreal and commodious counterfeits? I can imagine a day when, if we choose postmodern realism, people look back at us and wonder how we could have been confused about the eloquence of reality, for it will be so natural for them to think in these terms, having had much more experience with giving reality a say. I can imagine a day when we celebrate, as if rediscovered or discovered really for the first time, not only the union between mind and body but also the union between ourselves and our physical circumstances, the union between ourselves and things. For now, as the problematization of the physical becomes more evident and obvious, this potential significance of physical things and the quality of life they sponsor—correlational coexistence—calls on us, as philosophers and artists, to explore and articulate them as never before.

REFERENCES

Borgmann, Albert. 1979. "Freedom and Determinism in a Technological Setting." *Research in Philosophy and Technology* 2:79–90.
———. 1984. *Technology and the Character of Contemporary Life: A Philosophical Inquiry.* Chicago: University of Chicago Press.
———. 1992. *Crossing the Postmodern Divide.* Chicago: University of Chicago Press.
———. 1999. *Holding On to Reality: The Nature of Information at the Turn of the Millennium.* Chicago: University of Chicago Press.
Bugbee, Henry G., Jr. 1976. *The Inward Morning.* New York: Harper and Row.
Fletcher, Colin. 1984. *The Complete Walker III.* New York: Knopf.

Heidegger, Martin. 1977. *The Question Concerning Technology and Other Essays.* Trans. William Lovitt. New York: Harper and Row.

Rawls, John. 1971. *A Theory of Justice.* Cambridge: Harvard University Press.

Rilke, Rainer Maria. 1939. *Duino Elegies.* Trans. J. B. Leishman and Stephen Spender. New York: W. W. Norton.

Strong, David. 1995. *Crazy Mountains: Learning from Wilderness to Weigh Technology.* Albany: State University of New York Press.

PART FIVE

Postscript

EIGHTEEN

Reply to My Critics

Albert Borgmann

INTRODUCTION

Nearly twenty years have passed between the writing of *Technology and the Character of Contemporary Life (TCCL)* and the writing of this reply. During those years communism has collapsed, the threat of a nuclear holocaust has all but disappeared, environmental problems have been reduced if not eliminated, the culture of consumption is approaching a sustainable modus operandi, the rate of population growth is declining, democracy has been spreading not only in the formerly communist countries but also in Latin America, Africa, and Asia. Global warming and third-world misery remain grave challenges. But the goal, as Richard Rorty has put it, of life "as it might be lived on the sunlit uplands of global democracy and abundance" is at least conceivable now and likely reachable (Rorty 1995, 89).

This country particularly seems to be close to those sunny uplands. After a period of economic turmoil and political self-doubt in the late seventies and eighties, the economy has defied the supposedly iron law that unemployment and inflation cannot be low simultaneously and that federal deficit reduction must lead to a slowing economic growth. As we enter a new millennium, the United States finds itself the sole and unchallenged superpower and the model of the kind of open and enterprising democracy that is most hospitable to full employment and vigorous economic growth. The more regimented democracy of Japan is stumbling, those of Europe laboring; both were once thought to be more stable and productive than the United States. The United States, moreover, has recaptured and strengthened its leading position in the characteristic social and economic event of the moment—the information revolution.

The general rise of freedom, security, and prosperity is without doubt reason for relief and gratitude and should be an obligation for us to extend these blessings as far and as rapidly as possible. At the same time, there is under the surface of the general contentment a sense of uncertainty and stress and a feeling of being rushed and restless.

PHILOSOPHY AND TECHNOLOGY

One way of capturing the developments of these past two decades is to say that the rule of technology has been enormously expanded and solidified. Paul Durbin, Andrew Feenberg, Andrew Light, and David Strong have remarked that mainstream philosophy has had no interest in technology, and the events just noted have left hardly a trace in the prestigious journals of the guild, neither under the heading of technology nor any other.

TCCL came out of the concerns that had their beginning in the founding of the Society for Philosophy and Technology and the establishment of the journal *Research in Philosophy and Technology* in the late seventies. Paul Durbin and Carl Mitcham were the major movers of these endeavors; Don Ihde, Kristin Shrader-Frechette, and Langdon Winner have been among the prominent authors of this movement. We were then full of hopes and plans for the philosophy of technology. It seemed obvious that something like it just had to develop. In important part we have delivered on our share of the project. We have written our books, developed our positions, brought some sense and structure to the amorphous and confused beginnings. Yet we have to join in Durbin's disappointment (46–47). The philosophy of technology has remained marginal within the profession and all but inaudible in the national conversation. Within its limits, however, the philosophy of technology has become a scholarly enterprise one can be glad to be a part of. In it one can find the rigor of analytic philosophy, the flair of Continental philosophy, and, distinctively, an engagement of issues that actually matter. The vigor of the contributions in this volume, the variety of their approaches, and their ability to engage one another directly and indirectly all testify to the vitality of the philosophy of technology. And I must add that in what follows I will be unable to do justice to the richness and subtleties of my critics' essays.

As regards our position within academic philosophy, there is not much reason to lament insignificance within an enterprise that is itself insignificant. Analytic philosophy, still dominant at all major universities, is something that some philosophers should do and all philosophers can benefit from. But culturally it is barren, and due to its excessive extent, it has had a deadening effect on academic philosophy (Borgmann 1995). Rorty has largely given up on it and we, the older, published, and reasonably well situated members of the philosophy of technology community may be tempted to do the same.

But this is not an option for the young men and women who are intrigued by both philosophy and technology. The profession controls who gets to philosophize professionally. In the sixties, when the older philosophers of technology obtained their positions, the rapid growth of higher education and a national spirit of greater openness and idealism allowed us to slip into the academy. Today every opening is hotly contested. We must see to it that philosophers of technology have a chance in these contests.

I am afraid there is no persuading the analytic philosophers or the remaining more or less marginal schools of philosophy that technology is a problem that cries out for philosophical reflection and has occasioned sound and helpful scholarship. I quite agree with Durbin (47–48) that in response we should resist small-minded, unrealistic, or fatalistic versions of the philosophy of technology. Engaging and changing the public sphere must remain our goal. Saying this may not amount to a full agreement with Durbin's proposal as I will try to show on pages 360–61 of this reply. What matters for the moment is the suggestion that a reform of professional philosophy is unlikely to come from within anytime soon though we certainly should not despair of it. If philosophy of technology obtains a public hearing, administrators will follow and so finally will professional philosophers.

As Durbin has pointed out, however, public attention is no easier to get than professional acceptance. But we must nevertheless do what we ought to do, viz., reach out to the public to change things for the better. As for success, our attitude should be the one Kant recommends toward happiness. We must do everything to be worthy of it. Whether we find it is beyond our control.

The Origin of Technology

Of course what matters in the end are not complaints about the state of philosophy or the receptiveness of the public nor the general claim that technology requires more thought, but what light philosophy actually sheds on contemporary life. More particularly the question is about what insights have come and can be derived from *TCCL*. And to begin at the beginning, the question is about the nature and origin of technology.

Feenberg believes that *TCCL* has answered this question with Heidegger and Habermas along essentialist lines, a concern he shares with Larry Hickman (93). In general, to say that technology has an essence is to claim (1) that it has a definite structure and (2) that it is the same everywhere and all times. To these standard features of essentialism Feenberg adds for the instance of technology (3) that it is part of the essence of technology to shape reality irresistibly. This third feature is sometimes singled out under the heading of substantivism.

TCCL entirely agrees with Feenberg on two counts, and so his worries are largely unfounded. Regarding Feenberg's concern that I reduce technology to "constant determinations" and "a technological a priori" (304; see also 311–12), one needs to remember this proviso (Borgmann 1984, 12): "To avoid misunderstanding, let me repeat that my concern is with modern technology and its character. I will at times use the appropriate qualifier as a reminder. But often, in what follows, I will simply speak of technology when I mean modern technology." As for Feenberg's worry that I remain "faithful to the determinist premises of an earlier generation of founders of the field" (294), one will find comfort in the rejection of determinism in the discussion of the substantive sense of technology (Borgmann 1984, 9–10) and of responsibility vis-à-vis technology (Borgmann 1984, 102–5), summarized in the remark: "The rule of technology is not the reign of a substantive force people would bear with resentment or resistance" (Borgmann 1984, 105). The question that remains is about the character of modern technology. It will concern us later in this section and in the next section.

Power calls for an account of "why technology developed as it did" (271). The answer, he suggests, comes from a consideration of "the role of social institutions in constraining and guiding the choices individuals make" (271). There is then a distinction, if not an antagonism, to be noted between institutions and individuals. What are those institutions? Chief among them appears to be the "capitalist economy" (272). That capitalism rather than technology is the driving force of contemporary culture is a concern Power shares with Douglas Kellner (251) and Feenberg (310–12). But Power also refers to the market as a shaping and constraining institution (278).

It seems then that we have three paradigms vying for the best explanation of the technological society: the market, capitalism, and the device paradigm. By their very nature, paradigms never explain anything fully. They highlight certain features of a phenomenon and obscure others (Borgmann 1984, 75). Not surprisingly then the three paradigms bring different features of an advanced industrial society into relief and are largely compatible with one another. They differ, however, when taken in their usual senses, as regards social justice. The market paradigm implies that the distribution of power is naturally and wisely effected as by an invisible hand. Capitalism, when critically used, denotes an economy where a minority class, the capitalists, has unjustly arrogated power unto itself. The device paradigm requires and is required by liberal democracy and its limited inequalities.

Power and I agree that the standard justice claims of the market and of capitalism are bogus. The market, as Power shows (290–91), is a particularly incomplete paradigm and needs the support of moral and cultural institutions to work beneficially. And, of course, Power shows in some detail how the thinness of the market paradigm has allowed it to be taken over by the device paradigm.

Capitalism rightly implies that there is a minority class holding a disproportionate share of power and affluence. But so far capitalism is compatible with liberal democracy. The crucial question is whether capitalists hold their advantages against the will of a majority of people. Power explicitly rejects the standard socialist construal of market institutions. Where then does the responsibility for our state of affairs rest?

At its very first mention (272), Power calls the capitalist economy "our ideology," and not much later (273) notes that the regimen the economy inflicts on us is something "we collectively believe" to be necessary. Finally and most clearly he says (281): "We purposely choose to be ruled by a device." With the possible substitution of "implicitly" for "purposely," this is the thesis of *TCCL*. The constraints Power notes at the beginning of his piece are acknowledged in *TCCL:* "Members of the technological society are largely impotent vis à vis corporations and government agencies when they are called upon to act as consumers or taxpayers" (Borgmann 1984, 108). The institutions that constrain us, however, are subject to our rule as citizens. In that sense they are self-imposed as is implied by the sentence that follows the one just quoted: "But as citizens they have a scope of action that is undeniably wide and genuine" (Borgmann 1984, 108).

Power is mistaken, I think, when he says the subsequent proposals in *TCCL* and *Crossing the Postmodern Divide (CPD)* require saintliness for their adoption (278, 279). What I have learned from Power in the many years we have spent here at the University of Montana is that reform cannot defy the economy and economics. It must take a constructive view of these institutions and recognize their positive potential. Had it not been for Power, I would have talked far less confidently about economic matters in *TCCL* and *CPD*.

In any event, it becomes clear in Power's pursuit of the issue that capitalism, the market, and the device paradigm are compatible aspects of one and the same state of affairs and that they are not the antagonists of individuals but the ways in which we as social beings agree, implicitly at times and explicitly at others, to order our fundamental relations. Few generations have had the privilege or the burden, as did the founding fathers of this country, of shaping these institutions nearly from the ground up. We are typically born into a world of existing and interlocking institutions. But as Power and I agree, we are still responsible for these institutions and able, albeit with difficulty, as Carl Mitcham shows (131), to change them for the better. Power is right, however, in urging that more needs to be said on just how the economy came to be shaped along the lines of the device paradigm and also on what openings the economy provides if reform is to succeed. Whether philosophers are the ones to say all that, I am not sure.

But what about the question of which paradigm—the market, capitalism, or the device—best explains technology? The market and capitalism explain much in the economy, less in politics, and, if not erroneously, next to nothing as regards responsibility and the good life. The chief function of the device paradigm, however, is to provide a perspective on these ethical issues (Borgmann 1984, 76–77). The moral problem of social justice is of course a widely and deservedly discussed concern among liberal economists like Power. But underneath that problem there is a still deeper issue. What finally is it that we want to see more equally distributed? A life of hectic and distracting affluence? Power has less to say on that issue, but certainly more than what we get to hear from mainstream social scientists, and what he says (287–88, 290–92) is compatible with the focal things and practices that *TCCL* moves to the center of the good life.

To understand social institutions in this way, however, is not to answer the why-question Power raises at the start. But then, showing how things came to pass is all we can do in historical or social analysis. In prephilosophical circumstances, as Carl Mitcham has reminded us (131–32), such questions were finally answered by reference to the will of an unsearchably powerful agent, a god or destiny. So to invoke technique or technology as an explanation today is unacceptable, as Feenberg rightly points out. That leaves us with modern science, where a why-question is properly answered by showing how an event had to follow from certain conditions according to scientific laws. But recent developments in the theories of deterministic chaos have shown vividly, if proof were needed, that no laws worthy of the name are to be had for history and society (Kellert 1993). Description of the how rather than explanation of the why is what we need to embrace.

But could we not say that the technological promise of liberty and prosperity and the growing recognition that it is only realizable through the combination of scientific research and technological devices constitute the origin of it all and the answer to the question why technology developed the way it did (see Power, page 272)? Does not this seductive promise connect uniquely well with human inclinations, as Mitcham suggests (140–42)? There is, I believe, expository and didactic value in introducing and adumbrating technology by way of its guiding promise. But the promise is still a manifestation rather than an explanation of technology. The historically specific and effective power of the promise is something prior cultures would not have understood or embraced. The rise of the promise is not the transhistorical cause of technology but its first epiphany.

The Device Paradigm

If the promise of technology sets the stage, the device paradigm is its chief character. Mitcham has shown that character is precariously poised between essence and accident, between the universal and the particular (133). It tends to show too much and too little, highlighting certain features sharply, obscuring features that fall outside of the template (134). But in the end we are left with Mitcham's dictum: "Character cannot be avoided" (135). The world of human acting and shaping is always characteristic. If we want to be fully equal to it, we have to discern its characteristic shape.

Given its breadth, the pattern of technology is bound to have qualifications and exceptions. There is of course a broad class of exceptions by definition as it were. In the Aristotelian language Mitcham employs, focal things and practices are substantial (though not ahistorically so) and fall outside of the scope of technology.

No one denies that there are technological phenomena that conform to the pattern of the device. But there is a legitimate question whether there are not objects or procedures that are clearly significant, modern, and technological in an ordinary sense and yet fail to fall under the device paradigm. My critics, it seems to me, have uncovered two instances that are clear exceptions and one case that is divided.

Film clearly fits the device paradigm. It is an entertaining commodity that exhibits the typical instantaneity, ubiquity, safety, and ease of consumption. It rests on a sophisticated machinery of production, distribution, and display that is invisible and often impenetrable to the viewer. As Phillip Fandozzi points out, much of the content of film agrees with its paradigmatic form, and he finds this particularly to be so when the technique of montage is used in making the film. Remarkably, however, in a certain kind of film, where the mise en scène technique is used, the commodity is transfigured into a kind of thing, something that gathers, discloses, and informs the viewer's world. Fandozzi's opening sentences are a fine evocation of such an event.

The other clear exception is Jesse Tatum's, the mirror image of Fandozzi's. While the latter shows how a commodity can in important regards become a thing, the former shows the same for a certain kind of machinery. In the home power movement, people resolutely break through the opaqueness of the standard electrical generating machinery and find engagement and insight where typically disengagement and ignorance prevail.

Gordon Brittan calls for a similar transformation of devices (85–86). As I will try to show on pages 364 below, an expert system may not be a good illustration. As it happens, a magnificent instance is the Windjammer, an electricity-generating windmill, that Brittan has been instrumental in devising and developing. In a recent essay (Brittan, forthcoming) he has shown that wind machines can be sited on either side of the device-thing divide. The typical wind generator today is an archetypical device—inaccessibly mounted on an 80- to 120-foot tower, the machinery concealed in a metal housing, the technology forbiddingly complex and unrepairable but by an expert, the entire device nearly devoid of any ties to its location, the local population, or the tradition of windmills. Brittan's Windjammer, to the contrary, sits on the ground, is intelligible to and repairable by a farmer or rancher, recalls the sail technology with its various branches and ancient roots, and can be adapted to the local weather and wind.

Technological machineries do not have to be opaque as the device paradigm has it. Consequently Feenberg's apprehension (311) that the device paradigm is substantially applicable throughout history is not warranted. In fact, if the notion of understandable and maintainable machineries were to spread across the entire culture and engineering ingenuity were directed toward perfecting Brittan-Tatum machines, we would witness the decline and fall of the device paradigm. But this is neither likely nor desirable, as I will try to show on pages 364–66 below. Meanwhile, as Tatum further shows, machineries that conform to the paradigm are not all of one kind. They can reasonably be divided into enabling and disengaging machineries.

An important if complex case has been presented by Diana Michelfelder. She follows up the well-known thesis of cultural feminism that women are bearers of a distinctive moral culture, and she has a persuasive illustration in the way the women in the Midwestern town of Prospect use a technological device, viz., the telephone. I share the intuition that underlies Michelfelder's claims and find it supported in many instances I witness firsthand. Moreover, there seems to be at first blush a close parallel between the device paradigm and the standard male morality as portrayed by, say, Carol Gilligan (Gilligan 1982). Both promote such traits as individualism, power, and control. If technological objects can be used in a relational and female as well as in an individual and male way and only the latter comports with the device paradigm, the latter's domain is much reduced. In any event, it would be good to discover a broad and well-articulated technological alternative to the pattern of the device.

Clearly the women of Prospect have been using the telephone in a caring and communal way that goes counter to what one would expect from the use of a technological device. Hopeful and inspiring as this case is, how typical is it? A rural community of fewer than a thousand people that can be characterized as "a predominantly white, Christian town" is surely marginal in this country (Rakow 1992, 17). There is no guarantee that this atypical setting and its admirable customs can endure against the general sweep of technology. That current keeps rising within and without. Externally, it takes the form of agribusiness that undermines and depopulates the rural towns. When the women, formerly of Prospect, no longer have a courthouse, schools, churches, and stores in common, the basis on which to use the telephone as an instrument of knitting the community together will be gone.

The intrinsic thrust of technology concerns the development of the internal structure of the telephone, the addition of caller identification, call screening, call waiting, and voice mail. All of these induce us to exert and accept more control on who gets talked to and who does not. E-mail is encroaching on telephoning and further enhances our control. Cell phones allow us to call and be called more instantly and ubiquitously. But typically such calls do not fall into an empty slot in our lives but compete with an ongoing event, often to the detriment of both the call and the event.

Such competitions should remind us that the use of a device is never a net addition to our lives but also and always the subtraction of something else. To the extent that telephoning displaces face-to-face encounters and letter writing, it disengages us from one another. To the extent that it replaces the drudgery of, say, washing clothes, obviated by washer and dryer, and connects scattered and distant people, it strengthens personal connections. But to avoid the traps of anecdotalism and instrumentalism, we must consider the typical ways in which women use technological devices. I share the intuition that by and large they do better than men, that they drive more safely, use computers more productively, and are more attuned to personal relations than to technical challenges.

This latter hunch seems to be supported by social science. Women, at least since 1965, have consistently spent more time on communication and organizations and less time on television. But the gaps between men and women have been closing between 1965 and 1985; and while women did more visiting in 1965, men did more in 1985 (Robinson and Godbey 1997, 199). In fact the overall trend in American culture is toward gender homogeneity (Robinson and Godbey 1997, 197–204). As the discussion of happiness on pages 357–59 below will show again, social philosophers constantly need to check and at times revise their intuitions against the data of the social sciences. As it turns out, the intuition that women drive more safely may well be in error (Li 1998). All this suggests that the case of the women in Prospect is neither typical nor a portent of the future.

But this does not diminish the importance of Michelfelder's central thesis. It does, however, shift its significance from the descriptive to the normative. We can capture the force of Michelfelder's point, first through the notion that caring is a three-term relation (one person cares for another in a certain setting) and second by noting that the quality of care depends on whether the setting is relational or not. The future of the ethics of care will depend in part on whether we take Michelfelder's point to heart. We can talk care until we are blue in the face if the material culture remains, or becomes even more, nonrelational.

How do we tell whether a setting is relational or not? If we are reduced to a case-by-case determination, the prospects of care are grim since the culture at large changes in large sweeps rather than in one instance here and in another there. Unless we can discern and appraise the large currents of the time, we will be swept off our feet before we know where to take a stand.

Currently the most vigorous cultural force is the information revolution. It indulges us with such marvels as virtual golf, the example Michelfelder mentions (232). The charm of virtual golf lies in its availability. An insomniac can play at two in the morning, a hyperactive executive between two appointments, a traveler in an airport. Not having to arrange things with a partner is part of virtual golf's convenience. But it could be relational. Assume an advanced stage of information and communication technology allows an executive traveling in France to arrange for a golf match with her son in California, she playing at six in the evening, he at nine in the morning.

But what a reduced experience this is compared with the real thing where mother and son share a time and a place that disclose the world to them and one to the other. In the latter case rising together, driving from home to the golf course, seeing the sun come up, feeling pride and gratitude to be citizens of this town and members of the golfing community—all this evaporates in virtual golf. You can try to share devices and commodities with your children in a caring way—the stereo, the television set, an exercise spa, Big Macs. But how strenuous and cramped such attempts will be.

Devices are highly relational but in the wrong way. The machineries of devices interlock more tightly and widely than the artifacts of premodern cultures ever did. But what meets us in the sphere of consumption is a commodity that has detached itself tangibly or experientially from all ties and encumbrances and is freely and smoothly available. Hence devices typically obviate and even repel engagement. Their commodities invite unencumbered consumption; unencumbered by, among other burdens, the demands of other people.

Focal things, in contrast, have a highly articulated and ramified structure that challenges us in several ways at once and so provokes engagement. The things, moreover, we are good at and love to do we want to share with our loved ones. This is caring of the highest order, the kind that opens up once basic necessities and urgencies are taken care of.

Consider then the rich and at times daunting opportunities of careful engagement that present themselves when you teach your daughter to ski, take your son to the theater, run with your loved ones up a valley, or prepare a meal with your spouse or partner. There is a symmetry between humans and their world, a "correlational coexistence" as Strong calls it (329–30). Humans can unfold their richness as resourceful and capable creatures only in a setting that is rich enough to provoke and answer the fullness of their gifts. I would suggest then that as a rule the world of focal things and practices is relational and the universe of devices and commodities nonrelational.

There are counterinstances to the nonrelational character of devices. Clearly Brittan's and Tatum's world of accessible machineries is one, and so are, with qualifications, communication devices, just as Michelfelder has it. The technological culture scatters loved ones, often for the benefit of their development, to the far corners of the country if not the globe. Daily engagement in focal things and practices is not an option. Caring must rely on the telephone and e-mail. In such cases, communication devices have the transparence Michelfelder speaks of. And it would not surprise me if women do the lioness's share of such caring.

The tendency of our culture, of course, goes in the opposite direction, toward a gender-neutral society (Robinson and Godbey 1997, 107, 198, 200–204, 296). More generally, to the extent that we are making headway against the injustices of class, race, and gender, we are moving toward cultural sameness. A Native American lawyer who together with her spouse makes $200,000 a year lives a life that is much closer to that of her white neighbors than that of her relatives on the reservation. Diversity can assert itself against the homogeneity of affluent consumption only if it comes to reside in a variety of focal things and practices.

Something like this point holds for the distinctiveness of gender as well. One focal if temporary practice is the conception, the bearing, the giving birth, and the nursing of human life. These are uniquely female privileges that in the future may have to be defended against a technological takeover (Selzer 1987, 95). Right now they need to be secured against suppression by the leading forms of contemporary life. Today it is nearly impossible to excel both as a mother and as a citizen of the dominant culture. To shine in the latter role a woman needs long and arduous training and then an unconditional commitment to her calling of excellence—the law, medicine, politics, business, research and development, scholarship, whatever. Many young women have done this and bravely excelled at motherhood too, but at a forbidding cost to their own well-being. Mora Campbell gives us a vivid sketch of this predicament (257–58).

In general women work no more than men though the components of their work differ. Women do more housework than paid work, men the reverse (Robinson and Godbey 1997, 119). Where such a division and balance rests on a considered and voluntary agreement between woman and man, it is unobjectionable. For gifted and ambitious young women, however, such an arrangement is not a welcome option. It implies a denial of their professional excellence. What is needed in the personal and private sphere is a fair division of domestic labor. In the public and political realm, it requires support by way of extended and paid parental leave and through measures that secure and ease reentry into work. Such measures should be framed in a gender-neutral way to avoid the charge of gender bias and accommodate fathers who want to stay or parents who want to share staying home.

All this leaves us with the question whether, regardless of gender, communication devices can be the instruments of a focal practice. We need to consider the concrete particulars. Certainly when parents on a weekend talk to their children, those are moments of pleasure and engagement. But notice the phenomenology of the occasion. Here are the parents, receiver between cocked head and raised shoulder, doing the dishes perhaps or sorting newspapers, listening now and exclaiming then. Is this what the richness of reality and the capacities of humans have come to? Is this what gives meaning and coherence to the life of the parents? To raise such questions is anything but a condemnation of what those parents do. But their activity is best described, it seems to me, as the beneficial use of a device that supports, but could not be at the center of, a life worth living.

INFORMATION TECHNOLOGY

Information technology is currently the prominent and most influential version of the device paradigm. It is sweeping everything before it and has so captivated the economy, politics from left to right, the president of Harvard as well as Montana's commissioner of higher education, medicine and MTV, the scientist no less than the Nintendo player, that one is sorely tempted to embrace the determinist and substantive vision of technology. There is no serious public discussion of the deeper implications of this most recent technology. Once more the liberating and enriching promise of technology is pressed into service. The occasional skeptics or critics are told that their fate will be irrelevance and oblivion.

As a work of engineering, computer technology deserves admiration for its ingenuity, energy, and cooperation. It has released astounding creativity and devotion in the construction of hardware and software. The high ethos of intelligence and diligence is alive and well in the instrumental uses of information technology too. Astounding work has been done in mathematics and the natural and social sciences with the crucial help of computers. The benefits of computers in business and industry are less clear (Landauer 1995). Yet the computerization of the economy is by now irreversible, on the whole for better than worse.

All these developments have subtly changed the background of our lives and our background assumptions about what is real and what is artificial, what is focal, and what is peripheral (Borgmann 1999). Philosophers of technology have been especially concerned with the effects of information technology on the foreground of life, on the area where we move as a matter of course and at times for no ulterior purpose but for some sort of final involvement, an experience that is an end in itself. If these developments in the background and foreground of contemporary culture are as radical as they are made out to be by Kellner and Feenberg, there is a legitimate question whether they do not transcend the schema of technological devices and focal things. In particular, one may expect the new experiences to rival or surpass focal things and practices—but in doing what?

It is hard to find criteria by which to judge this rivalry. We seem to be faced with incommensurable notions of the good life. The traditional notions of a coherent and centered life are countered by the postmodern norms of a polymorphous and decentered life. Cyberspace is embraced as the realm where one can live and celebrate the latter kind of life. To oppose it seems mere prejudice for an outdated ideal. So is the charge that life in cyberspace is inauthentic and unreal. As Kellner reminds us, "for postmodern theory, 'reality' is a construct and notions of the 'authentic' or really real are regularly deconstructed" (239). Consistently, Kellner proposes to "argue that new technological modes of experience and interaction are just as real and life enhancing as conversation, gardening, taking a hike in the wilds, or caring for animals" (242). These new modes are the ones that information technology has made possible and take place in a sphere we have learned to call cyberspace. In an engagingly fair-minded plea, Kellner urges not the rejection or dismantling of traditional practices but an appreciation of technologically mediated communication and the realization that it can be as positive as focal things and practices (243–44).

Kellner agreeably urges that cyberspace communication should supplement real-life connections and interactions (244–45). I would include under this heading the works of justice and compassion that are supported by the Internet (247–48). Cyberspace did not beget these causes but has amplified their endeavors electronically. It has amplified, as Kellner well knows, the works of oppression and hate as well (248–49). All these battles have at any rate failed to rectify the distribution of power. To be sure, cyberspace has elevated formerly poor if brilliant nerds to the ranks of business moguls. But the overall hierarchy of prestige and affluence has hardly changed.

Kellner appreciates, moreover, the inevitable and the potential losses of communication when it is electronically mediated (244, 245, 246). But again and again he professes the belief that cyberspace can be the medium of positive experiences sui generis (243, 250), and he puts this most explicitly in urging that a dismissal of cyberspace "would prematurely close off potentially exciting and life-enhancing new realms of experience and expansions of 'reality' " (249).

There is no good evidence as yet on how computer technology has insinuated itself into the foreground of daily life. Home computers are used primarily by the well educated and affluent. Hence the distinctive habits of computer users may be traceable more to class than technology (Robinson and Godbey 1997, 154–66). Sherry Turkle has given a helpful, if largely impressionist, account of the impact of computers on our notions of self and world (Turkle 1995). These, together with a phenomenology of computer use and the cultural trends of this century, allow for a tentative appraisal of Kellner's hopes.

Perhaps the distinctive feature of experiences that are centered in cyberspace is a peculiar kind of ambiguity. It has a more spectacular and more speculative counterpart in the ambiguity of the cyborg Campbell discusses and criticizes within the frame of time. The instability and inconsistency Campbell finds in cyborg ambiguity can also be uncovered in what I will call *virtual ambiguity*. It characterizes to various degrees an acquaintance that is established entirely within cyberspace, be it in a MUD, a MOO, through e-mail, a list, a bulletin board, whatever. To know someone only in a mediated way is always to encounter ambiguities. Pen pals of the olden days were will-o'-the-wisps until one met one's correspondent in the flesh. Authors may have a distinctive voice, but finally to meet them in person is usually a surprising resolution of one's vague anticipations.

While such traditional ambiguities are straightforwardly reducible to the austerity of one's information about the person in question, encounters in cyberspace can be both ambiguous and rich. In fact virtual ambiguity seems to have the positive function of creating a space for intense experiences.

In cyberspace one is liberated from the otherwise defining and confining burdens of life and can give free rein to desire and gratification. Turkle has telling illustrations of the rush of pleasure that people feel when they first immerse themselves in cyberspace. But the larger lesson of Turkle's book is to the effect that unencumbered freedom and intense engagement cannot consist. At length a cyberdenizen must choose between cyberspace cum triviality or gravity cum real life. There is yet a third possibility as the history of technology teaches us. As in consumption, one can endlessly cycle through hopeful desire and dawning disappointment and drift from one cyberencounter to another, always hopeful, ever disappointed.

It is against this background that one should read Feenberg's remark that Teletel, the French national network, "quickly became the world's first and largest electronic singles bar" (301). Feenberg rightly points out that this was an appropriation of a system that, like the Internet, was "originally conceived by technocrats and engineers . . . for the distribution of data" (301), and he asks rhetorically, "Why is it not a liberation to break such limitations in the virtual world that now surrounds us?" Such appropriation, secondary instrumentalization as Feenberg calls it, does in fact produce liberties, constructive liberty when it is used for good causes and dubious liberty when it produces the virtual ambiguity of electronic single bars. Whether such liberation constitutes an event that transcends and invalidates the device paradigm seems doubtful. Given the precedent of telephony that underwent the same development and given the general orientation of technology toward consumption, we should have expected what happened to Teletel and the Internet.

Feenberg believes that the exfoliation of cyberspace discloses yet a deeper problem in the device paradigm. The pattern it outlines represents a structure, called primary instrumentalization by Feenberg, but fails to illuminate the function (the secondary instrumentalization), i.e., the way people adapt and use a particular technology (303–4, 306). There is clearly something like secondary instrumentalization, and it is Feenberg's accomplishment to have defined and illustrated it. But it is a feature within technology as patterned by the device paradigm. *TCCL* from the start defines modern technology not merely as a structure but as a function (procedure) that has a structure (pattern): "The pattern of which I have been speaking inheres in the dominant way in which we in the modern era have been taking up with the world; and that characteristic approach to reality I call (modern) technology" (Borgmann 1984 3). The approach has two sides, the construction of machineries, undertaken by "technocrats and engineers," as Feenberg has it, and the consumption of commodities by, as Feenberg again correctly points out, "users" (301). That users sometimes shift the fulcrum of the device from machinery to commodity is remarkable and worth noting. But what is more remarkable is the further advance of the device paradigm against things and practices and the invasion of the person by the ambiguity of the commodity.

So far we have been commodifying things primarily. In cyberspace we make persons the objects of consumption. But we can satisfy this consumptive desire only if we commodify ourselves. Self-commodification, however, is deeply troubled and troubling, lurching back and forth between its unequal cancellations, triviality in cyberspace and gravity in reality.

There are constructive uses of campers, of snowmobiles, of jet skis, of dirt bikes, of Walkmans, of cosmetic surgery, of fast food, of microwave ovens, of VCRs, and of television. And so there are of cyberspace. But there is a dominant and dubious character to the standard use of these devices and commodities that needs to be outlined and evaluated. When the structure of these phenomena is so generalized and attenuated and the variety and force of uses so heavily emphasized as they are by Feenberg, one's approach comes close to the correct but finally unrevealing instrumentalist view of technology that occasionally surfaces in Feenberg's essay (305, 310–11).

Information technology is so interesting because it is or soon will be at once as helpful and necessary as the telephone and as distracting and dispensable as television. But this two-sidedness, I fear, also portends the cultural shape cyberspace will settle into, an irreplaceable and instrumental use and an entertaining and stultifying use with an unhappily slippery slope between them—or so it will be if the rule of the device paradigm remains dominant.

FOCAL THINGS AND PRACTICES

The counterforce to the rule of technology is the dedication to focal things and practices. This, at any rate, is the argument of *TCCL*. But can focal concerns carry that burden? Larry Hickman wants to subordinate them to "a flexible functionalism" (93) that would cover all human artifacts and enterprises and establish their values and ranks through tests. Pragmatist that he is, Hickman is bound to determine without prejudice or dogmatism what works and what does not, and he rejects standards that are excused from testing. I admire and have learned from the constructive spirit of Hickman's and Durbin's pragmatism and their belief that all our standards are fallible. But Hickman is not satisfied with the profession that focal concerns as ultimate standards can be mistaken, are fair game for criticism, and require explanation. He insists on tests. If such are available, the philosophy of technology will be both simpler and more effective. It can then dispense with the device/thing distinction and address and redress all problems via testing.

Tests need standards, however, and these cannot be testable in turn on pain of a vicious circle or infinite regress in which cases the very notion of a test is undermined. Whatever the standards, they can at most be fallible, contestable, and attestable. So they are no better off than focal things and practices. In fact they are in a worse predicament because they will be vague if wide and unreasonable if narrow. Hickman's standards belong to the former kind. In one place he mentions relevance and fruitfulness (96). But relevance to what and fruitfulness what for? Elsewhere he mentions health and psychological well-being and says that family life, e.g., can be quantitatively measured against these standards. But to the extent that this is possible, such standards represent merely necessary conditions of the good life. Still, I agree with Hickman's underlying concern that to criticize the technological society we need evidence that there is trouble (99).

Philosophers cannot be the conscience of their society if they believe their society has no conscience at all. To assume that people at large have entirely lost their sense of moral or cultural excellence or that they had it totally deformed by the powers that be so that in the midst of a moral calamity they lead to every appearance a happy and untroubled life is to take an impossibly dark or condescending view of people. In *TCCL* I argue that there are two kinds of evidence that all is not right with the technological culture. One says that there is growing unhappiness in the advancing technological societies, the other concerns the generally low level of excellence in this country.

Brittan has rightly pointed out that the claim of declining happiness is mistaken (79 and note 15). The data that in the 1970s seemed to indicate so were in error. But the larger point "that in technological societies happiness is not simply thought to be higher consumption" still holds, and the stubborn fact that "people remain both enthralled and unsatisfied" by technology is still with us (Borgmann 1984, 130). As for excellence, Brittan doubts it has been declining and questions whether society can determine what a life of excellence should be.

To begin with traditional excellence, the argument in *TCCL* is not that it has declined, but that it is low (Borgmann 1984, 126–27, 129). We stand condemned not by our predecessors but by our possibilities. In civilizations past, the majority of people have frequently been barred from the pursuit of excellence by oppression and poverty. Why do we condemn the former and deplore the latter? Surely the modern rhetoric of liberation and enrichment aims at something higher than unfettered and affluent consumption. Brittan is right that excellence is in part a matter of luck and (if the possession of willpower is not the same thing as luck) of individual resolve. Instructions of excellence, however, must always carry an other-things-being-equal rider. Of course a paraplegic cannot become an excellent runner. That leaves a large area of athletic excellence to the wheelchair bound, and some of them are exemplars of strength and endurance. As regards the interplay of social institutions and individual choice, I agree with Power that the former will constrain the latter for worse or better.

Happiness, Brittan argues, is indeterminate, subjective, and inconstant, and hence there is not only no evidence of its decline, there cannot possibly be any. This is in part an empirical question. Ruut Veenhoven, whom Brittan refers to as having found little consistency in happiness research and certainly no decline of it (Veenhoven 1984), has on further research found·that satisfaction is likely a universal human aspiration and that it is a state that can be validly and reliably measured (Veenhoven 1996a, 2–3). Contrary to what was suggested in *TCCL,* rich societies are happy, poor societies are not.

The social science news is not all good, however. Affluence yields a diminishing return of satisfaction. From a certain point on, greater affluence yields no more satisfaction (Schor 1998, 11–19, 165). This country must have reached that point soon after the second world war. The standard of living has doubled since. Satisfaction has remained roughly the same (Veenhoven 1996a, 20–21, 24, 30; 1996b, 38). And more evidence has surfaced for Power's thesis that there is a mismatch between our social arrangements and our deepest aspirations. Robert Wuthnow has titled the first chapter of his recent book "Having It All—And Wanting More: The Social Symptoms of Cultural Distress" and found that people both pursue material goods and disdain materialism (Wuthnow 1996, 17, 271–76). Juliet Schor has gathered much evidence to support that thesis in her *The Overspent American* (Schor 1998). Most particularly and convincingly, John Robinson and Geoffrey Godbey have shown that Americans have gained three hours per week of free time between 1965 and 1985 and yet feel more rushed and harried than before (Robinson and Godbey 1997, 81– 120, 229–40). Labor-saving domestic technologies fail to save them time (Robinson and Godbey 1997, 257–60). They spend the greatest amount of their free time on television (in fact, more time than ever and more than half of their free time if secondary as well as primary watching is counted), yet they rate it as average or below average in enjoyment (Robinson and Godbey 1997, 136–53, 241–51).

That seems to support Brittan's view that an incurable restlessness resides in the heart of humans and will frustrate any attempt to find rest in happiness. And if there is no good on earth that we know to bring us happiness, doing the right thing unconditionally and striving after moral dignity is the only moral way out. Aristotle must yield to Kant. Brittan invokes Goethe's *Faust* to support his point. In Goethe's telling there is no Faustian bargain (shortsighted gain for eternal loss) but a bet that Faust offers the devil—never to find rest in pleasure and to surrender his soul to the devil should Faust ever say to some moment: "Abide, you are so beautiful" (lines 1699–1706).

As it turns out, Faust, having engaged in many dissolute and dissatisfying pleasures, at length applies himself to constructive work and, seeing it close to completion, speaks the fatal words, albeit in a subjunctive mood, yet clearly enough for the devil who to his utter chagrin is after all deprived of his prize when the angels announce that they can save "whoever strives unceasingly"—restlessness vindicated and rewarded (lines 11581–93 and 11937–38).

There is more than a little irony, however, in the kind of work that Faust beholds with some satisfaction. Were it undertaken today, we would call it the destruction and development of wetlands involving the forcible and in fact fatal removal of elderly residents (lines 11559–80). Faust finds satisfaction not in seeing what is morally conclusive and good in itself but in the kind of penultimate and instrumental goods where we moderns typically like to rest our case—land wrested from the wilderness, building lots, the freedom to own the means of production.

Such Faustian restlessness was a deeply and honorably felt response to the challenges of a new era, the modern era. But today it must strike us as destructive. The challenge is no longer to conquer and dominate nature. Thus restlessness may be a specifically modern affliction, and the postmodern task is finally to come to terms with reality and seek out those moments of transfiguration that draw from us the words "It is good for us to be here" (Matthew 17.4, Mark 9.5, Luke 9.33).

For Paul Thompson the place of restful transfiguration is not the wilderness of biblical Mount Tabor but rather the fields and pastures of agriculture. For him, agriculture is "something like the quintessential focal practice" (166–67). It is obviously premodern or early modern agriculture that he has in mind. And he is right in complaining that agriculture has received too little philosophical attention. But it has not been as universally scorned as he makes it out to be. Martin Heidegger has repeatedly and affectionately referred to agricultural practices and just like Thompson he refers to them at critical junctures as the counterforce and cure of technology (Heidegger 1969, 44; 1971, 32–34, 147; 1977, 14–15). And again like

Thompson (173, 174) he finally despairs of the rural setting as the focal place of reform (n.d., 33–34).

That leaves us with Thompson's larger point—the need *to place* focal practices. Andrew Light has shown that *TCCL* is not inhospitable to places, and it briefly makes the case for cities as places (Borgmann 1984, 243–44). But what about country and nature as places? One thing is clear. Agriculture would have to be saved or restored as a focal practice before it could serve as the site for the reform of technology, and so agriculture is an enormously difficult problem rather more difficult than that of our best answer to the question of technology. Agronomy, scientific husbandry, agribusiness, and the globalization of the agricultural commodities market are conspiring to purge agriculture of everything focal and familial. It is the integration of agriculture into the technological and global economy that makes its focal restoration such a daunting task.

REFORM

Durbin's sense that our work must have practical consequences is one I share entirely, provided we read in a particular way Durbin's remark that in actually doing something "about the technosocial evils that motivated us in the first place . . . we will abandon any privileged place for philosophy" (47). If "privileged" means something like "superior" or "imperial," I am with Durbin, but not if it means "distinctive" or "special." There is a strong position in the Western tradition that reflection clarifies one's vision and aids one's action. It is a position we all share as scholars and one we need to defend against fundamentalists and technocrats. As citizens, of course, we have no privileged task but are bound to join with others in the mundane enterprises of social and environmental reform, and even as philosophers we have no monopoly on reflection but must welcome contributions from all quarters. But if we give up on the Aristotelian notion of *theoria,* we *eo ipso* have abandoned the philosophy of technology. The foregoing remarks must not be misunderstood—none of them are to be taken as an a priori brief for the value or validity of the philosophy of technology. Here again a pragmatic spirit should prevail: let the philosophy of technology be judged by its fruits.

The cultural climate of today is not hospitable to ambitious reforms. The triumph of democracy and technology seems obvious, and its fruits are palpable, particularly in this country that prides itself on its democratic principles and technological prowess. Refinements and adjustments are needed, it seems, not fundamental reform. The three issues that nonetheless command some reformatory zeal are social justice, the welfare of the environment, and the quality of life. All three are closely connected with technology. The first is intractable without a reform of the device paradigm. The second is normally feasible *within* the device paradigm. The third is as close as we come in conventional discourse to a concern with a genuine reform of technology. As *TCCL* has it, social justice gets entangled in the way technology comes to specify liberal democracy. The leading idea is that individual choice is realizable only in something like a realm of consumption and commodities. The tendency of technology is both to make consumption widely available and to do so in an order of inequality so as to keep the delivery of commodities dynamic and captivating (Borgmann 1984, 107–13). Thus technology aids and undergirds liberal democracy. What *TCCL* missed is the converse relation, the fact that liberal democracy is uniquely hospitable to the flourishing of technology. Economists like John Kenneth Galbraith (Galbraith 1967, 107–8, 389–91) and Robert Heilbroner (Heilbroner 1974, 61–95) had assumed that technology would assert itself and prosper under communism or socialism as well or as poorly as in a liberal democracy. But the events of 1989 have taught us otherwise (Borgmann 1991).

It has turned out that when the design and direction of technology are arrogated by a vanguard or a party, economic complexities vastly overtax the administrative resources of such a relatively tiny group. The mass of people feels both disfranchised from the common enterprise and frustrated by the bungling of bureaucracy. What is needed for technology to prosper is a distribution of power and an allocation of benefits that are in principle open to everyone and in practice to everyone's benefit if not equally so. It has also turned out that in the conjunction of democratic liberty and technological prosperity one or the other can take the lead, but neither can greatly exceed the other without jeopardizing itself. In western and now in eastern Europe and Russia liberty came first; in Chile, Taiwan, Singapore, and presumably China, prosperity has been or is forging ahead. But liberty will have to catch up with prosperity in China and prosperity with liberty in Russia.

Though I agree with Durbin that we should use every opening and occasion to promote greater social justice, constrained inequality within and indifference to suffering without will remain the hallmarks of the advanced industrial countries until the more concealed ingredient of our technosocial arrangement, namely technology, is explicated in reflection and reformed in practice.

If social justice is one cause that still provokes ire and effort, the environment is the other. But as Easterbrook (1995) and Sagoff (1997) have shown, much environmental concern has been overtaken by successful environmental reforms. Resources and wastes, the alpha and omega of technology, have been brought into a roughly sustainable relation with production and consumption in the technologically advanced countries. There will be some anxious decades before the rest of the world has risen to a similar level, and concerted reaction to global warming is still tentative. But if the spreading of democracy manifests the expansive power of technology, environmental improvements testify to its self-regulating tendency (Borgmann 1984, 145–48).

An environmentally sustainable regime, however, is compatible with the draining of wetlands, mining in roadless areas, and losing the grizzly bear. As Eric Higgs has shown, it is misleading to challenge purely prudential environmentalism this way. The crucial issue of a respectful environmentalism can no longer be framed as the clash between preservation and destruction or development. The power of technology has touched every last bit of nature in some way, and hence environmentalism, no matter how high-minded, is unavoidably a matter of interaction and negotiation with nature and one another; and hence, as Higgs argues convincingly, a morally sustainable environmentalism, no less than the struggle for social justice, requires a reform of technology.

As *TCCL* and *CPD* have it, a reform of technology amounts to making the world more conducive to focal things and practices and communal celebrations and giving these a central place in our lives. Some of my readers find this reform weak in substance or procedure. It is not always clear what a strong reform would look like though the implications of the criticisms allow one to guess. A procedurally weak reform is one that lacks an effective procedure of implementing the substantive goal. Though Brittan is drawn to a life of focal things and practices, he knows "of no way in which to make more than a 'hortatory' case for it" (85). The implication, I take it, is that a reform proposal requires a rationally compelling argument. But such arguments are never stronger than their premises, and in cases of disagreement nothing more than exhortation is available to make one's premises prevail. Moreover, even when the premises are culturally fundamental and unassuming and inferences are rigorous, the arguments can fail as often as they succeed as the case of smoking shows.

Higgs is forthright in stating what he considers a procedure more adequate than that of *TCCL*. He sees an opening for reform in "the interest in local and bioregional economics coupled with the development of left ecological politics, especially variations on a theme of libertarian socialism or communal forms of anarchism" (215). But how do we kindle an effective concern in such movements? We can be sure that Higgs would not think of trying the didactic measures of Baader and Meinhoff. But that leaves us again with nothing more than pleading and exhorting.

Power thinks that the proposals of *TCCL* require the moral heroism of saints and are therefore doomed (278–79), and evidently he believes that the road to reform proceeds via a change of our social institutions. Agreeably, Power wants greater equality, stronger welfare institutions, and more ennobling work. But people will not hand us the keys to institutional changes. They elect politicians who promise affluence rather than equality and well-paying rather than intrinsically rewarding jobs. Once more we are reduced to persuasion and to asking people to be, if not saints, better citizens, better stewards of the land, and less venal in their choice of work. As it happens, Power does more than his share of pleading and persuading as a public speaker and radio commentator.

In all these cases of exhortation, pleading does not come merely to emotional appeals or persistent haranguing. There are facts that people need to be reminded of, connections and implications that must be clarified, deceptions and fallacies that have to be exposed. But none of this will do much good as long as most people remain enthralled by consumption. *To dislodge them from that persuasion, they need to be presented with or reminded of an alternative vision of life.*

Hickman finds the substance of the vision in *TCCL*, distorted as he sees it by the countervailing device paradigm, to be narrow and rigid (90–91, 96). And surely, other things being equal, a broader and more flexible view of reform is preferable. But Hickman's functionalism gains flexibility at the cost of definition. Feenberg similarly rejects the characterization of technology through the device paradigm and calls for a conception of technology that is of itself morally, aesthetically, and socially beneficial (304). This again, ceteris paribus, is a desirable goal. Feenberg's proposal is twofold. The first and more distinctive part is the notion of secondary instrumentalization (308–10). This is a significant feature of how people on occasion appropriate technology, but as far as Feenberg's illustrations go (301), it is not a feature that contravenes the pattern of the device. The second is a reminder of the beneficial uses of technology. That technology has such uses is evident and acknowledged (Borgmann 1984, 36, 139–40, 246, 248). But even when it does, technology serves us in a particular style. For people suffering from Lou Gehrig's disease (302), it makes communication ubiquitous, instantaneous, safe, and easy, i.e., available, and thus gives it the presence that is characteristic of a commodity. This is clearly a positive result. But even here technology subtly and distinctively transforms and patterns communication.

PRACTICING PHILOSOPHY OF TECHNOLOGY

All these criticisms and rejoinders, however, overstate our disagreements. The authors of this book obviously do not occupy one and the same position. But within the universe of academics and social critics they are close neighbors. The boundaries of that neighborhood can be roughly traced by a common diagnosis of contemporary culture and a shared view of a cure. There is something like an agreement that this society is shaped by a social system that (1) constrains individual choice, (2) is largely implicit, (3) is in important part realized in the material culture, and (4) has failed to conduce to the good life. We therefore need to (1) place the social system on the agenda of the national conversation through scholarship, education, and the media and (2) bring about a closer alignment of our best aspirations with our cultural resources, particularly in material culture.

We should follow the example Light has marked out for environmental pragmatism that seeks consensus among scholars in the service of concerted efforts of reform (Light and Katz 1996). Philosophers of technology should similarly look for common ground and shared proposals. The outline of such joint concerns will always be rough, and there will be different ways of filling in details. What follows is one of those ways.

The device paradigm requires neither intrinsic reform nor global replacement. To be sure, the proposals of Brittan and Tatum constitute a genuine and desirable alternative to both devices and traditional things. But only local replacements of devices by Brittan-Tatum machines are desirable and possible. The general infrastructure, e.g., of communication, transportation, and health should provide the instantaneity, ubiquity, safety, and ease that only advanced technological devices can provide. We cannot responsibly, to illustrate the point, risk human health or life in order to save a medical skill from obsolescence. Having the best of both worlds, computerized expert systems enhancing rather than displacing a skill, is an unlikely prospect. Assume the rural doctor in Brittan's essay (85–86) takes an electrocardiogram to determine whether her patient had a heart attack or not. Interpreting the record correctly is difficult, yet to get the right reading is crucial because a false reading may lead either to the omission of life-saving surgery or the needless administration of a traumatic operation. An expert system is apparently accurate more often than the most skillful human (Gawande 1998). Thus the expert system does in one regard diminish the scope of medical practice.

It would be foolish to try and construct an a priori argument to the effect that advanced computerized devices can never have the gathering and engaging force of a thing, an instrument, a hand tool, or a Brittan-Tatum machine. But for the moment consider this. The enormous information storage and processing capacity of today's computers has to have some function. Broadly these functions fall into two classes: those that deal with a complex contingency of the real world and those that construct a virtual reality of some sort. In the first case they obviate our engagement with the contingency in question, in the second they detach us temporarily from all real contingencies in exchange for a contingent world we have the power to enter or leave at will. In either case a highly computerized device comes between us and our immediate and primary world.

There is normally no need or possibility to bridge technologically the distance between ourselves and our surrounding world. It appears that the arts of dozens of millennia have pretty well discovered and crafted the tools and instruments that best engage humans and disclose the nearness of their world. It is in scientific research that information technology truly comes into its own. There are mathematical structures, natural phenomena, and social patterns that would have remained unreachable and unknowable without computers. Thus in addition to the Brittan-Tatum machines, devices that furnish a productive commodity, as it is called in *TCCL* (Borgmann 1984, 139), fall in part outside of the device paradigm. Computers do disburden researchers from prohibitive computational tasks through machineries that are forbiddingly complex and quite opaque. At the same time they open up demanding and engaging possibilities of exploration and investigation. Here too the device paradigm is perfect within the scope of serving researchers. Computational power should be made as instantaneous, ubiquitous, safe, and easy as possible. To stick to the "engagement" that is enforced by having to work with a centrally located mainframe computer tractable only in machine language would have been stultifying.

There are then areas of contemporary life that have irretrievably lost their engaging character—telecommunication, long-distance travel with the bed and board that come along with it, the utilities in the urban areas, the technologized parts of medicine, and many of the appliances and utensils of daily life. If even a small fraction of the machinery of our culture were in some way rendered accessible and engaging again, engagement would quickly turn into numbingly ceaseless and utterly exhausting labor. The machineries of life in a technological society have been configured into an interlocking system, and if we want to hold on to anything like the physical welfare and cognitive scope it affords, its basic machinery must be kept intact.

The legitimate scope of the device paradigm furnishes outer limits to constructive reforms but leaves us with the task of sketching a more definite and detailed outline of an excellent life within those boundaries. The pragmatic goal should be to circumscribe a commonwealth of the good life. There is broad agreement that skilled practices and precious things are the two landmarks of such a commonwealth and that both require more institutional support than they are typically getting. Larry Haworth's singularly lucid and helpful essay has shown that we can think of one landmark as implying the other. The care of precious things can lead to appropriate practice, and certain practices can generate precious things.

Demanding practices without the complement of valuable things are diminished and vice versa. A museum is a collection of valuable things. It can be the reminder or inspiration of a vigorous culture but does not constitute one by itself. Programming is an admirably exacting and inventive practice. But its resonance is limited by the instrumental and generally invisible status of its object. Consider also Brittan's "range of basic skills" (84), including "hunting and fishing, gardening, and raising and training animals" (87). Surely their nobility depends on whether they are centered on the appropriate things or are practiced by way of computer simulations.

Two kinds of focal things and practices are basic somewhat in Brittan's sense—the culture of the table and the culture of the word. The former relates to the sustenance of our very existence, the latter to the way we comprehend the wider world. But there will be a commonwealth of the good life only if a life of excellence is within everyone's reach. Durbin more than anyone has stressed this point, and I agree entirely.

The modern era has enough illustrations to show that people who lack the pretensions and accoutrements of the rich are capable of a fine kitchen and a well-set table and can be consummate tellers of tales and writers of letters and comprehending and discriminating readers of fiction, poetry, and social commentary. Obviously reform must include a guarantee of the means such competence requires. Similarly there is no doubt, if assurance is needed, that common folk can as competently participate in most communal celebrations as their presumed social betters.

Let me summarize these remarks on reform by adverting again to its substance and procedure by asking, What sort of difference would a commonwealth of the good life make? and How can we hope to build such a commonwealth? A world that is devoted to focal things and practices and communal celebrations would be quite different from ours, materially and practically. The hypertrophic utilities of consumption, the expressways, high-rises, shopping malls, and theme parks would shrink, or at any rate cease to expand. The focal points of a city would be its concert halls, theaters, parks, playing fields, public squares, and houses of education or worship. Cities would be livable and enjoyable for pedestrians. People would spend their free time in communal engagements, large and small. Houses would be built to favor dining, music making, and conversation or reading. It would be a world where life would come to rest in celebration more often and more regularly.

Of course, there should be no prohibitions on a life of hyperactive work and restless consumption. The idea is to structure house, city, and the ferial and festal patterns of life in such a way that the default case is not an inducement to turn on the TV and open the refrigerator. Obviously also, the suggestions that have been made for urban life have natural analogues in Higgs's practices of restoration and in other diverse and kindred practices. And before one dismisses all this as weak or reactionary, one should ask if the generally grand and gauzy gestures of reform we usually get are stronger or more progressive.

What is the prospect of coming closer to a commonwealth of the good life? Light is right in stressing that the idea is not fashionable. But it is unfashionable, I believe, more in the official rhetoric of the culture and in the discourse of social theory than on the ground where telling signs of discontent with the current rule of technology and encouraging movements of reform can be found. As for concrete steps that philosophers should take, I join the pragmatism of Durbin, Hickman, and Light. I take from Durbin the commitment to social justice and social activism, from Hickman the diversity of approaches, and from Light the call for a measure of cooperation. Philosophers of technology will leave no impression if each of them stresses his or her differences from all the others. Unless people hear something like the same message, albeit from different directions and in various keys, they will be left with confusion rather than enlightenment and encouragement.

A distinctive and shared concern cannot be cast into the form of a political or economic program. The economic measures proposed in *TCCL* (Borgmann 1984, 237–42) are no more than illustrations. We need something like a broad goal that is stated in terms of Light's "prepolitical conditions" (113–20). Such conditions have to be as close to sufficient for the good life as is tolerantly possible. When they are presented by way of concrete examples, they are sufficient for the good life at that moment. As guidelines for the future they can and should be nearly sufficient. Politically such illustrations and norms are far from neutral, yet they certainly underdetermine particular political programs. Philosophers, in their attempts to find a bridge between theory and practice, need to seek the conversation with activists like Richard Sclove and economists like Power and defer to their expertise in advocating concrete steps.

THE COMPLETION OF THE PHILOSOPHY OF TECHNOLOGY

Whatever else the philosophy of technology may be, it *is* philosophy and should recognize the standards of its guild and tradition. It is the merit of David Strong to have made that case clearly and vigorously. The task has a professional and a substantial aspect and an intensive and extensive dimension. Philosophies that we as professional thinkers admire and emulate have never been specialized. The great moral doctrines, e.g., have invariably been of a piece with an ontology or metaphysics, a psychology or epistemology, and a cosmology or theology, and these professional achievements reflect the fact that the world is of one piece or at any rate has one structure. Hence not much light can be shed on any one part to the exclusion or in ignorance of all others.

One honorable and helpful way of meeting this requirement of comprehensiveness is to draw on a great thinker or tradition, on pragmatism, phenomenology, the Frankfurt School, analytic philosophy, on Kant or Heidegger. But one can also take up the suggestion Strong finds in *TCCL* and *CPD* and has been developing in his own work. The immersion in technology may give a philosopher access to a strand of reality that, when fully traced, reveals a new vision of the fabric of reality.

The pursuit of this hunch has an intensive and an extensive dimension. The former directs us to the core convictions that, as Durbin has it (47), "motivated us in the first place" and that, in a sort of reflective equilibrium, are being tested and refined by our philosophical work. Strong is quite on the mark in stressing that the claim of things and the devotion to those things are pivotal to *TCCL* and *CPD*. To this I should add the point that Pieter Tijmes has urged on me in conversation time and again—among things, the sacramental ones that prefigure the Kingdom of God are central.

One task that flows immediately from the statement of such a conviction is some kind of explanation of how today strong and incompatible convictions are possible and can consist with the cooperative pragmatism proposed above. That issue further opens up on the extensive dimension of a complete philosophy. Formally this comes to outlining a cosmology at least and a metaphysics and natural theology if obtainable. That enterprise also includes the theories of time and space that Campbell and Thompson have been calling for.

These challenges go far beyond my space if not also my sangfroid. Let me conclude with a few surmises. Science makes reality ever more transparent, and technology makes it more and more controllable. But at the end of our inquiries and manipulations there is always something that reflects rather than yields to our searchlight and presents itself as given to us rather than constructed by us. It is intelligible not because we have seen through it or designed it but because it speaks to us from within the continuities of history and nature. Thus the task of cosmology is to understand the interconnection of lawfulness and contingency, of human construction and objective givenness. There is then the possibility that at the far end of scientific transparency and technological control an unforethinkable and uncontrollable reality newly presents itself and will suggest a resolution of contemporary ambiguities.

In particular we may come to answer the question of the status of focal things. At times they look like driftwood from a once-flourishing grove, now uprooted, worn and bleached, drifting along on the supporting flood of technology. Within a new cosmology, however, we may learn to recognize focal things as islands, once the high country of an ancient continent and still anchored and connected with one another beneath the surface of technology.

If these places are firm and inhabitable, they can provide those points of orientation and restful celebration that lend life dignity and pleasure. Such points and periods of rest are compatible, however, with a higher kind of restlessness, the one Augustine had in mind when at the beginning of the *Confessions* he said, "[R]estless is our heart till it may rest in thee."

REFERENCES

Borgmann, Albert. 1984. *Technology and the Character of Contemporary Life: A Philosophical Inquiry*. Chicago: University of Chicago Press.

———. 1991. "The Development of Technology in Eastern and Western Europe." In *Europe, America, and Technology*. Ed. Paul Durbin. Dordrecht: Kluwer.

———. 1992. *Crossing the Postmodern Divide*. Chicago: University of Chicago Press.

———. 1995. "Does Philosophy Matter?" *Technology in Society* 17:295–309.

———. 1999. *Holding On to Reality: The Nature of Information at the Turn of the Millennium*. Chicago: University of Chicago Press.

Brittan, Gordon G., Jr. 1998. "The Wind in One's Sails: Wind Turbines, Comprehension, and Community." Unpublished manuscript.

Easterbrook, Gregg. 1995. "Here Comes the Sun." *New Yorker*, April 10, 38–43.

Galbraith, John Kenneth. 1967. *The New Industrial State*. Boston: Houghton Mifflin.

Gawande, Atul. 1998. "No Mistake." *New Yorker*, March 3, 74–81.

Gilligan, Carol. 1982. *In a Different Voice*. Cambridge: Harvard University Press.

Heidegger, Martin. 1969. "Ein Wort des Dankes." In *Martin Heidegger zum 80. Geburtstag*. Frankfurt: Klostermann.

———. 1971. *Poetry, Language, Thought*. Trans. Albert Hofstadter. New York: Harper and Row.

———. 1977. *The Question Concerning Technology*. Trans. William Lovitt. New York: Harper and Row.

———. N.d. "Danksprache von Professor Martin Heidegger." In *Ansprachen zum 80. Geburtstag*. Messkirch: Messkirch.

Heilbroner, Robert. 1974. *An Inquiry into the Human Prospect*. New York: Norton.

Kellert, Stephen H. 1993. *In the Wake of Chaos: Unpredictable Order in Dynamical Systems.* Chicago: University of Chicago Press.

Landauer, Thomas K. 1995. *The Trouble with Computers.* Cambridge: MIT Press.

Li, Guohua, Susan P. Blake, Jean A. Langlois, and Gabor D. Kelen. 1998. "Are Female Drivers Safer? An Application of the Decomposition Method." *Epidemiology* 9:379–84.

Light, Andrew, and Eric Katz, eds. 1996. *Environmental Pragmatism.* London: Routledge.

Rakow, Lana F. 1992. *Gender on the Line.* Urbana: University of Illinois Press.

Robinson, John P., and Geoffrey Godbey. 1997. *Time for Life.* University Park: Pennsylvania State University Press.

Rorty, Richard. 1995. "Two Cheers for Élitism." *New Yorker,* January 30, 86–89.

Sagoff, Mark. 1997. "Do We Consume Too Much?" *Atlantic,* June, 80–96.

Schor, Juliet B. 1998. *The Overspent American.* New York: Basic Books.

Selzer, Richard. 1987. Contribution to Marion Long, "The Seers' Catalog." *Omni,* January, 36–40, 94–100.

Turkle, Sherry. 1995. *Life on the Screen.* New York: Simon and Schuster.

Veenhoven, Ruut. 1984. *Conditions of Happiness.* Dordrecht: Reidel.

———. 1996a. "Developments in Satisfaction-Research." *Social Indicators Research* 37:1–46.

———. 1996b. "Happy Life–Expectancy." *Social Indicators Research* 39:1–58.

Wuthnow, Robert. 1996. *Poor Richard's Principle.* Princeton: Princeton University Press.

On April 1, 1999, the *New York Times* published its first article on philosophy of technology since it had covered and published the infamous *Unabomber Manifesto* several years earlier. It was the first story on the academic field in the paper's history. The article by Katie Hafner—featuring interviews with Albert Borgmann, Hubert Dreyfus, Andrew Light, and Langdon Winner—soberly characterized philosophy of technology as a form of "scholarly skepticism" between the excesses of the Unabomber's revolutionary Luddism and the "gee-whiz" technophilia of *Wired* magazine. Though certainly welcome as one of the more prominent pieces of publicity for the field, the story did reveal one of the less encouraging aspects of philosophy of technology today: the fact that as a philosophical subfield, philosophy of technology has been comparatively stagnant when compared to other important areas like environmental philosophy and medical ethics.

What accounts for such differences? Surely the problems facing us on the technological front are as pressing as those concerning the environment (if these arenas of concern can be usefully distinguished at all). Surely the need to investigate these problems is as worthy of philosophical scrutiny. But still a gulf remains between the development of philosophy of technology and other fields. Explanations are difficult and can range from the precarious dynamics of the different philosophical societies formed around these interests to the larger cultural context in which these conversations have developed. One cultural studies scholar interviewed for the *Times* piece (though later cut from the article) opined that philosophy of technology was less popular than other fields today because younger generations of academics have grown up in a world so thoroughly saturated with technology that they are unable (or unwilling) to find space in their work to critique it. Critique of technology on such a view must be too countercultural even for the postmodernists of the current academic wave. But certainly the same could be said of environmental problems such as our relations with other animals. Ethical vegetarianism is as much on the fringe of our society as intentional simplifications of lifestyle. And yet many more younger scholars are working on questions of animal welfare than philosophy of technology.

Most curious of all, however, is that when more mainstream philosophers do turn to technological issues they sometimes willfully ignore the field of philosophy of technology. This problem was made particularly manifest when a volume was published in the prestigious United Kingdom Royal Institute of Philosophy supplemental series in 1995, supposedly on our field. The volume is titled *Philosophy and Technology,* and is edited by Roger Fellows for Cambridge University Press. But as Fellows makes clear in the first line of his introduction, the essays in the volume do not represent philosophy of technology in the sense that, as Fellows puts it, "Don Ihde requires." True to this promise, other than this reference to Ihde, there is not one reference in any of the papers in the volume to any of the prominent members of the Society for Philosophy and Technology, and thus, we can assume, to any of the prominent philosophers who have considered themselves doing philosophy of technology. Further, there is not one reference to any paper published in any issue of either of the field's two traditional forums: *Research in Philosophy and Technology* or the old Kluwer series, *Philosophy and Technology.* When philosophy of technology hit the philosophical big time, as it were, those spending their energy on developing the field, apparently, were irrelevant to the conversation. Clearly, if philosophy of technology is to survive into the next century it must broaden its appeal so as not again to find itself left out of such conversations, at least those occurring among philosophers inclined to finally get around to discussing technology.

So what is to be done? After going through the work brought out in this book, we see at least one possible answer, or at least hope that it represents one response toward revitalization of the field. We hope that this volume, by critically and extensively engaging Borgmann's particular philosophy of technology, will help to initiate a new level of discussion for the philosophy of technology and technology studies. Hopefully, as philosophy of technology achieves a stronger sense of self-identification and professional recognition, it will join forces with other fields and subdisciplines to create widely shared, critical, and socially relevant perspectives on technology as well as better philosophical foundations for the field.

No doubt, however, many readers will appreciate this volume more narrowly as only a Borgmann book. Indeed, it does engage the key issues of Borgmann's theory and present background information in sufficient detail that it could serve as a "Borgmann reader" for general scholars and for courses where Borgmann's philosophy is addressed. Most of us who have used one or another of Borgmann's works in a course have wanted a book that helps grasp Borgmann's theory, place it in perspective, and critique it. Accordingly, careful inspection of this book shows that it advances Borgmann's work itself through dialogue with others. For example, Andrew Feenberg's and Tom Power's contributions have made Borgmann stretch for better clarity than he attained in *TCCL*. Gordon Brittan forces him to take back an argument he made about happiness and technology. Criticisms by other contributors, such as Larry Hickman and Diane Michelfelder, have made Borgmann reach for fresh and more resourceful arguments for his basic conclusions. In his response to Douglas Kellner, Borgmann shows how his position is in the process of development and further substantiation. Other critics, such as Mora Campbell, Eric Higgs, David Strong, and Paul Thompson, have pointed out new areas that require exploration. Finally, Borgmann now shows through his pragmatic goal "to circumscribe a commonwealth of the good life" that his differences with his critics are not nearly as important to him as the common ground he believes he shares with them. In doing this, Borgmann claims that he is responding to the pragmatic calls at the heart of the work of Paul Durbin, Larry Hickman, and Andrew Light.

Yet, as what one reviewer described as a "focal event" for the field, this volume serves a much broader purpose—it addresses the problems at work in philosophy of technology today. Through their differing interpretations of Borgmann's ideas, all the contributors have also introduced their own approach to the fields of philosophy of technology and technology studies more broadly constituted. They have done so by showing their positions in relation to, basic critiques of, and, at times, indebtedness to Borgmann. In this sense, this volume serves as an introduction to the contributors' works as well and to the varieties of positions in the field more generally. Readers may certainly find our contributors' theories as attractive and on the mark as Borgmann's, if not more so.

Yet as a focal event for the field, this volume is something more and other than a collection of essays introducing the various positions of the contributors, and it is also something more than an introduction to the ongoing conversation of the field of philosophy of technology. Rather, considered as a whole, it initiates a new kind of activity in the field and a new kind of conversation. Plenty of us in the field of philosophy of technology have had questions about Borgmann's philosophy of technology or about the theories of Don Ihde, Kristin Shrader-Frechette, or Langdon Winner, but never before have we come together as a community and discussed our concerns with each other and with the philosopher or theorist whose ideas prompted these concerns, questions, and critiques.

For the first time there exists an extended and varied critical discussion from an array of different perspectives on such matters as the problems of technology, the limitations of the technological device, the so-called framework of technology, focal things and practices, the possibilities of reform of technological systems, the politics of critique and reform of technology, and the difficulties confronted when raising questions about what is good, what is useful, and what is a good life. Unprecedented in our field, a community of thinkers is gathered to think philosophically about a particular philosophy of technology. In this sense this volume serves as a focal event for our concerns. We hope that it is soon joined by other works of a similar nature—critical in outlook, extensive in discussion, focused on one theorist, unified by a professional academic community, systematic in treatment, and varied in illuminating approaches. Only through more discussions like this one will the field as a whole flourish. Only with such flourishing will a philosophical voice on the technological problems facing humanity, in a technologically mediated world, have any hope of being heard.

Eric Higgs
Edmonton, Alberta

Andrew Light
New York City

David Strong
Urbana, Illinois

August 1999

INDEX

morality (*cont.*)
223–24; society and, 357; temporal
ambiguity and, 259. *See also* ethics
moral theory, feminist ethics and, 224–25
multiplicity, the body and, 267
Mumford, Lewis, 186n. 3
music. *See* performance

national park, 203. *See also* Jasper National
Park
natural law, 136–37
nature, 19; Borgmann on, 336; character
and, 130–33; vs. commodities, 285;
community and, 34; culture and, 201;
decontextualization and, 306; device
paradigm and, 89; Disney Corporation
and, 200, 202, 203, 204; domination
of, 235, 237; ecological restoration and,
199; focal practices and, 263; focal
restoration and, 211; Heidegger and,
297, 312–13, 330; Jasper National Park,
196; maturity and, 36; psychology and,
128; Strong and, 176. *See also*
environment; wilderness
Nazis, 113
networking: patients and, 302–3; politics
and, 248. *See also* computer
communication
New Left, 42; Marcuse and, 40, 44
Newton, Sir Isaac, 307
New York Times, 371
Nicomachean Ethics (Aristotle), 77, 127,
140
Nietzsche, Elisabeth, 113
Nietzsche, Friedrich, 124, 251; Borgmann
and, 108, 110–11, 113, 117–18;
Kaufmann and, 109; the political and,
52–53, 106–8, 113–17
"Ninth Elegy" (Rilke), 336
Noble, David, 42
Noddings, Nel, 225–26
nomos. See law
North America, 4, 17; Borgmann on
culture of, 119–20; identification cards
and, 1; psychology and, 128, 129
nostalgia, the nostalgic, 82; Borgmann
and, 70–71

One-Dimensional Man (Marcuse), 40
ontology, the ontological, 121; ambiguity
and, 264; art and, 155; authenticity
and, 167; Borgmann on, 316; devices
and, 175; peasant experience and, 173;

philosophical history of farming and,
177; reality and, 239. *See also* being;
human
ordinary life: the moral and, 223–24, 230;
reform of technology and, 233
Oregon, 102
organism, 264, 265
organizations: Internet and, 247; men vs.
women, 349
orientation: sites of, 262; traditions of,
263. *See also* disorientation
Overspent American, The (Schor), 358

Painters Painting (Antonio), 246
painting, Microsoft Art Gallery and, 10
paradeictic, the: argument and, 139;
knowledge and, 36; reform and, 140
paradigm shift, 239, 240
parallelism model, 56–59
parents, 351
Parmenides, 304
participation in technological design, 187,
188n. 4
passive consumption, 30, 324
past, technological culture and, 82
patient vigor, 242, 243
performance: computer communication
and, 221; excellence and, 61, 62,
63–64; focal practice and, 24; vs.
practices, 63; tradition and, 64. *See also*
engagement; focal practices; practices
performer (musical), 23–24
Perry, Jonathan, 205
personality, 128, 129
Person to Person (Graham-LaFollette), 223
phenomenology, the phenomenological:
Berger and, 45; Borgmann and, 11
philosophy: academic, 342; activism and,
47; of agriculture, 174, 178, 180;
analytic, 342; ancient moral, 75;
argumentation and, 325–27; Aristotle
and, 131; Borgmann and, 47, 217, 235,
316–31, 334–35; Borgmann on, 19;
Continental, 342; essence and, 304;
focal things and, 21, 51, 328; history of,
331–32; language and, 326–27; in the
light of things, 327–29; modernity and,
236; moral, 223, 225; nature and, 132;
postmodern divide and, 240;
production and, 179; in the service of
things, 316–22, 326–31, 333, 335;
Society for Philosophy and Technology
and, 4–5; substantivist, 305; in tandem